# LUDWIG BOLTZMANN
# HIS LATER LIFE AND PHILOSOPHY, 1900–1906

# BOSTON STUDIES IN THE PHILOSOPHY OF SCIENCE



VOLUME 168

To Ludwig Boltzmann's descendants, who have been so helpful in the preparation of this book and to my initial colleagues in this project: Karl von Meyenn and Willibald Pricha.

Boltzmann, Ludwig
Mathematician, Physicist
Photo Deutsches Museum München

# LUDWIG BOLTZMANN
# HIS LATER LIFE AND PHILOSOPHY, 1900–1906

## *Book One: A Documentary History*

*Edited by*

JOHN BLACKMORE

SPRINGER-SCIENCE+BUSINESS MEDIA, B.V.

Library of Congress Cataloging-in-Publication Data

Boltzmann, Ludwig, 1844-1906.
Ludwig Boltzmann : his later life and philosophy, 1900-1906 : book one: a documentary history / edited by John Blackmore.
p. cm. -- (Boston studies in the philosophy of science ; v. 168)
Includes bibliographical references and index.
Contents: Bk. 1. A documentary history.
ISBN 978-90-481-4492-1 ISBN 978-94-017-0489-2 (eBook)
DOI 10.1007/978-94-017-0489-2
1. Boltzmann, Ludwig, 1844-1906. 2. Physicists--Austria--Biography. I. Blackmore, John T. II. Title. III. Series.
QC16.B64A3 1995
530'.092--dc20
[B] 94-39753

ISBN 978-90-481-4492-1

*Printed on acid-free paper*

# ACKNOWLEDGMENTS

I take special pleasure in acknowledging the role of extrahelpers by name, men and women who did not just help make this book possible by carrying out their duty or felt obligation - though this has been important and I am also much indebted to them - but those people who voluntarily made an extra effort which resulted in an indispensable contribution. I would also like to describe their wonderful deeds in detail.

Many years ago, My doctor father at the University of California at Los Angeles, John Burke, repeatedly tried to get me to write my doctoral dissertation on Ernst Mach, whose philosophy I strongly opposed. Kicking and screaming I finally agreed, but through that experience I first learned about Ludwig Boltzmann and became enthusiastic about him. But Engelbert Broda in Vienna had already written his biography. Roman Sexl, however, in 1982 invited me to help edit an anthology on Boltzmann. The book also included much information about previously unpublished correspondence, so I suggested to Professor Sexl, then head of the Department of Theoretical Physics at the University of Vienna, that if we tried hard enough we might find enough letters *by* Boltzmann in scattered archives (it was then believed that all of his *Nachlass* had been lost in the war so that no letters *to* him could be found) to complete an additional book in Sexl's planned series on the complete works of Ludwig Boltzmann of which two had already been published in addition to the anthology.

Karl von Meyenn and Willibald Pricha joined the project and indeed we found many letters, typed them, and added numerous footnotes, but sadly Roman Sexl became ill and soon died, but not before we had found another colleague, Walter Höflechner, who was very skilled at researching academic documents and letters relating to Boltzmann's transfers from university to university. But without a sponsor we drifted apart into different projects and our Boltzmann material remained largely unpublished.

Two philosophers got me going again. Sir Karl Popper suggested that I write a book about Ludwig Boltzmann's life and philosophy in English and Roderick Chisholm made Boltzmann's correspondence with Franz Brentano available to me, the first letters with truly significant philosophical content. Meanwhile, Boltzmann's grandson, Dieter Flamm, and his greatgrandson, Gerhard Fasol, both physicists, made additional correspondence available, and for the first time a number of letters *to* Boltzmann thereby demonstrating that a Boltzmann *Nachlass* did actually exist and that more material might be in prospect, which Fasol's mother Ilse M. Fasol-Boltzmann largely published in 1990.

I was now determined to write a book in English on Boltzmann's later life and philosophy, while Walter Höflechner was committed to publishing all of Ludwig Boltzmann's correspondence and basic academic documents in the original German. But not only has cooperation among the above scholars and many others as well remained excellent, but a number of people who had known Boltzmann personally freely shared their memories. Lily Hahn, the wife of the mathematician Hans Hahn, told me many amusing stories as did L.B. Loeb the physicist son of Jacques Loeb.

A number of librarians also walked the extra mile. Ellen B. Wells at the Smithsonian worked hard to put otherwise unpublishable material at my disposal. Billie R. Bousman at the Department of Special Collections at the University of Stanford provided invaluable information on turn of the century Stanford professors who were interested in Boltzmann's lectures given in Berkeley in summer 1905. Dorothy E. Mosakowski, Coordinator of Archives and Special Collections at Clark University searched but was unable to find correspondence by Mach and Boltzmann in her archives, but she didn't give up, kept hunting, and finally found letters from both men to the President of Clark University in a rarely noticed "special events archival box". She also discovered a previously unknown publication by Ludwig Boltzmann.

And finally and most decisive, Professor Robert Cohen, Mrs. Annie Kuipers, and my well-informed wife Setsuko Tanaka who has long had an interest in Boltzmann's physics have all contributed innumerable ideas and suggestions. To all my helpers and extrahelpers: God bless you!

# CORRESPONDENCE

(Chapters 3, 4, 5, 6, & 8)

A. Institutional Repositories:

*Berkeley*. Bancroft Library, University of California at Berkeley.
Letter No. 53.[1]
*Berkeley*. The President's File, University of California Archives.
Letters No. 83, 98, 103, 104, 109, 110, 111, 112, 113, 114, 116, & 118.
*Berlin*. Ostwald Archiv, Akademie Archiv, Akademie der Wissenschaften.
Letters No. 8, 9, 11, 16, 18, 80, & 87.
*Berlin*. Staatsbibliothek Preussischer Kulturbesitz.
Letters No. 51, 57, 61, 63, & 72.
*Claremont*. The Honnold Library, The Claremont Colleges, California.
Letters No. 88 & 89.
*Cracow*. Uniwersytet Bibliothek.
Letter No. 15.
*Dresden*. Königlich Sächsisches Ministerium des Kultus und öffentlichen Unterrichts. Letters No. 10, 12, 13, & 14.
*Freiburg im Breisgau*. Ernst-Mach-Institut der Fraunhofer Gesellschaft.
Letters No. 131 &136.
*Göttingen*. Universitäts Bibliothek.
Letter No. 70.
*Leipzig*. Universitäts Bibliothek.
Letters No. 19, 20, 21, 23, 50, 55, 58, 79, 81, 82, 105, & 107.
*London*. Library of the Royal Society.
Letters No. 3, 5, & 22.
*Los Angeles*. Einstein Family Correspondence Trust.
Letters No. 32 & 41.
*Munich*. Deutsches Museum Bibliothek Handschriftenabteilung.
Letters No. 25 & 135.
*Paris*. Archiv Institut de France - Akademie des Sciences.
Letter No. 119.
*Paris*. Nationale Bibliotheque.
Letters No. 1 & 4.
*Stockholm*. K. Vetenskapsakademiens Nobelkommitteer.
Letters No. 71, 90, 121, 122, & 123.
*Stockholm*. Universitetsbibliotek.
Letters No. 2, 43, 54, 73, 85, 101, & 125.
*Vienna*. Stefan Meyer Archiv, Institut für Radium Forschung.
Letters No. 24, 26, 36, 37, 45, 47, 48, 67, 74, 100, 108, 115, & 126.
*Vienna*. K.K. Ministerium für Cultus und Unterricht, Österreichisches Staatsarchiv.
Letters No. 6, 7, 17, 27, 28, 30, 31, 35, 38, 39, 40, 44, 49, 62, & 95.
*Vienna*. Österreichische National Bibliothek, Handschrift Abteilung.
Letters No. 33 & 46.

---

[1] *Plain numbers* refer to original correspondence, or where it was understood original letters were located at the time copies were acquired. *Numbers in italics* refer to where copies are or were located, including typed or printed copies. The lists are for the benefit of readers and to acknowledge our appreciation for the assistance of archivists, librarians, and family members, but may not reflect where used copies were actually obtained.

B. Published Letters:

Blackmore, John, (ed.) *Ernst Mach: A Deeper Look, - Documents and new perspectives* Kluwer: Dordrecht, 1992. Letter No. *87*.

Blackmore, John & Klaus Hentschel (eds.), *Ernst Mach als Aussenseiter - Machs Briefwechsel über Philosophie und Relativitätstheorie mit Persönlichkeiten seiner Zeit*, Wilhelm Braumüller: Vienna, 1985. Letter No. *117*.

Brentano, Franz, *Über Ernst Machs 'Erkenntnis und Irrtum'*, edited by R. Chisholm & J.C. Marek, Rodopi: Amsterdam, 1988. Letter No. *136*.

Buchholz, Hugo, *Die Mechanische Potential nach Vorlesungen von Ludwig Boltzmann*, Leipzig, 1908. Letter No. *127*.

Dick, A. & G. Kerber, *Ludwig Boltzmann - Katalog zur Ausstellung an der Zentralbibliothek für Physik in Wien*, Vienna, 1982. Letters No. *127 &134*.

Fasol-Boltzmann, Ilse M. (ed.), *Ludwig Boltzmann Principien der Naturfilosofi - Lectures on Natural philosophy 1903-1906*, Springer Verlag: Berlin, 1990. Letters No. *76, 77, & 78*.

Flamm, Dieter, "Aus dem Leben Ludwig Boltzmanns", *Ludwig Boltzmann Gesamtausgabe*, Band **VIII**, edited by R. Sexl & J. Blackmore, Vieweg: Braunschweig, 1982, pp. 21-56. Letters No. *29, 56, 65, 66, & 68*.

Haller, Rudolf & Friedrich Stadler, (eds.), *Ernst Mach - Werk und Wirkung*, Verlag Hölder-Pichler-Tempsky: Vienna, 1988,. Letters No. *88 & 89*.

Hoffmann, Dieter & Hubert Laitko (eds.), *Ernst Mach - Studien und Dokumente zu Leben und Werk*, Deutscher Verlag der Wissenschaften: Berlin, 1991. Letter No. *87*.

Höflechner, Walter, "Ludwig Boltzmann: sein akademischer Werdegang im Österreich - dargestellt nach archivalischen Materialien", *Mitteilungen der Österreichischen Gesellschaft für Geschichte der Naturwissenschaften*, **2** (1982) 43-62. Letters No. *28, 30, 31, 39, 62, 95, & 128*.

Körber, Hans-Günther (ed.), *Aus dem wissenschaftlichen Briefwechsel Wilhelm Ostwalds*, Band **I**, Berlin, 1961. Letters No. *8. 9. 11, 16, 18, & 80*.

Kox, Anne, J., "The Correspondence Betwen Boltzmann and H.A. Lorentz" in *Ludwig Boltzmann Gesamtausgabe*, Band **VIII**, edited by R. Sexl & J. Blackmore, Vieweg: Braunschweig, 1982. Letter No. *133*.

Stachel, John (ed.), *The Collected Papers of Albert Einstein*, Volume **I**, *The Early Years 1879-1902*, Princeton University Press: Princeton, 1987. Letters No. *32, 34, & 41*.

Thiele, Joachim, "Briefe Hugo von Seeligers an Ernst Mach" in *Philosophia Naturalis*, **17** (1979) 391-399. Letter No. *130*.

C. Private Collections:[1]

Engelbert Broda (Vienna)
Letters No. *6, 7, 10, 12, 13, 14, 17, 27, 35, 38, 40, 43, 44, 49, 54, 62, 73, 85, 101, 105, & 124.*

Buchholz, Hugo[2] (Halle)
Letter No. 127.

Roderick Chisholm (Providence & Graz)
Letters No. 69, 84, 86, 91, 92, 93, 94, 96, 97, 99, 106, 124, & 132.

Gerhard Fasol (Cambridge)
Letters No. 42, 75, 102, 120, & 129.

Ilse M. Fasol-Boltzmann (Bochum)
Letters No. 76, 77, & 78.

Dieter Flamm (Vienna)
Letters No. 29, *36, 37, 45, 47, 48,* 52, 56, 59, *60,* 65, 66, 68, *108, 115, & 126.*

Milan Popovic (Belgrade)
Letter No. 34.

Hans Suess (Vienna)
Letter No. 64.

---

[1] Professor Broda allowed the editor of this book to make copies from his typed copies in 1982. He died two year's later. It is not clear where his typed copies currently are.
[2] Professor Buchholz died many years ago. The current whereabouts of the letter from Boltzmann to him is not known.

# BOLTZMANN FAMILY TREE

(Circa 1983)

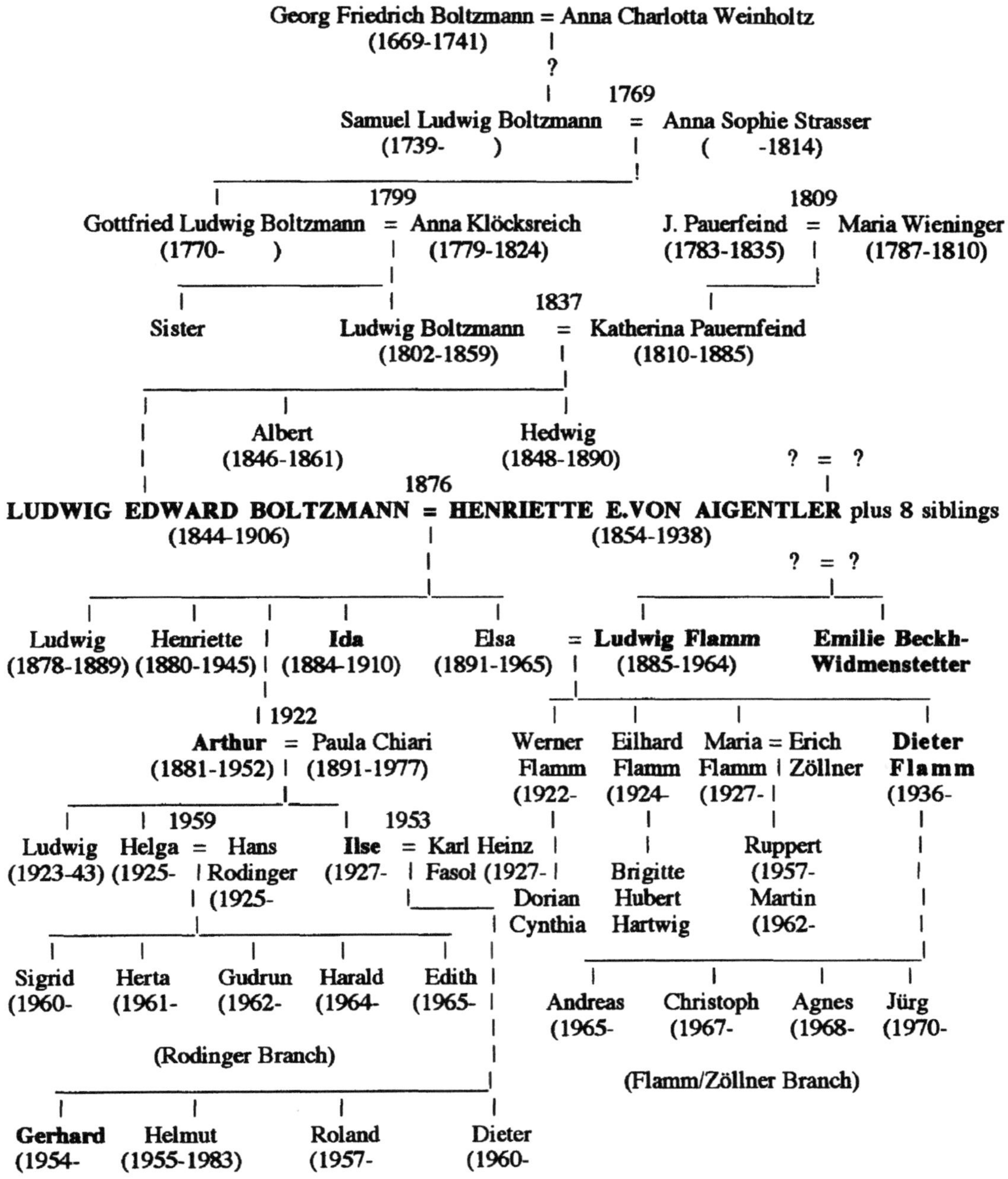


* The names in boldface represent authors or recipients of letters in this book or descendents who have published Boltzmann material or made it available for use.

ANALYTICAL TABLE OF CONTENTS

# Chapter 1

# INTRODUCTION

## Biography

Ludwig Boltzmann was born in the *Landstrasse* district of Vienna on the night of February 20, 1844, that is, between Shrove Tuesday and Ash Wednesday, a fact he later used to help explain his tendency to alternate between extreme emotional highs and lows. His father, an Austrian government official, was the son of a Protestant clockmaker from Berlin and his mother, born Catherina Pauernfeind, was Catholic. A second son, Albert, was born two years later in 1846 and a sister Hedwig in 1848. The family first moved to Salzburg and later to Linz where Ludwig entered the local Gymnasium and except for one year consistently received the highest marks in his class. But unfortunately his father died when he was fifteen and his brother a short time later. The widow did her best on a small pension to keep Ludwig in school.

He entered the University of Vienna in 1863 and specialized in mathematics and physics, though he also took a number of courses in philosophy. Among his older professors he was particularly impressed by the mathematician Petzval and the experimentalist von Ettinghausen. Among his younger teachers, Josef Stefan, a skilled theorist and experimentalist of Slovenian heritage, took precedence. But his closest helper and friend at that time was apparently Josef Loschmidt who in 1865 would make his famous calculation about the size of gas molecules. In fact, belief in atomism and use of the atomic theory was apparently so strong in Vienna that even Ernst Mach, who taught as a *Privatdozent* or unpaid lecturer in Vienna from 1861 dabbled in it until he received an appointment in Graz in 1864, where, being far away from his professors and colleagues, all mention of it stopped. He openly declared his presentist world view with its implied rejection of the reality of atoms in Prague in 1872. It is not yet clear whether Boltzmann attended any of Mach's lectures before the anti-atomist who would often cross Boltzmann's academic, scientific, and philosophical path left Vienna, but in any case Boltzmann, who became Stefan's assistant in order to help support himself, his mother, and his sister, rapidly threw himself into using the atomic theory to help understand thermodynamics more deeply, especially the second or entropy law.

Boltzmann published his first scientific article in 1865, received his PHD in 1866, and his *Habilitation* or permission to teach in a university in March 1868. He was appointed to a variation of Mach's erstwhile position in Graz in July 1869.[1] (Two professorships were now available instead of one.)

[1] For more exact dates see: Walter Höflechner, "Ludwig Boltzmann, sein akademischer Werdegang in Österreich - Dargestellt nach archivalischen Materialien", *Mitteilungen der Österreichischen Gesellschaft für Geschichte der Naturwissenschaften* **2** (1982), 43.

But already he had done important work on thermal equilibrium which helped generalize Maxwell's distribution law. Indeed, there is part of a letter by James Clerk Maxwell to Loschmidt from this period which runs: "I am very pleased over the outstanding work of your student; in England experimental physics is much neglected. Sir William Thomson has done the most in this connection, but you [in Austria] are ahead of us with your good example."[2] But while praise was fine, Boltzmann lusted after further travel. He wanted to know what other physicists were doing first hand.

In 1870 he attended lectures by Bunsen and Königsberger in Heidelberg, and in the same year went to Berlin only to scurry back to Vienna with the outbreak of the Franco-Prussian War, but Boltzmann was back in Berlin the next year attending lectures, visiting laboratories, and working on dielectricity more or less under the direction of Kirchhhoff and Helmholtz. Like many young scientists, Boltzmann was very conscious of gaining priority in submitting manuscripts for publication, and in a letter to Stefan[3] from Heidelberg, dated June 26, 1870, he was sending the latter a short notice about "Kirchhoff's problem" and hoping that it would be published by the Austrian Academy of Sciences "as quickly as possible". It would be followed later by a more extended account. And in the next year, the great Clausius himself would inadvertently fail to acknowledge that one of Boltzmann's papers had anticipated his own recent research. Boltzmann complained, and Clausius graciously acknowledged his oversight while still thinking that some of his own formulations were actually better.

But if there were still some immature edges to Boltzmann's behavior, and priority conflicts can reflect that, we must not let them obscure the truly important work he was undertaking. Just as the young mathematician Kepler used Tycho Brahé's calculations about Mars to discover that the actual orbit of that planet around the sun was elliptical, so Boltzmann, who was also very skilled at mathematics, had already by his early twenties used and gone beyond the work of Loschmidt, Clausius, and to some extent even Maxwell and thus made substantial steps toward giving an atomic and mechanical foundation to gas theory and in a somewhat more controversial sense to thermodynamics by means of his statistical interpretation of the second law. He was indeed a man of high promise. Fortunate are those whom training and circumstances favor so early in life.

Boltzmann's first appointment in Graz lasted only four years, minus his various trips to Germany. He returned to Vienna in 1873 as Professor of Mathematics. But he had left a dear friend behind. She was interested in mathematics but the University of Graz seems to have had difficulty making up its mind whether to allow her to attend lectures. Nevertheless, one source claims that she was the first female student there.[4] Boltzmann sympathized

---

[2] This fragment is included in Gustav Jäger, "Ludwig Boltzmann", *Neue Österreichische Biographie,* Band **II**, Amalthea Verlag: Wien-München-Zürich, 1925, [Kraus Reprint: Nendeln, Liechtenstein, 1970], pp. 152.

[3] The letter is located in the *Staatlicher Preussische Kulturbesitz* in Berlin.

[4] Höflechner, op. cit., p. 46.

with her plight and they began a correspondence which lasted for several years.[5] She was well-educated, of noble blood, and came from a large family but was an orphan. By 1876 Boltzmann wanted to return to Graz again and marry Henriette E. von Aigentler, but Ernst Mach also wanted to be reappointed to a university position in that attractive city, partly too because his own wife was also a Graz girl, and likewise an orphan. So the battle was on. Both Mach and Boltzmann had advocates among the university faculty, and both had published a great deal, though Boltzmann's work was closer to the main trends in physics and he had won a considerable reputation at home and abroad.

One of the more interesting and humorous parts of the fight was that the Graz mayor's family was in a position to transmit battle reports to both sides. Henriette was a protegé of the mayor and especially his wife and lived in their home, either in the *Paradeishof* as it was called in the city of Graz or in Stainz to the south. Meanwhile, a son of the mayor, Wilhelm Kienzl, the future opera composer, studied acoustics with Mach at the University of Prague. The Kienzl family was very musically oriented and both Mach and Boltzmann as music-lovers themselves were on very good terms with them.

On May 10, 1876, Henriette wrote to Ludwig that "Frau Kienzel tells me that she has heard from Professor Schönbach that [August] Toepler [the current physicist in Graz] has already signed a contract [to teach in Dresden]. The rumor has also been spread that Professor Mach from Prague will come here."[6]

Fifteen days later, Ludwig wrote to Henriette "Frau Kienzel told me [on her trip to Vienna] that according to a letter of Gulians [Wilhelm Kienzl] that Mach very much wants to go from Prague to Graz . . . ."[7]

But Henriette on June 11, 1876 wrote to Ludwig "that she [Frau Kienzl] has spoken with Mach and told him that you would also like very much to accept an appointment in Graz. Mach explained that although he had placed uncommonly much hope on coming to Graz, that if he had to make a proposal he would propose you as first choice."[8] And so it turned out. The very next month the Ministry decided to appoint Boltzmann to the Professorship in Experimental Physics. Did Frau Kienzl, the mayor's wife play the central role? Or was it Mach's reluctance to press the matter? Well, the Graz Faculty and Education Minister Stremayr, also played a part as did Kaiser Franz Josef who personally approved and signed all university appointments in Austria. As for the main participants. Mach stayed in Prague for eighteen more years, leaving in 1895, and Boltzmann married his

---

[5] According to Dr. Emilie Beckh-Widmannstetter (the sister of Ludwig Flamm who was the husband of Boltzmann's youngest daughter) and who collected the early correspondence between Boltzmann and his future wife, the first letter was sent on December 13, 1872 and the last one on July 12, 1876. See Dieter Flamm, "Aus dem Leben Ludwig Boltzmanns" in *Ludwig Boltzmann Gesamtausgabe*, Band **VIII**, edited by Roman Sexl & John Blackmore, Graz/Braunschweig, 1982, pp. 31-56.

[6] Ibid. p. 47.

[7] Ibid. p. 49.

[8] Ibid. P. 50.

sweetheart, had five children and remained in Graz for thirteen years until 1890, which overall were probably the happiest of his life.

But strange as it may seem, if one identifies Boltzmann's basic work with his contributions toward the Maxwell-Boltzmann law (1868), equilibrium and transport processes (1872), dielectrics (1872), explanation of the "reversibility paradox" (1877) and his famous relation between entropy and probability (1877), then one might conclude that the 33 year old scientist was already passing his peak in Graz, and the rest of his life was primarily one of meeting and overcoming physical and philosophical challenges to the above work and elucidating ingenious refinements which have continued to thwart most criticisms even up to the present day. But actually he would still make a number of important contributions to physics not even mentioning his valiant defence of the atomic theory, and he had not even begun to spread his wings as a philosopher since the first article with significant philosophical content was only published in 1886. He still had a big future in front of him, and this book would help to describe by means of letters and notes what from a philosophical point of view could be regarded as the climax of that development.

It was in Graz that Boltzmann first became recognized as a brilliant lecturer, one who fascinated a wide range of students and colleagues. Nor should one overlook that all but one of the many books he published under his own name during the 1890's on gas theory, mechanics, and Maxwell's theory of electricity and light[9] were based on or taken directly from these and later lectures. Among his students one would find the future Nobel Prize winners Svante Arrhenius and Walther Nernst as well as many others who would make a mark in their discipline.

But dark clouds began to come from the personal side during the late 1880's. His mother died in 1885, his eldest son in 1889, and his sister in 1890. Nor can one overlook the strange Berlin interlude (1888) where Boltzmann first seemed to accept a position in the Prussian capital, then seemed to turn it down, then tried to reconsider, and finally most reluctantly let the professorship slip through his hands.[10] It was not just a matter of indecision but as he called it a case of "neurasthenia". It would be a prelude to many similar attacks in the future. His emotions had never been fully stable, but from the mid-1890's the peaks and canyons would often be extreme.

Though his fellow professors and the University of Graz had rewarded him with the highest honor and responsiblity they could offer, Rectorship for the year 1887-1888, the death of his eldest son from appendicitus appar-

---

[9] Ludwig Boltzmann, *Vorlesungen über Maxwells Theorie der Ekekricität und des Lichtes,* 2 Volumes, J.A. Barth: Leipzig, 1891 & 1893; *Vorlesungen über Gastheorie,* 2 Volumes, J.A. Barth: Leipzig: Leipzig, 1896 & 1898; *Vorlesungen über die Principe der Mechanik,* 2 Volumes, 1897 & 1904. [A third volumes would appear in 1920 based on lecture notes taken by Hugo Buchholz.]

[10] See Herbert Hörz & Andreas Laass, *Ludwig Boltzmanns Wege nach Berlin,* Akademie-Verlag: Berlin, 1989.

ently struck him heavily and he seems to have felt that he had to leave for somewhere. Many universities would have been happy to have had him, but he chose the University of Munich in Bavaria. And there he spent four fine years from 1890 to 1894. Among his students he had in addition to Bavarians, Americans and even Japanese. Below is part of a letter from Hantaro Nagaoka to Aikitsu Tanakadate dated April 23, 1894:

> Professor Boltzmann is a man of bushy beard as you know. Students are much impressed by his features. But his lectures are surprisingly clear in contrast with those of Professor Helmholtz. He appears to have an excellent brain, for he explained Hamiltonian functions or six-fold integrals without seeing any notes.
>
> The number of students to the Gas Theory is seven, four to Hamilton's Principle, two to the *Oberseminar* and four to the *Unterseminar*. It is very fortunate for me that there is a small audience to these lectures. Professor Boltzmann is gentle and honest, and has a personality to be loved by everyone rather in contrast with his features.
>
> A laboratory building is under construction; it is about half the size of that in Berlin, but appears to shake less easily. Professor Boltzmann guides only one student in experimental physics . . . .[11]

This letter suggests that one reason that Boltzmann may have enjoyed his stay in Munich was that he seems to have taught small classes and not been overworked. He also found a group of congenial colleagues who met during the week in the *Hofbräuhaus*. They included the chemist Bayer, the mathematician Dyk, the physicists Sohnke & Lommel, the mathematician Pringsheim, and the astronomer Seeliger. Munich was also more centrally located for most conferences, and closer to England and France.

Doubts have been raised about whether he was really a materialist in an ontological sense, but concerning food and good times there was no doubt. Though he rose early and worked hard, there was a strong "materialistic" element. He loved to eat and drink. The heavy man become corpulent. He was also given to uttering strange combinations of words such as "immensely small" and "vastly trivial".[12] On his good days, he was a born conversationalist, who could string jokes together like firecrackers. But on the other side, he was also an "idealist" in the sense of being a man of high ideals and a strong believer in scientific progress. He loved his Schiller and Goethe and especially Beethoven. He was an outstanding pianist who had taken lessons

---

[11] My thanks to Setsuko Tanaka for translating this letter from Japanese and making it available for this book.

[12] Karl Przibram, "Erinnerungen an Boltzmanns Vorlesungen", *The Boltzmann Equation*, edited by E.G.D Cohen & W. Thirring, Springer-Verlag: Wien/New York, 1973, p. 642.

from Brückner, and later in Vienna the family held reserved seats for what was then called the *Hofoper*.

During his time in the Bavarian capital, he corresponded with Arrhenius, F. Exner, Fitzgerald, Graetz, Groth, Helmholtz, Lord Kelvin, Kienzl (the composer), Lilienthal, Lorentz, Loschmidt, Mach, Ostwald, Rayleigh, Runge, Schwarzschild, Sommerfeld, C.L. Weber, and many others. Though unfortunately, like many great scientists including Einstein in his younger days, he could be careless about keeping letters, such that only a relative handful sent *to* Boltzmann appear to be extant. Those which exist, apart from carbon copies not sent, are almost entirely *by* Boltzmann and located in scattered libraries and archives. Boltzmann's descendents do have a reasonable amount of family correspondence, but the descendants have become so numerous (See the family tree in this book) that many unsuspected documents may turn up in the future. Until about fifteen years ago it was widely believed that the entire Boltzmann Nachlass had been lost during World War II. But as has become increasingly evident this is far from being the case.

In this book we have included all the known correspondence by Boltzmann from 1900 to 1906 seen by this editor plus virtually all of the extant letters to Boltzmann from this period plus many letters about him written by third parties. We have included a few of the letters between government and academic officials about Boltzmann but only what seem to be the more important ones.

Many theorists who understandably focus on natural laws and relations between them sometimes seem to think that correspondence and the biography of a scientist while often interesting have nothing to do with his scientific or philosophical work and hence can be safely ignored. But in what most historians call "the real world" we normally learn about a person's hopes, plans, expectations, and basically how and why he did his scientific work by studying his life in detail. We learn basic historical causes, his challenges and how he responded to them. Theorists often like to think that it is they who understand "the real world" or at least its physical essentials, but with all due respect, most people would probably agree on reflection that history is reality and that most historians tend to know more about it than most theorists, intelligent and valuable as they often are in their largely abstract, idealized way.

But Boltzmann, a theorist par excellence, and most research historians have shared at least three beliefs, namely, first, that people have obtained "sway over the laws of nature",[13] second, that we should be cautious about accepting artificial meaning, and third, that there is no need to accept the existence of what philosophers call "universals". Boltzmann knew that laws are abstractions; they do not force anything to happen. Also, he took his opponents and what they were driving at seriously. He realized that valid

[13] Ludwig Boltzmann, "Acceptance Speech on Becoming an Honorary Doctor of Law", *Clark University Decennial Celebration*, Worcester, Mass., 1899, p. 20.

refutation requires understanding what one's opponent means, intends, or is driving at, not just formally deducing a strawman from "objective" or "convention-based" meaning. As for universals and their abstract "quiddity", Boltzmann consistently denied that they were real. Only particulars existed. Though Boltzmann thought largely in mathematical terms when doing science, he was also capable of quite competent, detailed-focused, historical description expressed in a highly amusing or artistic way, as his article "On the Trip of a German Professor into El Dorado" demonstrates (and which is included in Chapter seven of this book with a new translation).

But biography or writing about *other* people, especially someone as multi-faceted and controversial as Boltzmann, is not just a matter of historical description; it often includes the interpretations of biographers, which admittedly not everyone accepts, nor in many cases should they, though people perhaps ought to agree at least with some if the work is well-researched and thorough and the intentions and manner of proceeeding of the writer are reasonable. Nevertheless, because a few periods of Boltzmann's life are still a virtual blank and there is still relevant material which has not yet been made accessible to this editor, he would prefer at this point not to write a full-scale biography but to present *a documentary history* about the period for which we have the most information. The purpose of this book is not to persuade but to inform, to add new material, especially biographical and philosophical, which will make a better, richer understanding of Boltzmann more likely in the future.

Our purpose in this introduction, on the other hand, is to refresh the reader's mind with biographical information about Boltzmann's life *before* 1900, give an overview of his major contributions to physics, attempt an explanation of how and why he became interested in philosophy, and acquaint readers with a variety of philosophical positions which have been attributed to Boltzmann by different scholars.

Returning to Boltzmann, unlike his older colleagues Stefan and Loschmidt, he loved to attend scientific conferences which he regularly did, but partly because Boltzmann found that scientists had greater interest in his contributions to the kinetic theory of gases in general and his statistical interpretation of the second law of thermodynamics in particular in England than in Germany, he seems to have enjoyed most his visits to conferences in that country. Martin Klein writes about their concern with some of the problems connected with Boltzmann's theory:

> There was enough interest in these difficulies in England so that the British Association for the Advancement of Science appointed a committee to report on 'The Present State of Our Knowledge of Thermodynamics'. The committee consisted of Joseph Larmor and George Bryan; Bryan's lengthy report was submitted in two parts, appearing in 1891 and 1894. The first report covered the various attempts to relate the

> second law to the principles of dynamics, which included the early work of Boltzmann and Clausius as well as the later statistical theories. Bryan also analysed the monocycle analogies introduced by Helmholtz in 1884 and discussed and generalized at length by Boltzmann. Bryan's second report, on "The Laws of Distribution of Energy and Their Limitations" was presented to the British Association at its annual meeting which, in 1894, was held at Oxford. Boltzmann himself was there, presenting two short papers and taking part in the discussion which followed Bryan's report. He must have enjoyed himself immensely. He had an audience that was really interested in the kinetic theory of gases and generally sympathetic to his point of view. Many of it s members were well prepared to argue technical details as well as general principles, and generally to carry on a vigorous debate. No wonder Boltzmann referred to it as 'The unforgettable meeting of the British Association at Oxford.'[14]

Furthermore, the meeting helped spur a vast contribution of letters to *Nature* magazine about his theories during 1894 and 1895. But many criticisms were sharp, and as soon as they eased off, a very talented mathematician by the name of Ernst Zermelo, Planck's assistant, severely attacked Boltzmann's famous *H - Theorem,* insisting that "physics would have to choose between the second law of thermodynamics and the mechanical interpretation of nature."[15] Planck himself soon entered the lists but unlike Zermelo began to see increased merit in some of Boltzmann's ideas, and when forced to use some of them in his attempt to understand blackbody radiation which resulted in the earliest version of quantum theory in 1900, soon became a strong supporter of Boltzmann. But while Boltzmann had been involved in controversies virtually all of his adult life, even with his good friend Josef Loschmidt in 1877, somehow as he grew older they seemed to increase in frequency and ferocity.

Partly encouraged by his wife and the realization that Bavaria offered no pensions for retired professors in those days (The famous physicist Ohm had died in poverty there), when Stefan died in Vienna in 1893 Boltzmann in on and off fashion announced that he was a candidate for his former teacher's position. But the faculty was divided, and the Education Ministry thought Boltzmann's financial demands were too high.[16] Some members

---

[14] Martin J. Klein, "The Development of Boltzmann's Statistical Ideas", *The Boltzmann Equation,* op. cit., pp. 86-87.
[15] Ibid., p. 90.
[16] Höflechner, op. cit., 49-52.

supported Ernst Mach,[17] but once again when Mach saw that Boltzmann stood a strong chance of being in the race he discouraged his supporters and Austria's most famous theoretical physicist was finally persuaded to return once more to Vienna.

But Boltzmann was not happy in the Ring City. He thought there was an oversupply of *Privatdozenten* who wanted academic posts and a shortage of good undergraduate students. (Also, while many students chose Boltzmann as second advisor for their doctoral examinations, it seems that unlike Graz and even Munich few or none selected him as first advisor, at least for this period, i.e. 1894-1900. See Appendix A for more details.) Furthermore, a year after Boltzmann's return to the University of Vienna, Ernst Mach was appointed to Franz Brentano's old chair in philosophy in 1895 which was given a new title as "The History and Philosophy of the Inductive Sciences", and he began to attract or distract some of Boltzmann's students and assistants such as Anton Lampa as well as interest young philosophers like Wilhelm Jerusalem and Heinrich Gomperz. Ernst Mach's support of "presuppositionless science" as if there were no need for the atomic "hypothesis" and very little for other theories, brought criticism of Boltzmann's ideas from the pages of journals directly into the university itself in the form of Mach's widely appreciated lectures. To many, it seemed as if Boltzmann represented the old "mechanistic science" while Mach's call for "modern physics " was the harbinger of the future. With the advantage of hindsight the situation may look to us like the reverse, but Boltzmann allowed a negative interpretation of the Vienna situation to depress him deeply, and once again he dreamed of packing his bags and looking for a better position. But the Ministry of Education was not about to let him leave the Fatherland again without a fight, though it didn't mind him teaching summer school or special lectures abroad such as at Clark University in Massachusetts (1899). Section chief, von Hartel, who would be appointed Minister a few years later, argued in a letter dated November 4, 1896 that among the theoretical physicists "after the death of Helmholtz Boltzmann incontestably comes first".[18] They would struggle hard to keep him.

Nevertheless, under the reasonable impression of having bested Ostwald and his cohorts in a famous debate between "atomism" and "energeticism" at Lübeck in September 1895 and also being on good personal terms with this influential co-founder of physical chemistry, Boltzmann in between articles replying to Zermelo, defending atomism, bashing energeticism again, and criticizing Mach, and while completing his two volume book on gas theory, began during the late 1890's to inquire if it might be possible to find a position for himself in Leipzig, far away from dull students and dangerous philosophers. Ostwald was cooperative, and after Mach rejected his offer to come to Leipzig, he did his best to bring Boltzmann to Saxony. It is at this

---

[17] See Gustav Tschermak to Ernst Mach, Vienna, March 2 & 20, 1893.

[18] This was also the opinion of of the physicist Woldemar Voigt. See his article, "Ludwig Boltzmann", *Physikalische Zeitschrift*, 7 (October 1, 1906), 650.

point that the correspondence section of Chapter Three in our documentary history begins, namely, with the last month of 1899.

## Physics

The development of chemistry into a generally or increasingly reliable science during the 19th century under the influence of Lavoisier's theory of elements, Dalton's theory of atoms, and Avogadro's law (when it was finally understood and accepted) was naturally supported by many chemists and some physicists, but the emergence of thermodynamics during the midddle of the 19th century as based on the first and second laws which could be supported by macroscopic observation appealed to other scientists who were reluctant to make hypotheses about entities which were too small to be observed.

Furthermore, some physicists under the influence of German or Positivist philosophy thought that all inquiry or hypotheses beyond the appearances seemed "metaphysical" or "unscientific", hence for a number of them as well, thermodynamics appeared to offer a wonderful opportunity to reconstruct science on the basis of direct observation which would either refute the "mechanistic" science of Newton and the "atomistic hypotheses" of Dalton or subordinate them to thermodynamics. Indeed, at least in German the first two laws of thermodynamics were not simply called laws (*Gesetze*) but *Hauptsätze* or fundamental propositions as if as the ultimate foundation of all physical science they were more important than mere laws.

But Rudolf Clausius, the very discoverer of the second *Hauptsatz*, began investigating the possibility that the atomic theory might be helpful in understanding why entropy always increases. This of course was anathema to many who preferred description of conscious experience to "speculation" about what they regarded as non-existent physical causes. People experienced heat and it was relatively easy to understand how heat in steam engines could produce work, but both the caloric theory as if heat were an invisible liquid and the kinetic definition as if heat were merely the movement of tiny particles seemed abstract and removed from experience. Also, some experimentalists were reluctant to accept that mind and matter really were different and that heat in a mental or experiential sense was quite different from heat in a physical or trans-conscious sense. If heat consisted of the movement of particles why couldn't we see or detect them?

When Ludwig Boltzmann's much older friend, Josef Loschmidt, calculated the size of gas molecules in 1865, it clearly seemed to the younger scientist that atomism was here to stay and that one as gifted in mathematics as he himself could make a significant contribution to physics by relating thermodynamics more closely to the atomic theory. Boltzmann was also strongly influenced by the work of Rudolf Clausius who had thought of what we now call the second law as early as 1850 but only fully worked out his concept of entropy in 1865.

These two remarkable developments spurred young Boltzmann to be ambitious. According to Martin Klein's translation, Boltzmann announced in an 1866 paper: "It is the purpose of this article to give a purely analytical, completely general proof of the second law of thermodynamics."[19] This would remain Boltzmann's primary concern in physics for the rest of his life,[19a] though as Klein mentions his attitude toward the entropy law would undergo a measure of evolution.

At first glance the entropy law gave a very simple relation such that "if one took the ratio of the heat content of a system and its absolute temperature, this ratio would always increase in any process taking place in a closed system". But of course in reality there are no closed systems except the universe as a whole.

It seems that Boltzmann's attempt to establish a direct connection between the entropy law and the mechanical principle of least action ran into difficulties until Boltzmann studied Maxwell's statistical approach which was published in 1868. Below is Martin Klein's account of the well-known story:

> When Boltzmann made his original attempt to understand the second law in mechanical terms, he had apparently not yet read James Clerk Maxwell's early papers on the theory of gases, but his ignorance of Maxwell's work did not last long. Boltzmann has described how, during his student years and before he had learned any English, his teacher, Josef Stefan, handed him Maxwell's papers on electromagnetism and an English grammar book. And so by the time Maxwell's articles "On the Dynamic Theory of Gases" appeared, Boltzmann was prepared to appreciate them. His study of Maxwell led him directly to a new approach to the study of gases, one which the molecular distribution function rather than the complete set of molecular variables played a central part."[20]

Boltzmann's first subsequent paper which was also published that year using a statistical approach "was a new exponential formula for molecular distribution which is now known as the 'Boltzmann factor' and is basic to all modern calculations in statistical mechanics."[21] Maxwell had suggested that "most of the observable properties of a gas could be calculated if one knew,

---

[19] Klein, op. cit., 57.

[19a] Boltzmann's scientific articles have been collected into *Wissenschaftliche Abhandlungen*, 3 Volumes, edited by F. Hasenöhrl, J.A. Barth: Leipzig, 1908-1909 [reprinted by Chelsea Pub. Co., 1968].

[20] Ibid., p. 61.

[21] Stephen Brush, "Ludwig Boltzmann", *Dictionary of Scientific Biography*, Vol. **II**, edited by C.C. Gillispie, Charles Scribner's Sons: New York, 1970, p. 261. The major articles by Maxwell and Boltzmann referred to by Brush can be found in his book *Kinetic Theory*, Volumes **1** & **2**, Pergamon Press: Oxford, 1965 & 1966.

instead of the positions and velocities of all the molecules at any given time, only the average number of molecules having various positions and velocities."[22] This and subsequent papers helped to clarify what has become known as the Maxwell-Boltzmann distribution law.

It would be hard to overemphasize the importance of this 1868 paper. Most of the seeds of his later work can be found in it. Boltzmann employed two derivations one based on molecular collisions and one independent of collisions. Professor Brush writes about the latter: "He simply assumed that there is a fixed total amount of energy to be distributed among a finite number of molecules, in such a way that all combinations of energies are equally probable."[23]

This meant that while the atomic theory was basic to Boltzmann's work that not only had he found a way to determine the characteristics of a gas without having specific knowledge of individual molecules, not to mention atoms, but that the less those characteristics were specified the more freedom he might have in developing his theories about ensembles, equilibrium states, average conditions, disorder, and the like. He identified molecules with material points, circles, and spheres, especially the latter when theorizing about collisions. Such idealization was forced both by ignorance of actual characteristics and to attain simplicity. He was an atomist of a sort, but was somewhat like Daddy Longlegs in the story and movie of that name who helped his adopted daughter with money and education but had little or no interest in actually seeing her.

In fact, as the years passed he would not only talk more and more about idealized models or *Bilder* as if they were the atoms or molecules themselves but may have gradually or fitfully even begun to believe it, but the manner of doing so would be by mentally trivializing the distinction between molecules and models or pictures of them, as if the difference "were merely linguistic". But one must not assume that Boltzmann was ever fully consistent in employing this kind of reductionism or in imagining that it constituted a final answer. But he did increasingly nurture the idea, especially after he became acquainted with the theories of Heinrich Hertz during the late 1880's and early 1890's. A possible motivation behind his approach was to try to win favor for his kinetic views even among people who rejected the reality of atoms and molecules. But let us go back in time and return to our main story about the kinetic theory of gases.

In addition to "finding the correct distribution laws by assuming that the gas is in an equilibrium state"[24] Maxwell and Boltzmann wanted to take the next step as well, namely, "to show that the kinetic theory should also be able to show that a gas will actually tend toward an equilibrium state if it is not already there".[25] This might seem like forcing matters, but in 1872, only four years later, he published a paper indicating just that together with a

[22] Ibid.
[23] Ibid., p. 262.
[24] Ibid.
[25] Ibid.

description of transport processes in gases as if the time rate of change of particle density were equal to the rate of production minus the rate of leakage and the rate of absorbtion. The transport equation is now frequently called the *Boltzmann Equation.*

Professor Brush has concluded largely from the 1868 and 1872 papers that "It was Boltzmann's achievement to show in detail how thermodynamic entropy is related to the statistical distribution of molecular configurations and how increasing entropy corresponds to spreading randomness on the molecular level."[26]

The next step was to generalize the approach such that it applied not just to equilibrium states of gases but to non-equilibrium states as well. (A system is said to be in thermal equilibrium if the temperature is uniform throughout the system and is the same as that of the surroundings.) This was accomplished through the introduction of Boltzmann's notorious *H - Theorem,* (though it was generally called "Boltzmann's minimum theorem" in the 19th century and another symbol was used in place of *H* before 1890), which provided "an extension of the definition of entropy to nonequilibrium states not covered by the thermodynamic definition".[27] The minimum theorem held that states far from thermal equilibrium were so improbable, though not impossible, that it was always likely that the next significant fluctuation would be in the direction of equilibrium.

But the distribution law and its extension to nonequalibrium states as well as the effort to maintain that gases tend toward equilibrium made it more important than ever to try to resolve what appeared to be a manifest contradiction between the traditional reversibility of all purely mechanical motions at least in principle and the irreversible increase of entropy. For many experimentalists it was precisely this contradiction which seemed to prove that empirical "presuppositionless" thermodynamics was sound and Newtonian mechanics had only limited reliable application. They couldn't both be true, and many physicists were sure that the two *Hauptsätze,* minus atomistic speculation were part of the wave of the future.

The first major scientist to publically raise the objection was apparently Lord Kelvin (1874). But it was the objection as expressed by Boltzmann's friend Josef Loschmidt which he chose to answer in 1877. Loschmidt pointed out an apparent contradiction between "one of the basic premises of Boltzmann's derivation - the reversibility of individual collisions - and the irreversibility predicted by the theorem itself for a system of many molecules."[28] In fact, however, this objection was not Loschmidt's main point, but while Boltzmann understood what Loschmidt was driving at, he decided to strengthen a secondary feature of Loschmidt's argument, made it more general, and tried to put it in as compelling a manner as possible, aware that

---

[26] Ibid.
[27] Ibid., p. 263.
[28] Ibid.

refutation of this objection in its strongest form should discourage subsequent criticisms along the same line.[29]

According to Stephen Brush: "Boltzmann's initial response (1877) to the reversibility paradox was the suggestion that the irreversibility of processes in the real world is not a consequence of the equations of motion and the form of the intermolecular force law but, rather, seems to be a result of the initial conditions. For some unusual initial conditions the system might in fact decrease its entropy (increase the value of $H$) as time progresses; such initial conditions could be constructed simply by reversing all the velocities of the molecules in an equilibrium state known to have evolved from a non-equilibrium state. But, Boltzmann asserts, there are infinitely more initial states that evolve with increasing entropy, simply because the great majority of all possible states are equilibrium states."[30]

But one must not imagine that most other physicists accepted Boltzmann's explanation, especially in Germany. Most were experimental physicists who could follow observational work in thermodynamics but tended to be less interested in or knowledgeable about the highly mathematical work of Maxwell and Boltzmann. Experimental physics was still king, and while it was useful and slowly becoming fashionable to have a theoretical physicist around to help decode mathematical hieroglyphics for "real" physicists, they rarely attained top positions. Professor Klein also points out that "a more profound reason why Boltzmann's ideas were not widely understood was the way in which he changed his point of view without informing his readers."[31] Furthermore, many "theorists" actually did experimental work, perhaps in the hope that it would benefit their careers. Indeed, Boltzmann's own position in Graz was in experimental physics and he carried out a substantial amount of experimental research, particularly concerning electricity and electromagnetism. The shoe was on the other foot in those days, almost everywhere.

While Boltzmann was still primarily concerned with gas theory he could hardly avoid recognizing that the use of statistics and thinking in probability terms could be applied more generally to thermodynamics and perhaps even to other domains of physics as well. Thus was born especially with the help of J. Willard Gibbs what we now call "statistical mechanics" (a term which Gibbs introduced). On the other hand, Martin J. Klein insists

[29] Edward E. Daub in his article "Probability and Thermodynamics: The Reduction of the Second Law", *Isis*, **60** (1969) 319 writes the following about Loschmidt: "Boltzmann was referring to the first two parts of a series of articles by Josef Loschmidt, 'Über den Zustand des Wärmegleichgewichtes eines System von Korpern', *Die Sitzungsberichte der Akademie der Wissenschaften*, Wien, **3** (1876) 128-142 & 366-372. Loschmidt was not opposed to purely mechanical proofs of the second law; in fact, he favored them. His primary target was the general thermodynamic principle that no work can be obtained from the kinetic energy of particles at equilibrium, a principle which he called a 'terroristische Nimbus. that would make death the fate of the universe (cf. pp. 133-135). His [main] objection to Maxwell's and Boltzmann's treatment was their contention that the average temperature of a gas at equilibrium would be uniform throughout. . . ."

[30] Brush, op. cit., pp. 263-264.

[31] Klein, op. cit., p. 83.

that "it was Boltzmann, and not Maxwell or Gibbs, who worked out precisely *how* the second law was related to probability, creating the subject of statistical mechanics."[32] While many textbooks in the field simply start with the Maxwell-Boltzmann distribution function "as the basic postulate for calculating all the equilibrium properties of a system",[33] it seems that Boltzmann himself preferred to begin from a more general foundation.

Apparently, Boltzmann's answers to the recurrence paradox helped suggest to him a profitable path to follow. He mused: "One could even calculate from the relative numbers of the different distributions, their probabilities, which might lead to an interesting method for the calculation of thermal distribution."[34] In the same year (still 1877), Boltzmann put this remark into mathematical terms in a paper which related entropy and probability. Thus was born what is often called the *Boltzmann Principle*, which has been put on his gravestone in Vienna,

$$S = k \log W,$$

where "$W$ is the number of possible molecular configurations corresponding to a given macroscopic state of the system."

Using once again the clear description by Stephen Brush: "The new formula for entropy - from which formulas for all other thermodynamic quantities could be deduced - was based on the assumption of equal a priori probability of all microstates of the system (that is, all microstates that have the same total energy)."[35]

Nor did Boltzmann forget his *Doctorvater*. In 1883, while reading a paper on radiant heat as a possible exception to the second law of thermodynamics, he learned that an Italian physicist had worked on radiation pressure. Boltzmann studied the matter further and worked out "a theoretical derivation based on the second law and Maxwell's electromagnetic theory of the fourth-power law previously found experimentally by Stefan". Not only was the *Stefan-Boltzmann Law* further proof that "pencil pushers" (now called "computer punchers") could add something important to the work of experimentalists, but it also showed "a possible connection between thermodynamics and electromagnetism", one which would be exploited in the later quantum theory.

But increasingly Boltzmann began to feel uneasy about his assumption of "equal a priori probability of all microstates of the system", and both he and Maxwell began looking for support. After Maxwell's death, Boltzmann thought out in papers written from 1884 to 1887 what has become known as the "ergodic hypothesis", the assumption "that a single system will eventually pass through all possible microstates." If Boltzmann had already climbed out on a limb by stretching his work from equilibrium to nonequilibrium states

---

32 Ibid., p. 77.
33 Brush, op. cit., p. 264.
34 Ibid.
35 Ibid., p. 265.

and using his controversial minimal or *H -Theorem* to defend it, as many critics would maintain, then putting together and advocating the ergodic hypothesis appeared to some to be cutting off the limb behind him. It was taking an unnecessary risk.

But Professor Brush maintains that many or most critics misunderstood Boltzmann:

> There has been considerable confusion about what Maxwell and Boltzmann really meant by ergodic systems. It appears that they did not have in mind completely deterministic mechanical systems following a single trajectory unaffected by external conditions; the ergodic property was to be attributed to some random element, or at least to collisions with a boundary. In fact . . . he used them not for single systems but for collections of similar systems with the same energy but different initial conditions.[36]

Understandably, controversies have arisen among mathematicians about the possible existence of dynamical systems that pass through all possible configurations. In 1911, Boltzmann's gifted student, Paul Ehrenfest, with the help of his wife attempted to defuse part of the controversy by suggesting that even if the ergodic theory is not literally true that a "quasi-ergodic" system that passes "as close as one likes" to all possible states might still exist. But mathematicians soon found faults in that as well, but in the process seem to have opened up a new field for mathematical inquiry. But physicists have appeared to lose interest, since the demise of one assumption used to support a much more important assumption already supported on other grounds is rarely fatal to the latter or to the theory which apparently rests on it.

Nor was Boltzmann inclined to hold grudges when scientists disagreed with him or even philosophers. As is well-known, he was on good personal relations with both Ostwald and Mach, and even after his relations with his old friend Loschmidt may have cooled because of their debate in 1877 when Boltzmann first responded by calling Loschmidt's allegation "sophistry" among other bitter things, it was Boltzmann some years later who wrote a very considerate obituary after his death and it was Boltzmann who led the campaign to finance a statue of him. It was also Boltzmann who apparently broke the ice in restoring relations with the old scientist during the early 1890's. Here is part of the first of four letters sent by Boltzmann to Loschmidt, dated October 26, 1892:

> Our paths in life have drifted so far apart, that I would like to send you the latest news, and the first

[36] Ibid.

> item to be shared is that I am still alive. Nor am I better off than in dear Austria. And now I will utter my request. I have got to have [*brauche nothwendig*] your little book *Über die Natur des Äthers*, Gerold & Sohn: Wien, 1862 . . .
>
> My bookseller tells me it is out of print and that he can't obtain a copy for me . . . . I don't want to judge your book, I just want to read it, and the only way I can do that it seems is to turn to you for a copy. In the hope that you will not take my request in the wrong spirit, I have the same respect for you as of old . . .[37]

The most memorable attack on Boltzmann's statistical interpretation of the second law of thermodynamics has undoubtedly been "the recurrence objection". The philosopher Friedrich Nietzsche during the 1880's had alleged that finite reality moving in infinite time must eventually repeat itself. He felt sure that stalwart heroes would not mind being reborn and repeating previous actions to the letter. Real stoics are tough. They invite "Eternal Recurrence". But some mathematicians and physicists have thinner skins and have not been so sanguine, particularly since it seems inconsistent with the notion that entropy always increases. The objection appears to have first been put into rigorous form by the brilliant French mathematician, Henri Poincaré in 1890. "According to this theorem, any mechanical system constrained to move in a finite volume with fixed total energy must eventually return to any specified initial configuration."[38]

Unlike Nietzsche's formulation, the phrasing of Poincaré and Ernst Zermelo (1896) was specifically aimed at a mechanical system and hence "mechanistic science", saying in effect that given finite space and energy there would eventually have to be an exact recurrence of mechanical behavior if Newtonian mechanics were true, which would make it incompatible with the second law of thermodynamics. But since increase of entropy had been repeatedly observed and was an empirical fact, then the incompatibility with Newtonian mechanics, would suggest that it was the latter which was false.

Boltzmann's reply came in the form of three articles published in 1896-1897, two directed at Zermelo and one at Poincaré.[39] In the words of

---

[37] This and three other letters by Boltzmann to Loschmidt can be found in H. de Martin, *Josef Loschmidt,* Doctoral Dussertation, University of Vienna, circa 1950.

[38] Brush, op. cit., p. 264.

[39] See in order: Ernst Zermelo, "Über einen Satz der Dynamik und die mechanischen Wärmelehre", *Annalen der Physik,* **57** (1896) 485-494; Ludwig Boltzmann, "Entgegenung auf die wärmetheoretischen Betrachtungen des Hrn. Ernst Zermelo", *Annalen der Physik,* **57** (1896) 773-784; Ernst Zermelo, "Über mechanische Erklärungen irreversibler Vorgänge - Eine Antwort auf Hrn. Boltzmann's Entgegnung", *Annalen der Physik,* **59** (1896) 793-801; Ludwig Boltzmann, "Zu Hrn. Zermelos Abhandlung 'Über die mechanische Erklärung

Stephen Brush: "Boltzmann replied that the recurrence theorem does not contradict the *H - theorem,* but is completely in harmony with it. The equilibrium state is not a single configuration but, rather, a collection of the overwhelming majority of possible configurations, characterized by the Maxwell-Boltzmann distrbution. From the statistical viewpoint, the recurrence of some particular initial state is a fluctuation that is almost certain to occur if one waits long enough; the point is that the probability of such a fluctuation is so small that one would have to wait an immensely long time before observing a recurrence of the initial state. Thus the mechanical viewpoint does not lead to any consequences that are actually in disagreement with experience."[40]

Boltzmann was brutal: "All the objections raised against the mechanical world view [with respect to kinetic theory and statistical mechanics] are therefore aimless and are based on erroneous ideas. But anyone who is incapable of overcoming the difficulties involved in obtaining a clear understanding of the theorems of gas theory ought indeed to take Mr. Zermelo's advice and decide to give it up completely."[41]

But as with ergodic theory, Boltzmann had trouble leaving well enough alone. He couldn't resist adding an ingenious corolary, namely, that were the universe as a whole in thermal equilibrium, which from our perspective is at least imaginable, then there might be no distinction between forward and backward directions in time. There could be small parts of the universe such as a particular galaxy which contain ordered states on some planets where life may have developed. Where these unusual fluctuations occur they may identify time with an opposite direction. "In other words, the concept of 'direction of time' is statistical or even subjective, and is determined by the direction in which entropy happens to be increasing. Thus, the statement that 'Entropy increases with time' is a tautology."[42] One may also view this corolary as an attempt to support a traditional assumption in Newtonian mechanics: If all time is reversible and all entropy, at least in some regions of the universe, then perhaps all motion may somehow be reversible as well at least in principle, regardless of how improbable in fact.

Boltzmann gave people who enjoy cosmic speculation a lot to chew on, but most people would probably respond that his hypothesis about time is not unambiguous enough to clearly understand, and even if it were, there would almost certainly not be enough factual data available to us to be able to determine either the truth or the falsehood of this corolary, which seems to have been the most radical version of his *H - theorem.*

But not all criticisms of Boltzmann's theories and use of mathematics were public; many were carried on by correspondence. The recent publi-

---

irreversibler Vorgänge'", *Annalen der Physik,* **60** (1897) 392-398; Ludwig Boltzmann, "Über einen mechanischen Satz Poincaré's", *Wiener Ber.* **106** (1897) 12-20.

[40] Brush, op. cit., p. 264.

[41] This is Professor Klein's translation of Boltzmann's concluding remarks from Boltzmann's 1896 reply to Zermelo. See Martin J. Klein, op. cit., p. 92.

[42] Brush, op. cit., p. 264.

cation of Boltzmann's correspondence with Hendrik Antoon Lorentz indicates disagreements which are too technical to be discussed here, as are Boltzmann's unpublished letters to Rayleigh and Ramsay, but at least we can include parts of two letters, one by Boltzmann to Lord Kelvin and one by the latter. Boltzmann's letter was written in English, seems to have been partly corrected, and is dated January 6, 1893.

> I believe that only a misunderstanding of the true meaning of my theorem is the cause of all objections made to it. I do not pretend, to have rigorously demonstrated, that in all real bodies of equal temperature the average vis viva of each molecule is the same. I have only demonstrated a theorem of pure abstract dynamics, relating to mechanical systems, in the analogie of which to material bodies, has a certain degree of probability. This theorem does not maintain, that in every st[e]ady motion the average vis viva of each material point is the same. A stare with his trabant [a star with its satellite] (assumed to be alone in the world) or two perfect elastic spheres (moving in a straight line between t[w]o normal perfectly elastic screens, the velocities of which are in inverse proportion to its masses) A and C is not equal in every case. But my theorem postulates, that the mean kinetic energie is the same, if the masses are in mutual action with a perfect gas, and for the truth of that affirmation your test case [in May issue of *Phil. Mag.* 1892 or Lord Kelvin's non-extant letter of December 20, 1892] indeed is of a quite decisive kind . . . . [43]

We don't have Lord Kelvin's reply, but we do have a later letter, and here are two brief passages dated June 21, 1895:

> Lord Rayleigh has today put into my hands your letter to him of the 15th, in order that I might answer it for him . . . .
>
> I am sadly perplexed and disturbed by the great question of distribution of energy, originating in Clausius and taken up in so penetrating a manner by yourself and Maxwell. Whenever other occupations allow me to return to it, but alas! I make absolutely no progress towards comfort or happiness in regard to it.

[43] The letter is located in University Library, Cambridge University, Cambridge, England.

> This is very sad, as on it the whole of Thermodynamics hangs.[44]

But while Boltzmann, like the dogged Dutch boy desperately plugging holes with his fingers in the sea wall, was holding out against a flood of criticism, not only from Germans and Frenchmen but even from many English colleagues and friends, other fronts seemed to be collapsing. Mach in 1872 and Kirchhoff in 1874 argued that physics should be descriptive and not explanatory, and it should describe what could be observed. The rest was "metaphysics". Mach in 1883 declared in his book on the evolution of mechanics that Newton was wrong and that space, time, motion, rest, and place were always relative. And in 1892, Ostwald even published a chemistry book which had no reference to atoms or molecules while developing a theory of "energeticism" as if energy equations could be used to directly describe phenomena.

But it was developments in electromagnetism which seem to have shaken Boltzmann the most. He himself had worked in that field, knew the writings of Maxwell intimately, and became fascinated with the experimental breakthroughs of Heinrich Hertz starting in 1888 as well as with his book on how mechanics should be carried out in the future and his emphasis on the importance of what he called *Bilder* or pictures.

Just as Clausius, Maxwell, and Boltzmann had tried to give an atomistic and mechanical explanation of the second law of thermodynamics so Wilhelm Weber and other scientists had attempted to give a mechanical explanation of electricity, magnetism, and electromagnetism, but around 1870 Helmholtz through his analysis, Maxwell with his equations, and in the late 1880's Hertz's support of Maxwell, work toward clarifying the electromagnetic spectrum, discovery of radio waves and other unanticipated phenomena seemed more informative and reliable than Weber's work on a number of fundamental issues[45] such that even Boltzmann recognized that it might well be the case that a kind of electromagnetic world view might replace both atomism and "mechanistic science". Concerning philosophy, Boltzmann was struck by the apparent fact that Weber had wanted to understand physical reality but had largely failed at least with respect to electromagnetism, while Maxwell who often seemed forced to settle for mere equations as if without serious theory had prospered.

By the mid-1890's Boltzmann began to think of changing his strategy. He was not opposed to the new physics and in fact was contributing important understanding to it, but he wanted to make sure that what was of value from past physics, including many of his own older contributions, would not be lost. Perhaps anticipating his own change in attitude, he had already begun to study the works of his philosophically-minded fellow physicists more carefully so as to find a way of presenting classical views

---

[44] The letter is in the University Library of the University of Stockholm.

[45] See Christa Jungnickel and Russell McCormmach, *The Intellectual Mastery of Nature*, Volume II, University of Chicago Press: Chicago, 1986, pp. 22-24, 74-78, & 82-85,

and recent extensions of them which might be more acceptable to his opponents. But much of the terminology he employed was so unusual that clarification is in order.

As late as Boltzmann's introductory lecture in philosophy in Fall 1903, he was claiming that he had only published one article on philosophy in his life. But while from our perspective that was manifestly untrue, since he had published many articles on what we regard as philosophy of science by that time, the apparent explanation appears to be that Boltzmann considered those articles as contributions to *methodology* of science and not as philosophy at all.

Or to put the matter another way, like many German and Austrian scientists, Boltzmann often identified philosophy with metaphysics, which he thought should be avoided at all costs. Since he was not willing to consider his articles on methodology as metaphysics, he seemed reluctant to consider them philosophical. But while it would be risky to imagine that Boltzmann was consistent in his unusual definitions, let us for the moment follow his own habit of regarding most of his general articles as methodological rather than philosophical. In this manner we can probably tie his changing way of defending his contributions to science better. This methodological emphasis is most conspicuous from about 1895 to 1900. Before that his methodological articles were indeed primarily methodological, and after 1900 he was a philosopher, though still one who was adamantly opposed to "metaphysics" as he had begun to redefine it.

Let us turn now to Boltzmann's new methodological strategy.

## Methodology of Science

Once upon a time there was a young physicist of outstanding mathematical ability who used the atomic theory in order to give a mechanically-based statistical interpretation of the Second Law of Thermodynamics. The young physicist became very successful and famous at a relatively modest age but saw his later years hampered by increasing criticism from physicists, mathematicians, and philosophers. But in spite of being a very emotional person subject to extreme ups and downs, the now ageing and unwell physicist replied to his tormentors with brilliant effectiveness.

Unpopular as defending "old-fashioned mechanistic science" might seem to us and did seem to some of the more "advanced" physicists of the period, he did so in such a skilful manner as to consistently confound his opponents. His interpretation of the kinetic theory of gases as a statistical theory based on atomism and the recognition that probability does not exclude the possibility that entropy may decrease in some local situations surpassed the ability of his critics to clearly refute. They tried, kept trying, and still keep trying, but Boltzmann's *H -theorem* and of course the reality of atoms live on.

Concerning electromagnetism, this was Boltzmann's second major interest, and it was he who translated major work by Maxwell into German, arranged for the publication of part of Kirchhoff's collected papers, studied the work of Helmholtz and Hertz in detail and far from opposing what increasingly seemed incompatible with Newtonian mechanics and his own work in kinetic theory looked ahead to how the older mechanics should be treated given the overwhelming probability that a new "modern physics" was definitely on the way.

Boltzmann responded by arguing that atomism in some form was indispensable and that Newtonian mechanics was basically too practical to disregard. After Newton, it was no longer necessary to teach Aristotelian physics with its emphasis on "natural" and "unnatural" motion and insistence that the latter required constant force, but after Maxwell, Kirchhoff, and Hertz and possibly even more revolutionary developments in the future, as indeed would happen with the theories of Planck and Einstein, it would still be necessary to teach Newtonian or "mechanistic science". Where did its practicality lie?

According to Boltzmann, while theorists almost always attempt to find and use the most accurate data available, those natural laws which they discover which imply other laws or can be deduced from them are idealized, that is, they establish relationships between only a few of the vast number of actual variables which are or could be present in any real situation. In short, they are not literally true under real conditions but are at best approximations, and when the ignored variables are important as they sometimes are, they are not even that.

To be sure, the further physics advances new laws, empirical, statistical, and "deterministic" should be able to handle a larger number of variables, and ideally, the laws will become both simpler and more inclusive. But because a law can take more variables into account does not refute the law which takes fewer into account. As mentioned before, both are already idealized and thus at best merely approximations.

Perhaps the most authoritative opinion on Boltzmann's contributions to gas theory come from Stephen Brush. In his 1964 book he writes:

> The modern reader will rightly assume that some parts of Boltzmann's theory must have been rendered obsolete by later discoveries. In particular, we know that one cannot expect to develop an adequate theory of atomic phenomena without using quantum mechanics. However, it turns out that almost all the properties of gases at ordinary temperatures and densities can be described by the classical theory developed in the nineteenth century, and for this reason there has been a considerable revival of interest in classical kinetic theory during the last few years, in connection with rarified gas dynamics, plasma physics, and neutron transport theory. The reason why the classical theory works is that, while the internal

> structure of molecules must be described by quantum mechanics, the interaction between two molecules can be fairly well described by a classical model which ignores this structure and simply uses a postulated force law whose parameters can be chosen to fit experimental data. Aside from phenomena at very high densities or very low temperatures, the only property that the classical theory fails to account for is the ratio of specific heats. However, this failure was already well known at the time Boltzmann wrote his book, and he is therefore quite cautious on this point.[46]

What Boltzmann did was to place greater emphasis on an old criterion, practical utility. He did not abandon factual accuracy where it seemed important and possible to obtain, but he did attempt to make clear that where idealized laws which never literally correspond with nature are concerned another approach toward truth is needed. Wherever situations are found where "mechanistic laws" can handle the variables in a practical or sufficiently informative way, there is usually no reason to appeal to the newer, more inclusive laws. And since most situations which applied science encounters can normally still be handled by Newtonian mechanics and its various extensions we are stuck with it. In fact, all physics students have to master it before graduating to "modern physics", and many or most engineers and applied physicists learn no other kind. He thus attributed "relative truth" to Newtonian science to the extent it was practical, informative, or both.

But his point was much more fundamental. Newtonian mechanics is not alone in being idealized and hence not alone in being false in terms of correspondence with reality. All law-based science has idealized aspects, especially in the kind of work which theoretical as opposed to experimental, applied, and historically-focused scientists engage in. Thus where idealization is fundamental or unavoidable a non-correspondence theory of truth is needed. Boltzmann called it "relative truth", but at almost exactly the same time in America William James was advocating his variation of it under the label of "pragmatic truth".

Qualitative Aristotelian science does not have to be taught because part of it is naturally assumed by virtually everyone anyhow, much of it only holds under very special circumstances based on false understanding, and also because in a number of important circumstances it doesn't hold at all, but Newtonian mathematical physics including Boltzmann's own extensions of it does hold for virtually all circumstances which most people and applied scientists are likely to encounter and it also requires an understanding of calculus and other mathematics which almost always do have to be taught.

---

[46] Stephen Brush, "Translator's Introduction" to Ludwig Boltzmann, *Lectures on Gas Theory,* University of California Press: Berkeley & Los Angeles, 1964, p. 2.

Boltzmann won the reputation fairly early of being a staunch defender of atomism, but there was much disagreement among his colleagues and students as to just what that meant. For some it meant he believed in the reality of atoms; for others merely that he thought that the atomic theory was indispensable, and for still others that he identified atoms with mere pictures or models while agreeing with Ernst Mach (and hence disagreeing with Galileo, Descartes, and most leaders of the Scientific Revolution) that physics should not identify the physical world with what lay beyond sensory and conscious limits. Boltzmann's own words somewhat depended on his audience. In England, he generally favored the probable reality of atoms in a serious sense. But before a German-speaking audience which was supposedly more philosophically sophisticated he normally identified atoms with mere pictures or models, but not always. Before his own students, judging from their divided reactions, he apparently shifted from one approach to the other. To avoid the impression of inconsistency, he sometimes replied by alleging that both or by implication all three positions were basically the same. In short, he did not shy away from admitting a measure of relativism or loose approximation.

Once again Boltzmann appears to have outmaneuvered his critics and protected both atomism and the kinetic theory of gases, even if he couldn't manage to save "mechanistic science" except on the ground of indispensable utility while introducing a doctrine of relativistic or pragmatic truth to make practicality seem intellectually legitimate and primary. But the price was high. Once relativism about truth is permitted to enter the door, it often takes over the whole house, and many students become lazy. Why? Because: "If all truth is relative, why learn anything?"

Also, while the atomic theory or at least discontinuity of some sort may be indispensable in physics, when Boltzmann talked about atoms as mere pictures or models he seemed to be denying the existence of atoms and molecules as real entities possessing real energy and power, as if he were agreeing with his empiricist-foes that physics should restrict inquiry to saving the appearances or at least be modest in making forays beyond sensory experience.

Furthermore, Boltzmann appeared to be disregarding or downplaying the widely-held opinion that some idealized laws and theories were closer approximations to probable reality in terms of the variables normally taken into consideration than other laws and theories, that is, could be closer approximations in an absolute rather than merely a relative sense (disregarding other conditional factors for the moment).

But curiously, in Boltzmann's earlier highly successful scientific work it had been *he* who had emphasized the argument of "closer approximation to the truth", often understood as correspondence with reality, even if he was talking about ensembles or probability. Even while emphasizing difficulties in achieving greater approximation, he did not seem to be a relativist at all. How strange that in his later years, especially in terms of methodology, he should neglect or underemphasize his deep concern for exact truth. But when one is pressed hard enough by so many opponents as he was, perhaps one is not always as scrupulous about the type of ammunition employed to fire back. Nor was Boltzmann alone in appealing to relativistic notions.

But we cannot leave Boltzmann dying on the battlefield surrounded by victorious enemies. In most respects he was victorious but never knew it. "Mechanistic science" survived as Boltzmann predicted because it was too practical to abandon. No university would even think of teaching quantum and relativity theory without first teaching Newtonian mechanics. But more to the point, the major movements fighting against "the mechanistic world view" before the discoveries of Planck and Einstein including energeticism, phenomenalism, and the eletromagnetic world view would all come crashing down in much more dramatic fashion. Why? Because they were all incompatible with the reality of atoms.

Boltzmann was alive in 1900 when Planck's main work was first published and still active in 1905 in Einstein's *Wunderjahr,* but his eyesight was bad and philosophy already had its hooks in him. Decisive evidence has not yet surfaced that he was aware of these developments or had judged their importance wisely, though it could still be forthcoming in the future. It is known that Einstein corresponded with Boltzmann, but no letters appear to be extant on either side. The work on Brownian motion by Einstein, Smoluchowski, and Perrin from 1905 to 1908 ended the reign of much pseudo-modern physics which had seemed to bloom before 1900, and Ostwald at least, became willing to admit this. Mach of course remained obdurate.

Perhaps most ironically, what really compromised some parts of Boltzmann's work on the kinetic theory of gases and statistical mechanics was not the attacks on "mechanistic science" as reflected in Ostwald's rejection of atomism, Mach's philosophy, Zermelo's mathematics, or the anticipation that electromagneticism would become the new basis of physics, but increased understanding of atomic and subatomic reality. Boltzmann saved what he could by emphasizing physical behavior on the molecular level and even that on a statistical basis dealing with a vast quantity of particles as Professor Brush has pointed out, but it was Mach, Ostwald, and many supporters of the "electromagnetic world view" who were the big losers, even if energy still remains fundamental in physics as does electromagnetism and even though Mach's presentist philosophy when coupled with symbolic logic and some of Wittgenstein's ideas would have a rebirth with the Vienna Circle of the 1920's and the Logical Positivism and Analytic Philosophy of later decades.

If Boltzmann made a basic mistake, it may well have been in conceding too much to his opponents and being too pessimistic about the future of atomism. Indeed, there is a bit of evidence from his final year that he had begun to realize this, but it is still too fragmentary to clearly decipher. Nor has it been possible to include more than a relatively small amount of the new philosophical material in this book, though we have tried to translate into English what seems most understandable and important. Also, the new data about Boltzmann's life and philosophy has come too recently to fully understand. Hence the nature of this book as primarily a documentary history. Perhaps some readers can put the most important pieces together better and more quickly than the author of these lines.

But when our documentary history starts at the end of 1899, the omens still seemed very bleak for Boltzmann, kinetic theory, atomism, and mechanistic physics, at least in France, Germany, and much of Central Europe (though many chemists had never been pessimistic about atoms). That someone as beleaguered as Boltzmann should have looked to philosophy to help save the day should not unduly surprise us. Helmholtz, du Bois-Reymond, Kirchhoff, Mach, Ostwald, and Hertz had done it in their time, and Planck would become interested in philosophy as well. Physicist-philosophers were not rare, at least in Germany.

But one last point needs discussion about "methodology of science", or to be more exact, philosophizing about methodology or presupposing a philosophy which influences or helps determine methodological preference. While most unphilosophically-minded physicists probably were divided with respect to "mechanistic science" during the last quarter of the 19th century, some favoring and some opposing it, virtually all philosophers and physicists with a strong interest in philosophy from both the 18th and 19th centuries seem to have been almost fanatically opposed to it. One thinks of Leibniz, Berkeley, Kirchhoff, Hertz, and Mach above all. In fact, for many philosophers opposition to Newtonian mechanics or at least "the mechanistic world view" was almost a crusade. The question is why?

An initial answer might be that they were opposed to the reality of atoms which "mechanistic science" seemed to presuppose or at least favor. But while there was also that general opposition, it seems very likely that the major reason had to do with the suspicion that Newtonian science rested on the assumption shared by Galileo, Descartes, Locke, and many others that the real physical world as indicated by the representative theory of perception was beyond everything sensory and conscious. And the philosophers, especially those influenced by Berkeley, Hume, Kant, or Auguste Comte, could not accept that, and particularly the notion that science could or should try to understand that world.

In the main, there were two kinds of objections to indirect or representational realism about physical or external existence. First, was the religious objection emphasized by Berkeley, namely, that if matter was real (i.e. beyond the appearances) and had causal power, then man might be tempted to try to control it, which would make man more worldly and turn him away from God. The second line of objection, emphasized especially by Hume and Mach, was that human beings could be more certain of sensations than of what allegedly existed beyond them, hence science should restrict its concern to what at least in principle could be directly experienced.

By overcoming or replacing "the mechanistic world view" it was hoped above all that science could be returned to exclusive concern with sensations, immediate experience, or what could be directly "known", not back in every respect to the "save the appearances" Scholastic science before Galileo, but at least in the same general direction. What actually happened in the first two decades of the 20th century, however, was indeed a defeat for "mechanistic science" but a triumph for the underlying *assumptions* attributed to mechanistic science which the philosophers were against in a much more fundamental way, namely, the reality of atoms and an indirect world view.

But many philosophers and some scientists failed to see this. They falsely linked even the mature Einstein with Neo-Positivism and later Logical Positivism and falsely imagined that quantum theory and relativity theory favored their presentist or phenomenalistic world views. In name they won, but in fact they lost. Most scientists still think or at least normally assume that there is a physical world, that in accord with the representative theory of perception and the electro-magnetic spectrum it exists beyond everything sensory and conscious, and that it is the primary purpose of physical science to understand that physical (or in the words of their opponents "metaphysical") world.

Furthermore, fewer and fewer scientists believe in absolute, that is, unconditional certainty, i.e. epistemology in its most extreme or "philosophical" sense. Nor do they believe that knowledge of sensations or sensory experience is more certain in a public way than photographs and instrument readings in the absence of observers. "Eyeball observation" is subjective, and for other people it is often little more than hearsay. So much for phenomenalism and presentist philosophy.

## Boltzmann's Philosophy

Many people have recently commented on Boltzmann's philosophical views, but the diversity of opinion is startling, which is a major reason why this book is being published, namely, to add enough new information to help a consensus develop. First, there is praise by fellow Viennese thinkers for Boltzmann's philosophy without a clear description of what it was that deserved praise, but since these Austrian emigrés have become very prominent, even their casual remarks may deserve attention. For example, Sir Karl Popper has alleged that even though he is critical of Boltzmann's "subjective" theory of time: "Boltzmann is little known as a philosopher; until quite recently I too knew next to nothing about his philosophy, and I still know much less about it than I should. Yet with what I know I agree; more closely perhaps than with any other philosopher."* Paul Feyerabend, whose student years in Vienna are sometimes overlooked, has frequently commented in his publications on Boltzmann, for example: "It is unfortunate that Boltzmann's general philosophy, which is intimately connected with his physics, is practically unknown, for his ideas are still relevant to

* Karl Popper, *Unended Quest - An Intellectual Autobiography*, Open Court: La Salle, 1976, p. 156.

contemporary discussions and present a promising field for further study"** Presumably, Popper finds Boltzmann's defence of atomism attractive while Paul Feyerabend admires Boltzmann's willingness to experiment with new ideas as if nothing were sacrosanct in philosophy.

But basically, concerning Boltzmann's world view, upon which his philosophy of science apparently against his own will may be grounded, there are two schools, one which finds it impossible to reconcile all of the different strands in Boltzmann's outlook and another which thinks that the data can be reconciled in favor of a particular epistemological or ontological perspective. But the disagreements are much more varied than that. Within each camp there are serious differences of opinion. If Ludwig Boltzmann's philosophical remarks cannot be fully reconciled with each other, why is this so? Part of the problem may have been that Boltzmann wanted to consider Science primary, methodology of science secondary, and philosophy in all of its other forms superfluous. Given such an attitude, it is understandably difficult to pin down his epistemological and ontological assumptions. He normally wanted to deny that he had any!

But if we must eventually hack our way through Boltzmann's own occasional denials and wishful thinking, let us make matters a little easier for ourselves by beginning with a distinction between philosophy as philosophizing and as our deepest underlying assumptions or beliefs. If philosophy were only philosophizing, the thinking or talking about philosophical issues or problems, then one could speak of different opinions from the same person as simply reflecting opportunism, but since it is doubtful if any responsible adult can live without some assumptions and beliefs, that is, without having an underlying philosophy, incomplete or inconsistent as it may be, then what seems like opportunism may actually reflect an unwitting split between a person's philosophizing and his "real" philosophy.

In spite of his scorn for "metaphysics" and occasional support for a relativistic theory of truth, Boltzmann was such a conscientious person that it does not seem fair or plausible simply to consider him an opportunist, using one philosophy to solve one problem and a different or contrary one to help solve another. The tendency was there, but there is more to it than that. At times, he took philosophy very seriously indeed, especially in his later years.

What about responsible "irreconcilibilist" possibilities? Some might maintain that Boltzmann's philosophy simply kept evolving, such that consistency must be understood as a state only existing at a particular time or over a particular period with it always being possible that a few holdover ideas may have inadvertently survived from the past. Equally reasonable would be the supposition that his philosophy of nature and his philosophy of science were in conflict and that his philosophy of nature came closer to

** Paul Feyerabend, "Ludwig Boltzmann", *Encyclopedia of Philosophy,* Volume I, , New York, 1967, p. 336.

being his "real" world view.[47] A variation of this would be the opinion that much of his philosophy of science was primarily designed to persuade phenomenalists and presentists that even in terms of their own philosophy that atomism or at least some form of discontinuity theory was unavoidable, but that Boltzmann's "real" philosophy was quite different and made of sterner stuff, and included belief in the reality of atoms.[48] Perhaps the most initially plausible of the "incompatibilist" positions is that Boltzmann was a creature of moods and that in different moods he expressed different opinions.

Nor should we overlook the possibility that Boltzmann by the time he reached his sixties and was plagued by poor eyesight and illness simply found himself overwhelmed by the difficulties in the way of developing a philosophy which was both deep and internally consistent. Boltzmann had always profited from close study of the ideas of his scientific colleagues and opponents, but the same close attention does not always lead to mastery in philosophy, because many authors are not clear and perhaps a majority redefine basic terms, largely in order to place critcized perspectives in a more favorable light or in order to seem more original. When Boltzmann called metaphysics a "a migraine headache" he may have meant virtually all philosophy.

Among those who reject incompatibilism in favor of a unitary philosophy, as if consistency in Boltzmann's world view can be discovered, one is naturally tempted to start with mind-matter dualism combined with indirect or representational realism, but if that was Boltzmann's "real" philosophy and it may have been during the 1860's and 1870's for which we have little data, since his science articles almost never mention philosophy during those years, he kept it generally well hidden. To be sure, he often does seem to assume a type of mind-matter dualism, but the epistemology combined with it is direct rather than indirect, probably because of the German tendency to identify indirect realism with "metaphysics". In practical terms the difference between direct and indirect realism is that for the former if atoms were larger they could be directly sensed while for the latter atoms like everything physical are trans-conscious and hence can only be understand by inference or analogy.[49]

---

[47] See John T. Blackmore, "Mach vs. Boltzmann, Planck, Stumpf, and Külpe" in *Ernst Mach - His Life, Work, and Influence*, University of California Press: Berkeley & Los Angeles, 1972, pp. 214-216.

[48] This is the opinion of Professor Gerhard Fasol, a great-grandson of Boltzmann. See Gerhard Fasol, "Summary - Lectures on Natural Philosophy" in Ilse M. Fasol-Boltzmann (ed.), *Ludwig Boltzmann Principien der Naturfolosofi - Lectures on Natural Philosophy 1903-1906*, Springer-Verlag: Berlin, etc., 1990, p. 66.

[49] Martin Curd in his book, *Ludwig Boltzmann's Philosophy of Science*, UMI: Ann Aror, 1978, pp. 41-56, mentions several scientists and scholars who identified Boltzmann's philosophy with realism, especially an indirect world view such as representative realism. He mentions Boltzmann's successor Friedrich Hasenöhrl, Yehuda Elkana, and the earlier writings of John Blackmore. It also seems probable that Boltzmann's grandson, Dieter Flamm, also leans toward the position that Boltzmann was basically a realist. But some writers while calling Boltzmann a realist use the word in a somewhat unusual way. For

In this book, the editor defines epistemological "realism" itself in the standard or traditional sense as the position that there is a physical world and that it exists independently of consciousness of it. But he would also emphasize that since this definition existed *before* the term "epistemology" was introduced into English, it would be misleading to maintain that the definition has something to do with epistemology or theory of knowledge. If knowledge means something unconditionally certain or "immediate acquaintance", then it does not. It is more referential and ontological than epistemological, at least in the above extreme sense. The apparent reason why realism as basic opposition to the idealistic philosophies of Berkeley, Hume, and Kant became called "epistemological realism" almost certainly had to do with its secondary utility as a contrast to Medieval "realism", the Scholastic position that universals or only universals are real.

As for the definition of idealism [as idea-lism as opposed to both Platonic Idea-lism and ethical ideal-ism] it is simply the negation of the traditional definition of realism: Either there is no physical world or it is dependent on consciousness of it. This also means that logically there may be no third position between realism and idealism. One might also add that the traditional definitions have been present in most German as well as English dictionaries and encyclopedias of philosophy for almost two hundred years. This is, however, no guarantee that any particular philosopher including Boltzmann uses "realism" or "idealism" in a standard or traditional way. Nor is it a guarantee that a realist in his world view actually employs a realist methodology or philosophy of science, that is, makes it his primary goal to understand the real physical world.

Second, there are the many left-wing or Eastern European scholars who follow Vladimir Lenin[50] in considering Boltzmann a materialist,[51] but not

example, John Nyhof in his article "Philosophical Objections to the Kinetic Theory", *British Journal for the Philosophy of Science,* **39** (1988), 81-109 maintains that Boltzmann was a realist and never yielded to positivism or phenomenalism, but even though this editor sympathizes strongly with his wish to defend Boltzmann, he is somewhat surprised both by Nyhof's odd use of the word "realism" and his declaration based on an early reading of Sir Karl Popper that trying to understand physical reality must be rejected because it is "essentialistic". In short, the "moderate realism" which Nyhof's wants to attribute to Boltzmann looks suspiciously like a version of epistemological idealism as if there were no physical world or we should not try to understand it.

[50] See V.I. Lenin, *Materialism and Empirio-Criticism,* International Publishers: New York, pp. 295-298. Lenin wrote his original work in 1908 largely as an attempt to wean a number of Marxists away from supporting Ernst Mach, and it is easy to see how an author who was in a hurry and not very kean on distinguishing epistemology from ontology might be influenced by the titles of some of Boltzmann's articles into thinking that his "theory of knowledge was essentially materialistic". The titles in question in English "On the Indispensability of Atomism in Natural Science" (1897) and "On the Question of the Objective Existence of Processes in Inanimate Nature" (1897) do seem "realistic" and "materialistic", that is, until they are read carefully, when major doubts are almost certain to arise.

[51] Engelbert Broda, a pioneer in Boltzmann scholarship, who wrote the first biography of Boltzmann, *Ludwig Boltzmann - Mensch-Physiker-Philosoph,* Vienna, 1955, (which has been translated into English) as well as a great many articles about him, has alleged that Boltzmann was a materialist, but since he seems to feel that there is no difficulty in reconciling Boltzmann's relativism about truth and his picture theory about atoms with that

always in the traditional or Democritean sense of identifying the *external* physical world with what is trans-conscious.[52] In fact, there is some doubt whether anyone who identifies the physical world with appearances, and Boltzmann occasionally seemed to do that, can honestly and knowingly consider himself to be a materialist, since presentist ontology is often or normally identified with phenomenalism or idealism. In fact, it was precisely Lenin who criticized many Marxists for pretending to be materialists while actually holding a direct world view about the external world.[53]

Was Lenin correct in considering Boltzmann to be a materialist? There is no doubt that Boltzmann often expressed himself as if everything mental could be reduced to what is physical, but if he identified what was physical with what could be sensed or made conscious, then in traditional philosophical terms the reduction was from mind-matter dualism to phenomenalism, that is to a type of idealism (in Berkeley's sense), rather than to materialism (in the sense of Democritus or Lenin himself).

Phenomenalism is normally considered to be a type of idealism since in Berkeley's sense it identifies physical realisty with sensations, which Berkeley called "ideas". On the other hand, at least in Germany and Austria, phenomenalism or "neutral monism" was usually called "*Phänomenologie*" in the late 19th century. It was apparently only after an article by Alois Höfler in 1910 and especially Hans Kleinpeter's book *Der Phänomenalismus* in 1913 that the distinction seems to have been widely accepted in Central Europe. Nor was Ludwig Boltzmann's own use of terminology faultless. He himself tended to identify "*Phänomenologie*" not just with what we now call phenomenalism but with a phenomenalism which did not use models or pictures such as the position Mach allegedly held as contrasted with his own (unadmitted) phenomenalism or direct world view which did use them, which he often clothed in physicalistic language and which thereby may have confused some people into thinking that he actually was a materialist.

Several scholars[54] imagine that Boltzmann was a representist, but in saying this at least some of them are once again following Boltzmann's own

---

materialism, then even with great respect for Broda's pioneering efforts one can only suppose that what he means by "materialism" has little resemblance to what the term traditionally means in ontology.

[52] Wolfgang Stiller's recent book ***Ludwig Boltzmann -Altmeister der klassischen Physik - Wegbereiter der Quantenphysik und Evolutionstheorie,*** Verlag Harri Deutsch: Thun & Frankfurt am Main, 1989 is a very interesting, well-researched, and quite informative book about Boltzmann's life, work, and especially his colleagues, students, and opponents. Since the author was writing in East Germany before unification, he was not in a good position to question or even qualify the claim that Boltzmann was an ontological materialist, but the manner in which Stiller writes, and especially what he doesn't write about, makes clear that he is aware of the problems in trying to pin down Boltzmann's ontology and world view.

[53] Lenin's epistemological position is normally identified with a version of indirect realism called the "mirror theory". He was especially critical of Alexander Bogdanov, an influential Menshevik, whom he suspected of being a Machist who only gave lipservice to indirect epistemology and materialist ontology. See Lenin, op. cit., pp. 51-53, 119-125, 130-134, 229-236, 279-281, 333-338, and 340-343.

[54] See Ludwig Boltzmann, "Theories as Representations" in ***Philosophy of Science,*** edited by Arthur Danto & Sidney Morgenbesser, Meridian Books, Cleveland & New York, 1964

suspect terminology and ignoring the larger history of philosophy and history of terminology. Boltzmann at least in his conspicuous philosophy of science was clearly not a representist in the Cartesian or dualistic sense of ideas or images representing a transcendent physical world,[55] but merely in the presentist or phenomenalist sense of pictures[56] representing conscious experience. It can be quite misleading to fail to distinguish between these two senses of representism. Boltzmann, however, as a non-historian had the excuse of not being very familiar with the early representism, but most scholars today can scarcely pretend to possess the same ignorance.

Some thinkers believe that Boltzmann's interest in Darwin's theory of evolution as modifed by Lamarck's idea of the inheritance of acquired characteristics was the key.[57] They use this to explain Boltzmann's opposition to Kant's static a priori categories of the mind. If human beings inherit mental characteristics and if acquired characteristics can be inherited, then there ought to have been a number of changes in those categories over time, an evolution of mental categories. Boltzmann liked to argue that philosophy was metaphysics in the sense that it retained a number of old ways of thinking which either had been made obsolete by evolution or science or which were useless in a future struggle for survival. But his emphasis on this attitude seems to have been of rather brief duration and it may well have been partly caused by the deep depression which Boltzmann sufferered during the middle months of 1904 when he cancelled his philosophy lectures for a semester. After his correspondence with Brentano picked up and he was invited to lecture in California in the Summer of 1905,

---

[1899], pp. 245-252; Andrew D. Wilson, *Representing Reality: Ludwig Boltzmann and the Nature and Purpose of Theoretical Physics*, Cornell University Doctoral Dissertation & University Microfilms, 1990; and Blackmore, 1972, op. cit., p. 215.

55 See E.A. Burtt, *The Metaphysical Foundations of Modern Science*, Doubleday: Garden City, 1954.

56 There has been recent interest among historians and philosophers of science in Boltzmann's methodology of science, especially his picture or *Bild* theory and its possible connection with Heinrich Hertz. See: Martin V. Curd, *Ludwig Boltzmanns Philosophy of Science: Theories, Pictures, and Analogies*, University of Pittsburgh Doctoral Dissertation & University Microfilms, 1978; Friedrich Wallner, "Boltzmann, Hertz, and Wittgenstein" in *Ludwig Boltzmann Gesamtausgabe*, Volume **VIII**, edited by Roman Sexl & John Blackmore, Vieweg: Braunschweig, 1982, pp. 143-153; Andrew D. Wilson, "Hertz, Boltzmann, and Wittgenstein Reconsidered", *Studies in the History and Philosophy of Science*, **20** (1989) 245-263; S. D'Agostino, "Boltzmann and Hertz on the *Bild* - Conception of Physical Theory", *History of Science*, **28** (1990) 381-398. (Wilson in his paper mentions that Boltzmann had taken ten courses in philosophy while a student, p. 252, footnote 18.)

57 See Engelbert Broda, "Boltzmann and Darwin" in *Ludwig Boltzmann Gesamtausgabe*, Volume **VIII**, edited by R. Sexl & J. Blackmore, Vieweg: Braunschweig: 1982, pp. 129-142 and "Darwin & Boltzmann" in *8th Kühlingsborn Colloquim, Philosophical and Ethical Problems of Biosciences: Darwin Today*, edited by E. Geissler & G.W. Scheler, Akademie Verlag: Berlin, 1981, pp. 61-70 and "Boltzmann als Evolutionischer Philosoph", *Berichte zur Wissenschaftsgeschichte*, **6** (1983) 103-114. (This last article may also be found in Engelbert Broda, *Wissenschaft Verantwortung Frieden*, edited by Paul Broda et al, Deuticke: 1985, pp. 88-100.)

his spirits rose again and this sour attitude seemed to disappear at least for several months.

Many observers simply deplore the fact that Boltzmann appeared to succumb to a type of phenomenalism or "neutral monism", seeming to become a Machist in epistemology and ontology while only seriously disagreeing with him on methodological issues concerning hypotheses or "pictures".[58] Boltzmann's curious habit of identifying "*Phänomenologie*" with what we would would call a phenomenalism which does not use models or pictures, tended to obscure the fact that a phenomenalism *with* models and pictures is still phenomenalism as the word is normally understood today. The underlying position still identifies reality with sensations, ideas, or conscious experience. But one must keep in mind that so much German philosophy was presentist or phenomenalistic at the time that Boltzmann may simply have taken many aspects of it for granted. Furthermore, Boltzmann was looking for ways to persuade his colleagues to accept his view on physics and methodology of science, and appearing to acquiesce or even actually doing so in matters of epistemology or ontology might have made his intended audience more sympathetic to his message about physics.

Still other scholars have been intrigued by Boltzmann's linguistic philosophy.[59] They stress his late articles on statistical mechanics and Schopenhauer (1904-1905) as if he were not only a forerunner of Ludwig Wittgenstein but almost certainly must have exercized considerable influence on the young man. Comments on Boltzmann's lectures were much in the Vienna newspapers in those days. Furthermore, a number of *Gymnasium, Realschule*, or other middle school students sometimes visited university lectures and since Boltzmann held some philosophy classes at night in Spring 1905 it is even conceivable that Wittgenstein could have heard Boltzmann directly. Above all, It was Boltzmann who claimed that the proper purpose of philosophy, if it had one, was merely to examine language and eliminate all metaphysical expression, that is, expression which was no longer compatible with science or had lost its evolutionary value in the struggle for survival. It was also Boltzmann, clearly in one of his more relativistic moods, who seems to have been the first reputable thinker to parrot the age-old cry of frustrated students unable to penetrate to the depths of philosophy "that realism and idealism were only linguistically different".

---

[58] S.R. de Groot, "Foreword", in Ludwig Boltzmann, *Theoretical Physics and Philosophical Problems*, edited by Brian McGuinness and translated by Paul Foulkes, Reidel: Dordrecht, 1974, pp. ix-xiii; Erwin Hiebert, "Boltzmann's Conception of Theory Construction: The Promotion of Pluralism, Provisionalism, and Pragmatic Realism", in J. Hintikka et al. (eds.), *Probabilistic Thinking, Thermodynamics, and the Interaction of History and Philosophy of Science*, Reidel: Dordrecht & Boston, 1981, pp. 175-198; John Blackmore, "Boltzmann's Concessions to Mach's Philosophy of Science", *Ludwig Boltzmann Gesamtausgabe*, Volume VIII, edited by R. Sexl & J. Blackmore, Vieweg: Braunschweig, 1982, pp. 155-190.

[59] See Wallner, op. cit., pp. 143-153; Wilson, op. cit., pp. 245-263; and Girolamo Ramunni, "Peut-on analyser le langage scientifique sans se soucier de l'outile mathematique? Le cas Boltzmann confronté `a celui de Ostwald", *Documents pour l'histoire du vocabulaire scientifique*, **8** (1986) 121-132.

But this book provides information about Boltzmann's philosophy even *after* the linguistic philosophy outburst, that is during the middle and late months of 1905 and even into early 1906, the year of Boltzmann's tragic death. The discovery of Boltzmann's correspondence with Franz Brentano and the publication of Boltzmann's lecture notes for his courses on methodology of science and natural philosophy held between 1903 and 1906 have added so much new data, that virtually all of the earlier views may have to be revised. "There is gold in them thar hills!"

The letters to Brentano (and the one by him) are dated as follows:

By Boltzmann: November 19, 1903 and December 26, 1904.
" " January 4, 1905, and February 17 & 23, 1905.
By Brentano: undated (probably late February 1905).
By Boltzmann: March 6, 20, & 23, 1905.
" " April 29, 1905 and June 3, 1905.
" " January 16, 1906.

Boltzmann's lecture notes for his course on "Methodology of Science and Natural Philosophy" were transcribed from an obscure type of shorthand by his granddaughter, Ilse M. Fasol-Boltzmann. She has published a book, *Ludwig Boltzmann Principien der Naturfilosofi*, Springer-Verlag: Berlin, 1990., which is partly in German (the notes) with articles and other material in English. The title reflects Boltzmann's interest in phonetics and the fact that many of the words in his shorthand notes were written phonetically. The articles are by Stephen Brush and Gerhard Fasol. Concerning the lecture notes themselves, they are most complete for the first semester which Boltzmann taught, that is from October 1903 until March 1904. In fact, there are three different versions. The most complete which was written out in longhand was apparently written by an unknown student from the lectures themselves starting with Lecture #3. It is possible that Boltzmann might at one time have intended to publish them. All of the versions emphasize mathematics and philosophy of mathematics. Because the humber of students apparently dropped dramatically at the end of the first semester, Boltzmann apparently became so discouraged that he didn't teach the second semester philosophy course at all.

In October 1904, Boltzmann started once more, this time by lecturing on Schopenhauer's philosophy. These notes are also fairly extensive, but it is not always easy to determine what represents his own ideas and what are supposed to be descriptions of what he was reading or Schopenhauer's philosophy. The Spring semester 1905 was taught, but shortened. In this semester and in the Fall semester 1905-1906 Boltzmann focused more on basic philosophical problems and often expressed his own point of view on them, but unfortunately, the notes grow shorter and shorter as if he increasingly felt he could lecture effectively without them.

This editor has only translated a small proportion of what Frau Dr. Ilse Boltzmann-Fasol has with great difficulty made available in her very valuable book, but understandably, the emphasis is on those parts of the last two semesters for which we have notes. (Boltzmann did not teach the Spring 1906 course and may not have even completely finished the preceeding one.) Because the notes are so abbreviated, sometimes this editor has taken the liberty of putting some of them into complete sentences, hopefully without distorting the meaning. But serious scholars are recommended to read Frau Dr. Boltzmann-Fasol's transcriptions in German, but no version is without problems since Boltzmann's original shorthand was also abbreviated and had some idiosyncratic symbols. Below is a list of the periods taught.

| Fall Semester | Spring Semester |
|---|---|
| Oct. 1903 - March. 1904 (Philosophy of Mathematics) | cancelled |
| Oct. 1904 - March. 1905 (Schopenhauer's Philosophy) | May1905 - June 1905 (Philosophical Problems) [course shortened to allow Boltzmann to prepare for his teaching at Berkeley] |
| Oct. 1905 - Jan. 1906 (Philosophical Problems) [shortened by illness] | cancelled |

The courses have been described in terms of their actual content rather than how they were listed in the university catalogue.

Returning to our discussion of some of the possible philosophical positions which have already been or reasonably might be attributed to Boltzmann, a very late possibility might be that he suddenly began to take increasing evidence for the reality of atoms more seriously (we know that there had been some correspondence earlier from Einstein, though no letters appear to be extant), saw the futility of both linguistic reduction of basically different philosophical positions and trying to defeat phenomenalists on their own ground, grasped that it was no longer necessary to appease the philosophers, bitterly criticized Berkeley and Mach, abandoned the book he was writing on philosophy, despaired of obtaining peace of mind, went into severe depression, and one day committed suicide.

But while the many possible philosophical positions mentioned above should be kept in mind for the sake of initial orientation, hopefully, the reader will allow the documentary evidence he is about to read to *prevail* over any and all such preliminary opinions. The reader himself with the help of old and new evidence must assume responsibility for determining what is true. The old puzzle has some new pieces which fill in a number of old gaps. But the reader has a chance to beat the "experts" by finding additional pieces in Boltzmann's "real" philosophy, that is, assuming that he had a coherent one. Good Luck!

But while some readers will welcome the opportunity to make up their own minds, others used to being spoon-fed what to believe will be unhappy that the editor has decided to be "so frustratingly neutral" about Boltzmann's philosophy, even after receiving so much new data. But realistically, the particular writer of these lines has developed more and more reservations over the years. Nevertheless, if he must venture an opinion it would be in admiration of Robert J. Deltete's analysis of Boltzmann's philosophy in his book *The Energetics Controversy in Late Nineteenth-Century Germany: Helm, Ostwald, and Their Critics*, Volume I, UMI: Ann Arbor, 1983, pp. 49-72. Like other physicists of his time, Boltzmann often used the language of phenomenalists and positivists and even did his best to believe that he rejected an indirect world view about matter. But like most other physicists when actually doing physics he often seems to have taken for granted an indirect world view in general and the reality of atoms and molecules in particular, regardless of how tentative and guarded his words or even how overtly hostile his conscious attitude was to "metaphysical assumptions". In short, a split-position was largely the rule in those days. Boltzmann's tragedy was at least in part his effort to make the split even wider. Every "advance" in his struggle against "metaphysics" seemed to make him feel less and less happy.

But another group of observers as yet undiscussed fail to see any tension in Boltzmann's philosophy. Martin Curd, for example, in his book *Ludwig Boltzmann's Philosophy*, UMI: Ann Arbor, 1978, pp. 41-70, maintains that Boltzmann's views were basically the same as those of Ernst Mach (i.e. in epistemology and ontology). He even rejects what he calls the "conversion thesis", that Boltzmann was a realist before 1890 and was "converted: by Mach or developments within physics into becoming a phenomenalist after 1890. Curd alleges that there is no evidence for this, and that Boltzmann might have always been a phenomenalist with positivistic inclinations. Curd attacks all "hagiology" which would bring Boltzmann into line with the subsequent victory of atomism over positivism, as if Boltzmann would have been as shocked by proof that atoms exist in a serious sense as Mach himself, even though both men may have accepted the "reality" of atoms in the rather secondary or trivial sense of "pictures"appearing on a spinthariscope screen as displayed by Boltzmann's assistant Stefan Meyer or via "materialistic" or "physicalistic language" in terms of the occasional forays of both men into linguistic reductionism.

## Myths and Exaggerations

But while it would indeed seem imprudent for this editor to express a final or unqualified opinion on the nature of Boltzmann's philosophy especially in what is primarily a documentary history, and particularly before other philosophers, scholars, and even himself have even examined the new material with the extraordinarily fine-combed brush it will eventually require, one can at least voice a rather firm opinion about a number of myths, exaggerations, and half-truths. For example, one often hears that:

"Most physicists in the late 19th century thought their discipline was basically sound and only had a few more 'puzzles' to solve before everything would be neatly wrapped up."

This extraordinary opinion gains no support either from Boltzmann or his opponents. It may have been present among some physicists during the 1860's and 1870's but after the startling work of Maxwell and Hertz in electrodynamics plus radiation discoveries in chemistry revolution was in the air such that only specialists in remote branches of physical science could imagine that there was little of importance left to investigate.

"Boltzmann was the last great representative of mechanistic science."

This allegation may be true, but it is so misleading unless severely qualified that it should probably be treated as a myth. One expects a representative of "mechanistic science" to be opposed to modern developments, but Boltzmann favored them. In fact, he went too far by exaggerating the role of both electromagnetism and phenomenalistic "presuppositionless" science as likely replacements. Furthermore, many aspects of Boltzmann's work on gas theory are still of practical use today. Also, his work contributed to the development of Planck's quantum theory, and Einstein's theory on Brownian motion largely confirmed Boltzmann's suppositions about the behavior of molecules. Boltzmann was a leader in the changes which were taking place in physics. He was a forerunner and transition figure who contributed much to modern physics. He defended selected parts of older physics because they were informative and practical and feared that they might be foolishly abandoned, not because he regarded them as literally true to reality or superior to modern physics. He was not an opponent of the sound changes which were taking place.

"Boltzmann hated philosophy and had only written one philosophy article before 1903."

Boltzmann himself is largely to blame for these myths. He had written many philosophy articles before 1903, but he apparently considered them to be merely methodological. His tendency to identify philosophy with "metaphysics" and desire not to be tarred with such a pejorative expression was apparently behind his misleading claim. In connection with that label, in his later years he voiced the thought that philosophy created unnecessary mental turmoil by inquiring beyond the scope of sensations and science, citing Kant's antinomies and du Bois-Reymonds "Seven Riddles". In short, during part of 1904 and early 1905 Boltzmann seems to have "hated" philosophy in a sense later taken up by Wittgenstein, but if the discipline of philosophy would stick to correcting language by eliminating metaphysical expressions it might have some valid use as a kind of handmaiden to science. But later in 1905, Boltzmann in his correspondence with Brentano gradually began to take pride in thinking of himself as a philosopher, only to relapse again in early 1906. But regardless of Boltzmann's feelings he was a philosopher in two senses, first, he had underlying assumptions and beliefs of a general or philosophical kind regardless of how incompatible with each other, and second, he increasingly philosophized about recognized philosophical problems as one can detect from his notes for his philosophy lectures. For people who "hate" philosophy to do as much of it as the two philosophizing Ludwigs from Vienna demonstrates that in a very real sense they didn't hate it. What spooked them was a fata morgana called "metaphysics".[60]

"Boltzmann committed suicide because of opposition to atomism."

By 1906, most chemists and an increasing number of physicists accepted the reality of atoms and the indispensability of the atomic theory. If he was going to commit suicide because of opposition to atomism, he should have done it earlier, say from 1895 to 1900, when the tide in France and Germany really did seem negative. The work of Ramsay and Rayleigh in England, Becquerel and the Curies in France, use of the spinthariscope to measure atomic or ionic scintillations, and the early work of Einstein and Smoluchowski on Brownian motion all occurred before Boltzmann's death. It is more likely that self-criticism for taking the anti-atomists too seriously was a factor. At the end he may have accused himself of being a gullible dolt for listening to the philosophers. Nor is there significant discussion about atomism in his philosophy lecture notes, either pro or con. In short, this editor can find no support for the hypothesis above.

"Ernst Mach and his ideas played a role in Boltzmann's suicide."

This is almost certainly pure myth. There is no evidence from Boltzmann's correspondence or notes that he was thinking about Mach or his ideas during the last months of his life. Boltzmann did thank Mach for sending

---

[60] Albert Einstein in his article "Remarks on Bertrand Russell's Theory of Knowledge" in the Schilpp volume: *The Philosophy of Bertrand Russell,* Open Court: La Salle, 1989 [1944], page 289, commented on this unreasonable fear of metaphysics (i.e. of indirect epistemology and ontology) as follows:

> By his clear critique . . . Hume created a danger for philosophy in that, following his critique, a fateful 'fear of metaphysics' arose which has come to be a malady of contemporary empiristic philosophizing . . .

him a copy of Mach's book, *Erkenntnis und Irrtum*, after he returned from America in late summer 1905, and he presumably did visit the philosopher shortly afterwards since they were on friendly terms. Boltzmann's list of infirmities would appear to be enough by themselves to explain his death, but admittedly there is some evidence that philosophy did play a role in Boltzmann's last extended period of deep depression, but there is no evidence that it had anything to do with Mach or his world view. Mach is very rarely mentioned in the notes, though Gabrielle Rabel remembered what she thought were Boltzmann's harsh criticisms of Mach in Boltzmann's late philosophy lectures, i.e. from 1905 or early 1906.

"Boltzmann was part of a 'declining research program'."[61]

"The opposition to Boltzmann and his work has been exaggerated."

Both of these extreme opinions appear to be myths. He was not part of a declining research program. There were more scientists working on kinetic theory and especially statistical mechanics in his last years in Germany and Austria than earlier, and many of them had been his own students such as Smoluchowski and Ehrenfest. Nor should we forget that Einstein himself published many papers on kinetic theory and even the mathematician David Hilbert.[62] But on the other hand, most English and American scholars seem to have *underestimated* the opposition to Boltzmann's ideas in France and Germany, opposition generated partly on philosophical grounds.

Some of the confusion concerns the fact that, yes, Boltzmann was a much sought-after theoretical physicist for a number of academic positions, but on the other hand, most physicists circa 1890-1906 were still experimentalists and many of them shared Ernst Mach's opinion that "theories were like falling leaves" and would eventually be replaced by straight description

---

[61] This point of view has been represented by some of the followers of the Hungarian-English philosopher, Imre Lakatos, in particular, Peter Clark in his article "Atomism versus Thermodynamics", in *Method and Appraisal in the Physical Sciences*, edited by C. Howson, Cambridge University Press: Cambridge, 1976, pp. 41-105 and later in his 1982 review of Stephen Brush's excellent book *The Kind of Motion We Call Heat*, a review which appeared in *British Journal for the Philosophy of Science*, **33** (1982) 165-186. But serious research historians tend to be wary of sociological theories imposed on historical data, whether in terms of "declining research programs", "scientific revolutions", "paradigms", "puzzles", "normal science", or in fact any type of sociological or economic determinism. It seems preferable to do enough factual research until the evidence itself seems to suggest something new, unexpected, or important. Historians must fight the imposition of theories on their discipline. Thesis history is not sufficiently inductive. Evidence should not simply be trotted out as an example to support or refute a thesis, but used as a source from which to help infer what is true. For research historians finding new evidence is almost always more important than theories and conclusions. It is evidence which is primary. Why? Because weight of evidence determines how probable it is that an idea or thesis is true, and that weight and probability can change depending on new discoveries of evidence.

[62] See the letter by Boltzmann to Hilbert in this book and Clayton A. Gearhart, "Einstein Before 1905: The Early Papers on Statistical Mechanics", *American Journal of Physics*, **58** (1990) 68-80.

of the appearances by mathematical functions. It was theoretical physics itself which was under attack.

It may be hard for us today to appreciate a situation so contrary to our own. After the work of Planck, Einstein, and the quantum theorists, those physicists acknowledged to be leaders of the discipline have generally been theorists, but before 1920 almost no theorists received Nobel Prizes, many were considered primarily mathematicans (as Boltzmann was when he lectured in Saint Louis in 1904), and theoretical physicists in general were still widely considerd to be low men on the totem pole.[63] It was Boltzmann's very success as a theorist which may have disturbed many experimentalists of his time. Their philosopher was Ernst Mach, and he downplayed theory as something often useful on a temporary basis but which had no final place in physics. Nor should one take subsequent criticisns of Mach by Planck, Einstein, and others as indicative of what they thought prior to 1900. To paraphrase Einstein, they all drank from Mother Mach's milk. Love of "presuppositionless physics" implied opposition to theoretical physics.

In short, no one with as many outstanding students and assistants as Boltzmann had in Vienna during his last stay from 1902 to 1906 could be part of a "declining research program", but opposition to his support of atomism and theoretical physics was widespread in Central Europe. Part of it was grounded in technical difficulties in his science, but the vast majority had philosophical roots given the influence of Kant and professional philosophers on the one side and phenomenalistically-focused physicist-philosophers on the other such as Mach and Duhem plus lesser-known figures like Frank, Petzoldt, Lampa, Kleinpeter, Wald, and Jaumann.

Both philosophical factions opposed the mind-matter dualism and representational realism of Galileo, Descartes, and most of the leaders of the Scientific Revolution. Germany under the influence of Luther and the 30 Years' War (apart from Kepler and Guericke) pretty much missed the Scientific Revolution, generally preferring a direct or presentist philosophy about the physical world (i.e. Werner, Goethe, etc.). And even as late as the early 20th century retained that overall perspective. Science did begin to

---

[63] The very useful volumes by Christa Jungnickel and Russell McCormach titled *The Intellectual Mastery of Science*, University of Chicago Press: Chicago, 1986, follow Boltzmann's academic career closely and mention how sought-after Boltzmann was, but in the process the authors tend to underestimate the opposition to him. Part of the reason is that the authors of the volumes seem uninterested in philosophy and how it affected physics, but mostly they look at the rising importance of theoretical physics from a recent perspective instead of how it was generally understood in those days. In fact, Volume Two of their generally magnificent work is titled *The Now Mighty Theoretical Physics 1870-1925*. Granted, that much important theoretical physics was done from 1875 until Boltzmann's death in 1906, but from the perspective of most scientists and the general public *of the time*, experimental physics was still generally considered more basic and important, and most experimentalists in physics either ignored or were opposed to Boltzmann's work. Indeed, when Boltzmann died in 1906 a newspaper article in a leading Vienna Newspaper mentioned how amazing it was that a mere physicist could know as much mathematics as Boltzmann. Physicists were still generally thought of as experimentalists, and the situation was even more pro-experimental in the United States.

flourish, but the country as a whole or at least most of the universities still seemed not just philosophy-ridden, but dominated by particular philosophical currents incompatible with many of the basic philosophical assumptions underlying the Scientific Revolution.

## Conclusion

We have already admitted that inspite of efforts at neutrality that "deep down" this editor and his colleagues do have views about Boltzmann, but the reader is not required either to try to discover them or if he can find them to consider them any more seriously than any of the other "anticipations" we have mentioned above. Nevertheless, one opinion may be worth mentioning: If it is maintained that Boltzmann's attitude *toward* rather than *in* philosophy was systematic in that he consistently tried to persuade scientists to accept his scientific and methodological ideas regardless of which philosophy they held - a common hope among many scientists - and an attitude which he tried to further by minimizing philosophical differences, as if all of them were "merely linguistic" or represented "rather the same position", then there was almost certainly a big measure of misunderstanding or self-deception in his attitude which the closer study of philosophy conducted during his final years was almost bound to explode.

To be sure, much science does appear to progress while ignoring philosophical differences and even the fact that different scientists mean different things by different expressions such as "matter", "mind", "energy", "mass", "reality", "truth", "fact", and so forth, but at some stage mutual miscommunication becomes so severe that it does affect scientific progress as is evident when one compares different sciences with each other and the very different approaches toward causes and causal explanation which they often employ.

If philosophy in its primary sense means not mere speculating or "philosophizing", but our deepest and hence most influential general assumptions and beliefs, *then no differences are more real, important, or influential,* and scientists who overlook this, as Boltzmann seems to have long attempted to do, are in for a big shock. Even today, idealists who follow in the traditions of Berkeley, Hume, and Kant, indeed all presentists, *still* reject the reality of atoms, which they would disguise by employing the euphemism "theoretical entities", that is, by redefining terms, much as Boltzmann did by talking about atoms and molecules as pictures or "*Bilder*", even if they are much more serious in their presentism than he apparently was and less inclined to employ it as mere methodological propaganda to persuade recalcitrant philosophically-inclined scientists.

All scientists who deny the importance of philosophy or who try to restrict its scope to "linguistic clarification" do not commit suicide as did Boltzmann, but if they study philosophical differences long enough, they are almost bound to eventually recognize that many or most are real, and concerning the most fundamental logical, epistemological, ontological, and ethical differences, *that nothing in this world can matter more!*

But the time has come to end this introduction and turn to Boltzmann's own words. After a short section on preliminary quotations and much more material from correspondence about biographical, academic, and scientific issues, philosophical concerns increasingly came to the fore. There are still some gaps, especially given Boltzmann's extremely depressed moods when he was apparently reluctant to write anything. But there is also considerable humor. Boltzmann during his good moods loved to make jokes, especially puns, which are not always easy to translate into English, though we have tried. Ludwig Boltzmann's brain and personality were indeed unique, as was his remarkable and surprisingly influential thought. he was a serious and important thinker, even if a rather troubled one.

John Blackmore
Tokyo, April 1994

# Chapter 2

# PRELIMINARY QUOTATIONS

## (mostly from letters and scientific articles)

### (1886)[1]

The remarkable facts that chemists daily discover are as many new proofs of atomism. . . . .

In his comprehensive work on mechanics, [Gustav Robert] Kirchhoff very clearly sets himself as a task merely to describe natural phenomena as simply and perspicuously as possible, renouncing all explanation, and since then what in physics used to be called explanation has repeatedly been called a mere description of facts . . . .

Many problems are like those put to a painter about what picture he was hiding behind the curtain, to which he replied 'The curtain *is* the picture' . . . .

If one seeks to explain motions from forces and these from the essence of things, that is phenomena from things in themselves, one always seems to start from the view that explanation requires reducing the explicandum to some quite new principle external to it. This view is alien to natural science . . . .

[But] are we not here once more placed in front of that painted curtain mentioned earlier? Are we going to regard it as a limitation of our sense of sight that nobody can tell what picture is concealed behind the curtain? We shall be able to retain the word 'explain' if from the outset all such reservations are kept at a distance.

We infer the existence of things only from the impressions they make on our senses. It is thus one of the fairest triumphs of science if we succeed in inferring the existence of a large group of things that mostly escape our sense perception.

---

[1] This lecture, "Der zweite Hauptsatz der mechanischen Wärmetheorie", was given before a meeting of the Imperial Academy of Sciences in Vienna on May 29, 1886. It was included in Ludwig Boltzmann's *Populäre Schriften*, Leipzig, 1905, pp. 25-50 and in English in *Theoretical Physics and Philosophical Problems*, edited by Brian McGuinness and translated by Paul Foulkes, Reidel: Dordrecht, 1974, pp. 13-32. Our quotations are drawn from pages 14, 15, & 16.

(1892)[2]

About the dogma that nature can only be explained mechanically (out of the movement of atoms), *I* reject the opposite dogma, that it is impossible.

Also the application of the principle of the relativity of movement doesn't seem to me to be free from objection, since even the sources of various energies at a distance are part of the 'reference system' here, which could affect the movement of body *A*. Only the absolutely uniform motion of body *A* with respect to all sources of energy at a distance (my example would be with respect to the earth) is *unconstatirbar*.

(1892)[3]

It was above all Kirchhoff who doubted whether the privileged position given to forces viewed as causes of phenomena was justified: whether with Kepler, we specify the shape of a planetary orbit, indicating the velocity at every point, or, with Newton, we give the force at any point, both are really no more than different ways of describing the facts; Newton's merit is merely the discovery that the description of the motion of celestial bodies becomes especially simple, if we specify the second differential coefficient of their coordinates with respect to time (acceleration, force). In half a page, forces had been defined away and banished from nature and physics made into a descriptive science properly speaking. The edifice of mechanics was too solid for this external change to occasion any essential influence on its inner character.

(1893)[4]

This lecturer will refer to the recent theories of electromagnetism. He will first address himself to the views which [Heinrich] Hertz inaugurated according to which equations should be completely separated from every mechanical representation, [even those] which help clarify them or cont-

---

[2] The first quotation is from a letter from Boltzmann to Wilhelm Ostwald and is dated June 11, 1892. The second letter to Ostwald is undated but was sent from a lakeside resort during 1892, and hence was apparently sent during that summer. See Hans-Günther Körber, *Aus dem wissenschaftlichen Briefwechsel Wilhelm Ostwalds,* Volume I, Berlin, 1961.

[3] Ludwig Boltzmann, "Über die Methoden der theoretischen Physik", *Katalog mathematischer und mathem. - physikalischer Modelle Apparate und Instrumente,* edited by W. Dyck, Munich, 1892. Included in *Populäre Schriften,* op. cit., pp. 1-10 and in *Theoretical Physics and Philosophical Problems,* op. cit., pp. 5-12.

[4] Ludwig Boltzmann, "Über die neueren Theorien der Elektrizität und des Magnetismus", *Verhandlungen der Versammlung Deutscher Naturforscher und Ärzte,* Nürnberg, 1893. See Ludwig Boltzmann, *Wissenschaftliche Abhandlungen,* Volume III, edited by F. Hasenöhrl, Chelsea Publishing Company: New York, 1968, pp. 502-503.

ribute to their discovery. He will point out that there are certain difficulties connected with the definition of fundamental concepts, and thinks that while on the one side the complete separation of what is given in experience from our subjective mechanical hypotheses would be useful, that on the other side a reduction in eagerness to find mechanical hypotheses or to value them less highly would damage science.

(October 1, 1893)[5]

Do you [Ernst Mach] still hold the opinion that the law of the conservation of energy can be deduced from the causal law? . . . . I believe that the impossibility of perpetual motion is a *purely empirical statement,* which in not yet examined cases may at any moment be refuted . . . .

(October 25, 1893)

I am also happy that we now completely share the same opinion, and please forgive me for having caused you so much trouble. But on reading your publication *Die Geschichte und die Wurzel des Satzes von der Erhaltung der Arbeit,*[6] I really believed at first that you appealed to some other kind of evidence than plain experience. For the speculations with which I am currently occupied, the truth about the origin of the first law is enormously important, and therefore, I could find no better address than to turn to you. Now, I see that you understand the matter the same as me and therefore I have become completely satisfied.

(1895)[7]

[originally published in English]

I propose to answer two questions: -

1) Is the theory of gases a true physical theory as valuable as any other physical theory?

2) What can we demand from any physical theory?

[5] These are two of four letters which can be found in *Ernst Mach als Aussenseiter,* edited by J. Blackmore & K. Hentschel, Braumüller: Vienna, 1985, pp. 10-15.
[6] Mach's book was published by Calvé Verlag in Prague in 1872. For an English edition see: *History and Root of the Principle of the Conservation of Energy,* Chicago, 1911.
[7] Ludwig Boltzmann, "On Certain Questions of the Theory of Gases", *Nature,* **51** (1895) 413-415.

The first question I answer in the affirmative, but the second belongs not so much to ordinary physics (let us call it orthophysics) as to what we call in Germany metaphysics. For a long time the celebrated theory of Boscovich was the ideal of physicists. According to his theory, bodies as well as the ether are aggregates of material points, acting together with forces, which are simple functions of their distances. If this theory were to hold good for all phenomena, we should still be a long way from what Faust's famulus hoped to obtain, viz. to know everything. But the difficulty of enumerating all the material points of the universe, and of determining the law of mutual force for each pair, would be only a quantitative one; nature would be a difficult problem, but not a mystery for the human mind.

When Lord Salisbury says that nature is a mystery,[8] he means, it seems to me, that this simple conception of Boscovich is refuted almost in every branch of science, the Theory of Gases not excepted. The assumption that the gasmolecules are aggregates of material points, in the sense of Boscovich, does not agree with the facts. But what else are they? And what is the ether through which they move? Let us again hear Lord Salisbury. He says:

'What the atom of each element is, whether it is a movement or a thing or a vortex, or a point having inertia, all these questions are surrounded by profound darkness. I dare not use any less pedantic word than entity to designate the ether, for it would be a great exaggeration of our knowledge if I were to speak of it as a body, or even as a substance.'

If this is so - and hardly any physicist will contradict this - then neither the Theory of Gases nor any other physical theory can be quite a congruent account of [the] facts, and I cannot hope with Mr. Burbury,[9] that Mr. Bryan[10] will be able to deduce all of the phenomena of spectroscopy from the electromagnetic theory of light. Certainly, therefore, [Heinrich] Hertz[11] is right when he says:

---

[8]The reference is to Lord Salisbury's Address to the British Association at Oxford in 1894, a meeting which Boltzmann attended. Lord Robert Cecil (1830-1903) who became the 3rd Marquis of Salisbury had succeeded Disraeli as head of the Conservative Party in 1885 and became Prime Minister for the third time in 1895. He was well-known for his intererst in science.

[9] Samuel Hawkley Burbury of London (1831-1911) graduated from Cambridge in 1854, became a lawyer, but seems to have enjoyed most doing mathematics. He wrote "On the Second Law of Thermodynamics in Connection with the Kinetic Theory of gases" in *Philosophical Magazine* in 1876. He also wrote five letters to *Nature* during 1894-1895 period on Boltzmann's work.

[10] George Hartley Bryan (1864-1928), a Cambridge scientist at the time co-authored an article with Boltzmann in 1894, contributed six letters to *Nature* on Boltzmann's work in 1894-1895, and wrote an obituary on him in 1908. He was very interested in thermodynamics.

[11] Heinrich Hertz (1857-1894) by means of discovering that two metal balls generated sparks when a certain potential was reached (1888) helped demonstrate the existence of radio waves and the correctness of Maxwell's predictions about electromagnetic radiation at the speed of light. It also helped undermine trust in mechanical attempts to understand electromagnetism and optics. Hertz's work had practical application with the development of radiowaves. Hertz became profesor at Bonn in 1889 and wrote a somewhat philosophical

'The rigour of science requires, that we distinguish well the undraped figure of nature itself from the gay-coloured vesture with which we clothe it at our pleasure.' But I think that a predilection for nudity would be carried too far if we were to forgo every hypothesis. Only we must not demand too much from hypotheses.

It is curious to see that in Germany, where until lately the theory of action-at-a-distance was much more cultivated than in Newton's native land itself, where Maxwell's theory of electricity was not accepted, because it does not start from quite a precise hypothesis, at present every special theory is old-fashioned,[12] while in England interest in the Theory of gases is still active; vide, among others, the excellent papers of Mr. Tait,[13] of whose ingenious results I cannot speak too highly, though I have been forced to oppose them in certain points.

Every hypothesis must derive indubitable results from mechanically well-defined assumptions by mathematically correct methods. If the results agree with a large series of facts, we must be content, even if the true nature of facts is not revealed in every respect. No one hypothesis has hitherto attained the last end, the Theory of Gases not excepted. But this theory agrees in so many respects with the facts, that we can hardly doubt that in gases certain entities, the number and size of which can be roughly determined, fly about pell-mell. Can it be seriously expected that they will behave exactly as aggregates of Newtonian centres of force, or as the rigid bodies of our mechanics? And how awkward is the human mind in diving the nature of things, when forsaken by the analogy of what we can see and touch? . . . .

But how can the molecules of a gas behave as rigid bodies? Are they not composed of smaller atoms? Probably they are; but the *vis viva* of their internal vibrations is transformed into progressive and rotatory motion so slowly, that when a gas is brought to a lower temperature the molecules may retain for days, or even years, the higher *vis viva* of their internal vibrations corresponding to the original temperature. . . . .

What I proved in my papers is as follows: It is extremely probable that $H$ is very near to its minimum value;[14] if it is greater, it may increase or decrease, but the probability that it decreases is always greater. Thus, if I obtain a certain value for $dH/dt$, this result does not hold for every time-element $dt$, but is only an average value. But the greater the number of

---

book on mechanics which together with Hertz's physical discoveries would have considerable influence on Boltzmann's thinking, especially during the late 1890's. Hertz died prematurely on January 1, 1894.

[12] Some caution may be urged concerning Boltzmann's description of National tendencies in physics.

[13] Peter Guthrie Tait (1831-1901) was a mathematician and physicist who taught natural philosophy at Edinburgh from 1860 until his death. In mathematics he was interested in quaternions and in physics in thermodynamics and kinetic theory. He also worked on the compressability of water, glass, and mercury. Several of Boltzmann's papers focus on his work.

[14] Boltzmann's famous *H - Theorem*

molecules, the smaller is the time-interval *dt* for which the results hold good.

Mr. [Edward] Culverwell[15] says that my theorem cannot be true because if it were true every atom of the universe would have the same *vis viva,* and all energy would be dissipated. I find, on the contrary, that this argument only tends to confirm my theorem, which requires only that in the course of time the universe must tend to a state where the average *vis viva* of every atom is the same and all energy is dissapated, and that is indeed the case. But if we ask why the state is not yet reached, we again come to a "Salisburian mystery". I will conclude this paper with an idea of my old assistant, Dr. Schütz.[16]

We assume that the whole universe is, and rests forever, in thermal equilibrium. The probability that one (only one) part of the universe is in a certain state, is the smaller the further this state is from thermal equilibrium; but this probability is greater, the greater the universe itself is. If we assume the universe great enough we can make the probability of one relatively small part being in any given state (however far from the state of thermal equilibrium), as great as we please. We can also make the probability great that, though the whole universe is in thermal equilibrium, our world is in its present state. It may be said that the world is so far from thermal equilibrium that we cannot imagine the improbability of such a state. But can we imagine, on the other side, how small a part of the whole universe this world is? Assuming the universe great enough, the probability that such a small part of it as our world should be in its present state, is no longer small.

If this assumption were correct, our world would return more and more to thermal equilibrium; but because the whole universe is so great, it might be probable that at some future time some other world might deviate as far from thermal equilibrium as our world does at present. Then the aforementioned H-curve would form a representation of what takes place in the universe. The summits of the curve would represent the worlds where visible motion and life exist.

## (1895)*

One of the most important questions when Loschmidt was in his prime concerned the composition of matter, and indeed, it is still so today; only now one puts the question in a somewhat different style. While at that time one sought from matter itself the last elements of being [*Scienden* ], so today

---

[15] Edward Parnall Culverwell of Dublin (1855-1931) contributed five letters to *Nature* during 1894-1895 concerning Boltzmann's kinetic theory of gases.

[16] In extant correspondence, Boltzmann was rather critical of Dr. Schütz's abilities.

* Boltzmann gave an address honoring the memory and scientific work of his friend and colleague Josef Loschmidt before the Austrian Academy of Sciences on October 29, 1895. The address was included in *Populäre Schriften,* op. cit., pp. 228-252. Our quotation is from page 241.

one asks [following Hertz], out of which simple elements must one form mental pictures in order to acquire the best representation of the appearances [*Erscheinungen* ]. What one means in both cases is indeed therefore rather the same [*so ziemlich dasselbe* ].**

(1895)[17]

[Georg Helm writing to his wife]

The great action lies behind me [yesterday in the conference at Lübeck]. My lecture [in favor of energeticism and against atomism], I believe, went quite well and was praised, but in the discussion things went hard. Indeed, Boltzmann began with friendly acknowledging remarks, but . . . he and later [Felix] Klein then touched on things which during my preparation and in my correspondence I had not anticipated, and which seemed inappropriate, and it was hard for me to put in even a few words of clarification. Ostwald and Boltzmann went at each other very skilfully, but the discussion didn't come to an end, and is supposed to be continued this afternoon. The meeting lasted from 9 until 12 and took place in the large athletic gymnasium in which the general sitting had been held. The big hall was at least half-full, so that a couple of hundred people experienced what happened . . . .

I had already met Boltzmann yesterday before the talk and we had discussed things in a heated manner . . . . I absented myself from the evenings festivities in order to reflect once again . . .

(1896)[17a]

[by Boltzmann]

Who can see the future? Let us have free scope for all directions of research; away with all dogmatism, either atomistic or antiatomistic! In describing the theory of gases as a mechanical *analogy*, we have already indi-

** The passage is evidently of philosophical importance in attempting to clarify Boltzmann's theory of reality, nevertheless, it is not clear whether Boltzmann really thought matter and mental pictures were the same or was simply making a wry observation about the fashionable character of Hertz's philosophy of science as presented in the latter's recently published book on mechanics (1894) without in fact accepting Hertz's position. Also, how seriously is the qualification *so ziemlich* intended?

[17] The letter from Georg Helm (1851-1923), who was an ally of Ostwald and Professor of Analytic Geometry, Analytic Mechanics, and Mathematical Physics at the Technical University of Dresden, was directed to his wife, was written in Lübeck, and is dated September 17, 1895.

[17a] Ludwig Boltzmann, *Vorlesungen über Gastheorie*, Band I, Barth: Leipzig, 1896. The quotation is taken from Ludwig Boltzmann, *Lectures on Gas Theory*, translated by Stephen G. Brush, University of California Press: Berkeley & Los Angeles, 1964, p. 26.

cated, by the choice of this word, how far removed we are from that viewpoint which would see in visible matter the true properties of the smallest particles of the body.

(1897)[18]

Number is a pure concept of experience which cannot be defined by other concepts. One can only remember what experience has taught . . . . I find what Mach wrote in his [*Principien der*] *Wärmetheorie* [1896] to be very good: "Mathematics is economically-ordered experience of counting". The whole tendency and way of considering things which can be found in such [simple] words, is more preferable to me than a mountain of Scholasticism.

(1897)[19]

I have myself repeatedly warned that an extension of our thought-pictures [*Gedankenbilder*] beyond experience is too much to trust and reminded [everyone] that one must conceive and set forth the pictures [*Bilder*] of current mechanics and especially the conception of the smallest parts of bodies as provisional.

(1897)[20]

I have been pointing out for a long time[21] that we in no way think things themselves [*die Dinge selbst denken*][22] but construct representational

---

[18] This letter is from Boltzmann to Felix Klein and is dated August 14, 1897. See Karl von Meyenn, "Boltzmann als Kritiker und Rezensent" in *Ludwig Boltzmann Gesamtausgabe*, Volume VIII, edited by R. Sexl & J. Blackmore, Friedrich Vieweg und Sohn: Braunschweig/Wiesbaden, 1982, p. 119.

[19] Ludwig Boltzmann, "Zu Hrn. Zermelos Abhandlung 'Über die mechanische Erklärung irreversibler Vorgänge'", See *Annalen der Physik*, **60** (1897) 392-398 or Ludwig Boltzmann, *Wissenschaftliche Abhandlungen*, edited by F. Hasnöhrl, Chelsea Publishing Company: New York, 1968, p. 583.

[20] Ludwig Boltzmann, Über einige meiner weniger bekannten Abhandlungen über Gastheorie and deren Verhältnis zu derselben", *Verhandlungen der 69. Versammlung der Naturforscher und Arzte*, Braunschweig, 1897, pp. 19-26. Also in *Jahresbuch der Deutsche Mathematische Vereinigung*, **6** (1899) 130-138. and Ludwig Boltzmann, *Wissenschaftliche Abhandlungen*, ibid., pp. 598-608. Our translation was made from the latter, pages 598-600 & 607-608.

[21] Boltzmann makes reference to two articles, one published in 1896 and one in 1897. They were included in *Populäre Schriften* (1905) as "Über die Unentbehrlichkeit der Atomistik in der Naturwissenschaften", pp. 140-157 and "Nochmals über die Atomistik", pp. 158 to 161. For English translations see Ludwig Boltzmann, *Theoretical Physics and Philosophical Problems*, op. cit. pp. 41-56.

[22] Presumably, Boltzmann did not use the more natural expression "think about things themselves" (*denke an die Dinge selbst*), because it would suggest a type of representational understanding different from the one he wanted to emphasize, or rather, would have contrasted the two in a way which might have inadvertently seemed to trivialize his preferred form of representation.

pictures [*Vorstellungsbilder*] by which we represent how experience is related . . . .

The idea is now very widespread [that theoretical physicists] should begin with the writing down of differential equations and consider them as the most complete picture of [sensory] phenomena. One calls this method, since it attempts to describe phenomena without any concern for underlying causes, phenomenology. Its essence consists in the attempt by Maxwell and Hertz independently from each other to apply those equations which one favors directly to the naked [sensory] facts through formula without connecting them with the colorful mantle of any hypotheses.

I have not the slightest intention to oppose any justification by these great scientists for this preference for nakedness. The representation of the purely factual with the least arbitrary accompaniment is desirable and important at all times. But one must not therefore think that the representation by means of a continuum [picture] is simpler or goes less beyond the facts than atomic [pictures] in general, though I am of course not referring to special, highly artificial atomic models . . . .

Therefore, it seems to me that next to phenomenology that a theory would possess interest and importance which could be developed in a responsible way from the simplest possible and most clearly constructed individual entities . . . .Such a theory is the gas theory.

[To be sure] our representations of molecules are still quite raw and will always remain incomplete. But the theory agrees in so many of its various details with experience and is confirmed in so many respects . . . that its basic features [*Grundlinien*] will never disappear from science . . . .

I believe, therefore, that I am able to say with assurance about molecules:

And yet they still move![23]

(1898)[24]

It was precisely when I received your dear letter that I again suffered a neurasthenic attack, as I so often have now [in Vienna], while I was able to avoid having any in Munich. This time, I was overcome with fear that the whole H - curve was nonsense [*Unsinn*].

---

23 This famous line is one which was attributed to Galileo and referred to the movement of the earth around the sun. Concerning Boltzmann's philosophical position, his apparent ability to combine belief in molecules as pictures with belief in molecules as genuine entities, this has rarely been better demonstrated than by the first and last lines of this largely scientific article. He may have thought that these positions were "rather the same", but what did he mean in exact terms?

24 Ludwig Boltzmann to Felix Klein, Vienna, February 12, 1898. See Karl von Meyenn, op. cit., p. 121.

(1898)[25]

In my opinion it would be a great tragedy for science if the theory of gases were temporarily thrown into oblivion because of a momentary hostile attitude toward it, as was for example the wave theory because of Newton's authority.

I am conscious of being only an individual struggling weakly against the stream of time. But it still remains in my power to contribute in such a way that, when the theory of gases is again revived, not too much will have to be rediscovered. Thus in this book [this Part] I will now include the parts that are the most difficult and most subject to misunderstanding, and give (at least in outline) the most easily understood exposition of them. When consequently parts of the argument become somewhat complicated, I must of course plead that a precise presentation of these theories is not possible without a corresponding formal apparatus.

(1898)[26]

> Motto: "Boltzmann explains his interest concerning the analogies in the behavior of different forms of phenomena of energy to be so great, that he himself can be designated as a passionate energeticist."

I, and certainly everyone who stood at my side at Lübeck, are no less deeply persuaded than *Herr* Helm about the fundamental significance of energy and the energy principle. I am not opposed to the effort, to the attempt at a natural representation which would place energy foremost, I am only against the manner and the way in which *Herren* Helm, Ostwald, Meyerhoffer, and others have tried to do it . . . . It is precisely in the interest of the healthy further development of the consequences of the energy laws - therefore of energeticism according to the definition of *Hrn.* Helms - that I have tried to warn against this striving to win by the introduction of unclear concepts and incomprehensible formulas principles of genuine completeness and generality. These efforts on my part have not

---

[25] This quotation is from Stephen Brush's translation of the Foreword to Part II of Ludwig Boltzmann's, *Lectures on Gas Theory,* University of California Press: Berkeley & Los Angeles, 1964, p. 216. The first edition of Parts I & II of the original German volumes came out with the title: *Vorlesungen über Gastheorie,* J.A. Barth: Leipzig, 1896 & 1898. Boltzmann placed and dated the remarks above: Volosca, Villa Irenea, August, 1898.

[26] Ludwig Boltzmann, "Zur Energetik", *Verhandlungen der 70. Versammlung Deutsche Naturforscher und Ärzte,* Düsseldorf, 1898, pp. 65-68. See also Ludwig Boltzmann, *Wissenschaftliche Abhandlungen,* op. cit. p. 638.

been wholly without success, as *Hr.* Helm has himself acknowledged, and they have also not been without value for energeticism . . . .

(1899)[27]

We shall call an idea about nature false, if it misrepresents certain facts or if there are obviously simpler ideas that represent these facts more clearly and especially if the idea contradicts generally confirmed laws of thought; however, it is still possible to have theories that correctly represent a large number of facts but are incorrect in other aspects, so that they have a certain relative truth. Indeed, it is even possible that we can construct a system of pictures of experience in several different ways. These systems are not all equally simple, nor represent phenomena equally well. However, it may be doubtful and in a sense a matter of taste which representation satisfies us most . . . .

In using this, which we have called the purely deductive method we have of course not solved the question of the nature of matter, mass, and force, but we have circumvented it by making a prior consideration of it quite superfluous. In our conceptual scheme these concepts are quite definite numbers and directions for geometrical constructions which we know how to consider and execute in order to obtain a useful picture of the phenomenal [sensory] world. What might be the true cause for this world to run as it does, what is as it were concealed behind it and acts as its motor, these things we do not regard as the business of natural science to explore . . . .

The first inconvenience that meets us here is this: previously when setting up the basic principles we were dealing with purely mental constructs which we can mentally transform as we like and of which we can require that they always conform exactly to our demands; whereas now we wish to start from phenomena directly observed, which must always be very composite and complex. If from them we wish to obtain fundamental laws, we always must generalize and idealize the phenomena, so that we really are dealing no longer with what is quite exactly factual but with processes that are realized in nature only more or less approximately . . . .

What will be our attitude to absolute space and time, given our present mode of representation? We cannot put compass points on parts of absolute space, but only on material bodies. Hence we can determine only the motion

[27] Ludwig Boltzmann, "Über die Grundprincipien und Grundgleichungen der Mechanik", erste Vorlesung, *Clark University Dicennial*, Worcester, Mass., 1899, pp. 261-275. This article was also included in *Populäre Schriften*, op. cit., pp. 253 - 270. We have used, however, the translation by Paul Foulkes in *Theoretical Physics and Philosophical Problems*, op. cit., pp. 105-106, 109, 115-116, 120, & 125-126.

of material bodies relative to each other. We must not at this stage mix the real bodies we are now alone considering with the mental picture of a fictitious co-ordinate system, as we did with the deductive method . . . .

Here again we shall of course not begin with the laws for a material point, since these are an abstraction, although we shall likewise refrain from entertaining the illusion that we can do without all abstraction. In my view we cannot pronounce a single proposition that would really be merely a pure fact of experience. The simplest terms like yellow, sweet, sour, and so on which seem to indicate mere sensations already express concepts obtained by abstraction from a great many facts of experience. When Goethe says that only half of our experience is experience, what he intends to convey by this seemingly paradoxical dictum is surely that in every conceptual grasp of experience or verbal representation of it we must already go beyond experience. The frequently uttered requirement that natural science must never go beyond experience should therefore be reformulated thus: never go too far beyond experience and introduce only such abstractions as can soon be tested by experience . . . .

As we saw, we can never isolate a body from all external influence. If now there are such influences and each of them on its own would produce a motion but their joint effect produces rest in the reference frame, then we say that the several causes of relative motion cancel each other . . . .

Chapter 3

# CALL TO LEIPZIG

## A. Correspondence

#1

To Pierre Curie[1] December 20, 1899

Because Monsieurs Stefan Meyer[2] and Schweidler[3] are assistants and work in my laboratory, Monsieur le Professeur Suess[4] has given me your letter and requested that I answer it. We very much hope that you will have the goodness to send us this quantity of moderately radioactive barium of which you write in your letter as soon as possible, because it is of great interest for us to have as much radioactive substance as possible, even if your substance is inferior to that of Monsieur Giesel,[5] of which I am not persuaded. It is of great importance for us to have some from different sources.

If in some months it would also be possible to send us your new substances which are so admirable and strongly radioactive, we would be extremely obliged to you. With salutations I am most respectfully yours...[6]

---

[1] Pierre Curie (1859-1906) was professor of physics at the Sorbonne in Paris. He is well-known for his investigation of piezoelectricity and for the discovery that the magnetic properties of substances change at a certain temperature but he is best known for his collaboration with his wife Marie Curie (1867-1934) in the discovery of polonium and radium as radioactive substances in 1898 and the help he gave his wife in isolating the latter element by 1902, isolation which especially required separating radium from barium.

[2] Stefan Meyer (1872-1949) in late 1899 had co-discovered that radioactive rays were deflected when passed through a magnetic field, but naturally he needed as many different kinds of radioactive substances from as many different locations as possible in order to determine the significance of his discovery more accurately, Understandably, however, as a mere laboratory assistant without academic clout he occasionally needed men of major reputation such as Suess and Boltzmann to bolster his requests with their prestige. His industriousness at the time is indicated by his twenty-five publications between 1897 and 1904 including a paper co-authored with Schweidler: "Über das Verhalten von Radium und Polonium im magnet. Felde", *Physikalische Zeitschrift*, Volume 1, 1900. Meyer became the first director of the Radium Institute in Vienna, which was established in 1910.

[3] Egon von Schweidler (1873-1948) like Meyer and Smoluchowski received his *Habilitation* in 1899. He became professor of physics in Innsbruck in 1911. Beginning in 1899 most of his work was on atmospheric electricity.

[4] Eduard Suess (1831-1914) was a leader of Austrian political liberalism, a professor of geology, the major figure in providing a sufficient and hygenic water supply for Vienna, and longtime president of the Academy of Sciences. His major contribution to geology concerned his theory of uplift and shields to help explain geological evolution. His major book *Das*

*Antlitz der Erde* (The Surface of the Earth) was translated into several languages. He had helped the Curies receive radiocative pitchblende from Bohemia for their research.

[5] Friedrich Giesel (1852-1927), director of a German company which prepared quinine, also produced marketable quantities of impure radioactive material and like the Curies was reasonably generous in making his material available to other researchers.

[6] This letter and the followup on January 8, 1900 were written in French and signed "Louis Boltzmann".

#2

To Svante Arrhenius[1] [Vienna, January 1900]

Thank you very much for your friendly New Year's wishes which I heartily return. I am so very pleased at what you have accomplished during the year and in such an agreeable and stimulating way, plus how you have kept me in friendly remembrance. During the year which has just fled by, there has also been some personal compensation for the monotony of teaching activity in Vienna. After the conclusion of my lectures,* I made a short tour through America where I visited New York, Boston, Worcester, Montreal, Niagara, Buffalo, Pittsburgh-Allegheny, Washington, Baltimore, and Philadelphia and became acquainted with Hall,[2] Michelson,[3] Rowland,[4] etc.. I then lived in Abbagia by the Adriatic in order to regain my health, and finally, I visited Munich for the Natural Scientists' Convention. Under separate cover, I am sending you a copy of the lecture which I gave there.[5]

[1] Svante Arrhenius (1859-1927), who had studied with Boltzmann in Graz some years earlier, was a very personable Swedish physical chemist of considerable ability, winning a Nobel prize in 1903 for his contributions to our understanding of electrolytic conductivity. Since Arrhenius's letter to Boltzmann has not surfaced yet, a reasonable conjecture would be that Boltzmann might have been praising his former student for his work on a book which would appear in print in 1900 in Swedish and which could be translated as "Treatise on Theoretical Electrochemistry".

* In addition to his Vienna lectures, Boltzmann also gave several in summer 1899 at Clark University in Massachusetts. It is very possible that Boltzmann took his tour of eastern parts of America after the Clark experience.

[2] Boltzmann probably meant Edwin Herbert Hall (1829-1907), a physics professor at Harvard University, who was well-known at least within the discipline for discovering in 1879 what is still called the "Hall Effect", a deflection phenomenon which resulted when electricity was passed through gold in a magnetic field, and which Boltzmann had already written on. See "Zur Theorie des von Hall entdeckten elektromagnetischen Phänomens", *Sitzungsberichte der kaiserlichen Akademie der Wissenschaften.* Mathematisch-naturwissenschaftliche Classe. Wien, **94** (1886) 644-669.

[3] Albert Abraham Michelson (1852-1931) was professor of physics at Clark University before moving to the University of Chicago and is famous for his unsuccessful attempts to measure ether and for winning a Nobel prize in 1907.

[4] Henry August Rowland (1848-1901) was professor of physics at Johns Hopkins University in Baltimore. His work to make gratings more exact helped transform spectroscopy

into an exact science. He was widely considered America's leading physicist at the time, a view in which he himself appears to have concurred.

[5] The lecture was titled "Über die Entwicklung der Methoden der theoretischen Physik in neurer Zeit" and was given on September 22, 1899. It was included in Boltzmann's *Populäre Schriften*, Leipzig, 1905 and translated into English by Paul Foulkes in *Theoretical Physics and Philosophical Problems*, edited by Brian McGuinness, Reidel: Dordrecht, 1974, pp. 77-100.

#3

To Joseph Larmor[1] January 7, 1900

Before answering your letter, I had wanted to reflect on the matter more carefully, so that I would be able to write more exactly, but at that time the Natural Scientists' Convention in Munich required a lot of work as did the new semester in Vienna and to my dismay [*Schrecken*] a lot of time has passed.

I have now hurriedly looked around for information about models. I have looked at an apparatus which sculptors first use to form wax copies before attempting to work in marble, then what is called a "moulage" technique by means of which one imitates healthy and sick parts of bodies, and even a procedure which allows the plastic reconstruction of organic creatures with the help of a micrometer. I also went to the Vienna Polytechnicum in order to look at models of house, roof, and bridge construction as well as an endless variety of machines but came away persuaded that what we have on hand in Vienna is obsolete and miserable. How much better the article could be written were one exactly familiar with models in London and America!

I am also in the dark about how mathematical models should be handled. I don't even understand particularly much about algebraic surfaces of pure geometrical models. The wave surface of light in *zweiaxigen* crystals would be an example of more general interest. But how am I to manage bringing in something about the appropriate equation in order to discuss conic refraction more accurately? Also about models of electromagnetic light theory, I have no idea how much mathematical grounding should be provided. Could you send me an exemplary article from the already printed parts of the Encyclopaedia Britannica?

If I dare to write about every kind of model, I am very much afraid that in spite of all of my efforts that I will fail to come up to the Encyclopaedia Britannica's standards. I simply don't know the material well enough. For this reason it would be my dearest preference if it would not cause any great difficulties to take the preparation of the article away from me and give it to an Englishman. Otherwise, I would request as much exact instruction about and answers to the questions raised above as possible. But in any case, what I write about technical and machine models will not be complete and will have to be enlarged by an English technician. I hope that you understand my letter which has been written in German; It would be difficult for me to write everything correctly in English, but I can read English without any difficulty.

---

[1] Sir Joseph Larmor (1857-1942) taught mathematics and later physics at Cambridge University. He is perhaps best-known for his work on electromagnetic theory and in particular his attempt to accomodate ether with electrons. Indeed, his major book *Aether and Matter* came out in the same year in which Boltzmann sent his letter. Larmor himself had worked with mathematical and physical models and would have been as qualified as Boltzmann himself for writing the encyclopaedia article, with the possible exception of the philosophical dimension. Larmor's location in Cambridge where the Encyclopaedia was published at that time also helped make him seem suitable as a general editor for articles on physics.

#4

To Pierre Curie Vienna, January 8, 1900

I want to render my thanks for the sample of radioactive barium carbonate which we have quickly put to use for new experiments. If it would be possible to send us some samples of your new preparation, which as you mention is so extremely strong [*efficace*], I beseech you to oblige us by sending some.[1]

With the expression of my highest respect...

---

[1] See letter #1.

#5

To Joseph Larmor [Vienna, undated]

Since I promised to do it, therefore I am sending you an essay on models.[1] But I am not pleased with it at all. If you can use something from it, that will keep me happy.[2] I make no claims on an honorarium, which is a completely secondary consideration. With heartfelt greetings...

---

[1] See letter #3 for details on some of Boltzmann's reservations. The article was initially published in the *Encyclopaedia Britannica* in 1902 and again in the 11th edition, volume 18, Cambridge, England, 1911, pp. 638-640. It is perhaps most accessible today in the book by Ludwig Boltzmann, *Theoretical Physics and Philosophical Problems*, edited by Brian McGuinness, Reidel: Dordrecht, 1974, pp. 213-220.

[2] Since the original version of Boltzmann's essay does not appear to be extant, it is not clear to what extent Larmor altered Boltzmann's manuscript, but unless Boltzmann had grammatical corrections made by someone in Vienna concerning his use of English, Larmor would have almost certainly made at least some grammatical and stylistic changes. Whether he took Boltzmann's advice and asked an English technician to make additions is not known.

#6

Imperial Chamberlain
to the Education Ministry

Vienna, February 12, 1900

His Imperial and Royal Majesty [Franz Josef I] has signified his pleasure to most graciously bestow on the full professor of theoretical physics and director of the physical institute of the University of Vienna, Hofrath Dr. Ludwig Boltzmann, the Italian composer in Genoa, Giuseppe Verdi,[1] as well as the Russian [i.e. Polish] writer, Henryk Sienkiewicz,[2] the Imperial and Royal Austro-Hungarian Medal of Honor for art and science. .

[1] This award had some historical significance. Verdi had composed several operas during the middle of the 19th century directly or indirectly opposing Austrian domination of Northern Italy. The medal suggested that Franz Josef, who had been Austrian Emperor since 1848, had forgiven Verdi. The composer died the next year, however in 1901.

[2] Henryk Sienkiewicz (1846-1916) was a writer of historical novels of which the most popular was probably *Quo Vadis* (1896), a tale about Nero and Imperial Rome, and which later became a successful Hollywood movie. Most of what are generally acknowledged to be his best novels, however, were patriotic stories about Polish history. Since most of Poland was under Russian domination in 1900, it may have seemed politic to call Sienkiewicz "Russian".

#7

To Wilhelm von Hartel[1]

Vienna, February 16, 1900

I take the liberty of offering my sincere thanks for my share of the high decorations which have been awarded as medals for art and science and request that my appreciation also be made manifest in the very highest place.[2]

[1] Wilhelm von Hartel (1839-1907) was Austrian minister of education from 1900 to 1905, vice-president of the Austrian Academy of Sciences from 1900 to 1907, and a longtime professor of classical philology at the University of Vienna.

[2] Emperor Franz Josef I is meant. Boltzmann, however, perhaps in connection with his fervent admiration for Friedrich Schiller, liked to think that he himself was basically republican at heart and at least in principle opposed to monarchy.

#8

To Wilhelm Ostwald[1]

Vienna, March 3, 1900

My warmest thanks for your kind and friendly letter, and at the same time I would like to express my pleasure that the situation in Leipzig seems to have taken on a more favorable shape. I had feared that taking over the editorship of what had earlier been called Wiedemann's[2] *Annalen* would have been an important reason for Drude[3] not to leave Leipzig. Also, I cannot conceal from you that I have become more and more dissatisfied with work in Vienna. Students dedicated to higher things are completely lacking, which is also no surprise given the fact that in Vienna alone there are already seven *Privatdocenten* waiting for university appointments someplace, anyplace, which means that I only have potential teachers for *Gymnasium* as students, and worse, whose understanding of higher matters is almost non-existent and who often come from Croatian or Slovenian schools.[4]

On top of that, there is the shaky political situation in Austria, and no one knows where it will lead, plus the upsetting hordes of people and the lack of good conveyances for travel in order to get away from it all for a short time during the semester.

If there were really still a possibility which might yet offer itself to escape from this into genuinely fruitful activity which fits my inclinations like I had in Munich[5] and gave up in such a foolishly lighthearted way, then I would feel infinitely better. I would also be especially pleased to be able to directly thank you, since this would offer a striking example of how scientific differences of opinion[6] can be combined with the best personal friendship. Furthermore, perhaps in some respects such could also help us overcome those differences.

---

[1] Wilhelm Ostwald (1853-1932) was a co-founder of physical chemistry, Nobel prize winner in 1909, and director of a very active laboratory at the University of Leipzig from 1887 to 1909. Though he long defended what he called "Energeticism" against use of the atomic theory and had suffered mental anguish when Boltzmann severely criticized his stand at a conference in Lübeck in 1895, he remained on good personal relations with the Vienna physicist-philosopher. Ostwald had invited Ernst Mach to transfer from Vienna to Leipzig in late 1897, but Mach had declined to pursue the feeler. (See the letter from Mach to Ostwald dated November 18, 1897 in *Ernst Mach - A Deeper Look*, edited by John Blackmore, Kluwer: Dordrecht, 1992, pp. 73-74.)

[2] Gustav Wiedemann (1826-1899) was professor of experimental physics at Leipzig and editor of the journal *Annalen der Physik* during the last quarter of the 19th century. Hermann Helmholtz tended to be his advisor concerning theoretical physics until his death in 1894, a role which Max Planck then increasingly exercized.

[3] Paul Drude (1863-1906) became associate professor of physics at Leipzig and after Wiedemann's death in 1899 became editor of the *Annalen der Physik*. He taught in Giessen from 1901 to 1905 when he transferred to Berlin but unexpectedly died, presumably from suicide in 1906. It was Planck as advisor in charge of theoretical physics who gave the journal continuity, and after Drude's death he officially became co-editor with Wilhelm Wien.

[4] In the jargon of our own age, Boltzmann's remarks might seem like "racism" or "cultural prejudice", but Boltzmann's own main teacher, Josef Stefan, had come from a Slovenian background and Boltzmann's respect for him was virtually boundless. (See letter #15, footnote 2 for more information.)

[5] Boltzmann taught at the University of Munich from 1890 to 1894.

[6] The differences were primarily over the soundness of Ostwald's "Energeticism" and the indispensability of the atomic theory in physics. In an environment less dominated by Ernst Mach's opposition to "metaphysics", the debate would probably have centered more directly on the reality of atoms.

#9

From Wilhelm Ostwald Leipzig, March 10, 1900

*Strictly Confidential!*

The matter is going more favorably than I had dared to hope. Drude is definitely going to Giessen, and the prospect of winning you has stirred the liveliest interest in the faculty. What we can do we will do to bring home to the Ministry the urgency of bringing you here, and a favorable outcome may be designated as *very probable.*

The faculty has not yet made a decision, therefore, all that I am reporting to you is my supposition. If everything goes as we want and may hope, then you will receive the official *Anfrage* in *one or two weeks.* I don't know if you can make yourself free so quickly. If it goes through, you could even be here for the next semester.

I must repeat that these are only hopes. But I believe with good conscience that I am able to say that these hopes are well-grounded.

With best wishes...

#10

Beckmann[1] to Wislicenus[2] (telegram) Leipzig, [March 1900]

Drude has accepted Giessen. In the faculty meeting, Ostwald disclosed that *Boltzmann* would definitely accept. Concerning the professorship, the question will be put to the lower house [*Landtag*] and pushed through with acceleration. The commission meets at five o'clock on Monday, after that the faculty. Please answer whether you [*du*] approve of Boltzmann and expedited handling.[3]

---

[1] Ernst Otto Beckmann (1853-1923) was a professor of applied chemistry. He taught in Leipzig from 1897 to 1912.

[2] Johannes Wislicenus (1835-1902) had succeeded Kolbe in Leipzig in 1885 and been Beckmann's teacher. Later, he became professor of chemistry in Zürich. He is best-known for discovering two different forms of lactic acid (1863) which seemed to differ only by the way they reacted to polarized light. Van't Hoff's method for arranging organic molecules in three dimensions provided the key which Wisclenius was looking for to explain the two different forms in a deeper way.

It seems that Wislicenus may have arranged the schedule of speakers at the famous debate at Lübeck in 1895 in a way which favored the atomists over the energeticists. At least that was the subsequent opinion of Ostwald who was his colleague in Leipzig at the time. See Robert J. Deltete, *The Energetics Controversy in Late Nineteenth-Century Germany: Helm, Ostwald and Their Critics,* Volume II, UMI: Ann Arbor, 1983, pp. 414-416.

#11

To Wilhelm Ostwald Vienna, March 13, 1900

Let me express anew my best thanks for your second, so amiable letter. I cannot even utter how much it means to me in my current regrettably somewhat depressed state of mind that you and your estimable Leipzig colleagues have so good an opinion of me. For the rest, I wait with eager attention the further development of the matter.

With highest respect and hearty greetings...

#12

Sievers[1] to Waentig[2] Leipzig, March 13,1900

. . . . A possible quick resolution of the appointment would have the considerable advantage that it would allow Professor Boltzmann to be able to exercize his personal influence on that part of the future physical institute which will be arranged for theoretical physicists . . .

And finally, it should be mentioned that rarely have members of the philosophical [arts and science] faculty wished for the appointment of a new colleague in such a unanimous way and looked forward to having this outstanding man in their ranks with so much happy expectation as in the case before us. This attitude is shared above all by the colleagues who are nearest to his own field, particularly by the representatives of experimental physics;[3] but other members of the faculty more distant have also expressed themselves in the same manner.

I am adding Boltzmann's letter [to Ostwald?] in order to help make my sketch more understandable, but at appropriate times I will want it back.

---

[1] Eduard Sievers (1850-1932) was professor of philology and dean of the philosophy faculty in the University of Leipzig at that time. (But an undated document, presumably written a few weeks later, lists Otto Wiener as dean, but perhaps only temporarily since he writes "zur Zeit". See letter #19, footnote 1 for more information about Wiener.)

[2] Dr. K. Waentig was an official from the Saxon Ministry of Education and apparently the main representative concerned with Boltzmann's proposed transfer. The Minister himself it seems was Dr. Paul Sydewitz (1843-1910), whom Boltzmann later met in Leipzig on his way to Berkeley in 1905.

[3] While this comment may sound strange to people today very conscious of the fact that Boltzmann was primarily a theoretical rather than an experimental physicist, one may do well to keep in mind that at that time the distinction was generally little known or emphasized even by other academic people who happened to be outside of physics, and that when noted, experimentalists still tended to enjoy a higher reputation, hence their support was important. Also, a lot of experimental physics was going on in Boltzmann's Vienna laboratory at the time, and he had taught and contributed to experimental physics while in Graz during the 1880's.

#13

Waentig to Ostwald Leipzig, March 15, 1900

Since I know that you are especially interested in Boltzmann's appointment and also his good friend, I have allowed myself to turn to you in this matter and request, though in complete confidence, for full access to your opinion and for your full support.

#14

Ostwald to Waentig Leipzig, March 16, 1900

Boltzmann's salary in Vienna seems to have been about ten to twelve thousand Marks per year, and I believe that given his position in science - and he is incontestably the top man in his specialty, and not just in Germany - that he should not be offered less. I believe that I can guarantee that for that amount he is *certain* to come. His coming would be a great event for our university! - If you wish a more exact disclosure of his current salary, then I will request the information from Boltzmann (who I am also writing now) and send it to you. My personal relation to him is that of scientific opposition [*Gegnerschaft*] combined with the highest personal esteem . . . .

#15

From Marian von Smoluchowski[1] (draft) Cracow, March 29, 1900

May I express my heartful thanks for the good wishes which you have sent me by telegram, even if my reply is merely by letter.

I consider it a great honor that you have acknowledged my work by word of mouth on this and other occasions, [acknowledgement] which will [make] me [better known] to a wider [audience].

In theoretical physics, I am indebted first [of all] to my dead teacher [Jo-sef] Stefan,[2] and to you Herr Hofrath for almost everything else. You have exercized through your lectures and textbooks, which I have studied again and again with pleasure and profit, the greatest influence upon me, and I am proud of being able to count myself as having been one of your students. Please excuse me for allowing myself to say these things, but I do not want this moment to pass by, which is so important and pleasing to me, without sharing with you, Herr Hofrath, how much I am indebted to you.

In the hope that you will accept this assurance of my sincere thankfulness, I remain...

---

[1] Marian Smoluchowski Ritter von Smolan (1872-1917) was Polish and one of Boltzmann's most outstanding students. He became professor in Lemberg [Polish Lwów, Ukrainian Lviv] in 1900 and Cracow in 1913. He is best known for his collateral work with Einstein & Perrin on Brownian motion. His early death during World War I was a tragedy for science.
[2] Josef Stefan (1835-1893) was born in Klagenfurt, Styria of poor Slovenian parents. He taught physics in Vienna from the early 1860's until his death in 1893. He is well-known for the Stefan-Boltzmann law and for his work on rate of cooling of hot objects. Boltzmann was appointed to Stefan's chair in Vienna in 1894.

#16

From Wilhelm Ostwald Leipzig, April 10,1900

The newspapers announce that you have received a call from here, but that the Austrian government is making great efforts to keep you there.

If even after your repeated letters I am not to doubt that you will not allow yourself to persevere,[1] I would still do all I can in order to repeatedly make evident that an ample and successful appointment here just as you wish is certain to a degree like at no other university. I know the local conditions and the quality of the students, having been here for over twelve years. And I must say that both are outstanding. Only those who really want to work come to Leipzig, and even under the least likely circumstances one will always find a small band of loyal listeners [*Zuhörer*].

For me personally, it would be a hard blow if you didn't come. I have countered all of the objections which have been directed against your appointment that you wouldn't come, by arguing that in any case if the offer were good enough that you would. Let me add, that failure in this matter would undermine my whole reputation with the faculty and ministry. Therefore - if you vacillate [*schwanken*] on their account, come on mine. You won't regret it.

With best wishes...

---

[1] The translator admits that there may be a problem in catching the exact meaning in this passage in face of Ostwald's use of a double negative: "...nicht zweifeln will, dass Sie sich nicht halten lassen werden...".

#17

To Wilhelm von Hartel[1] Vienna, April 10, 1900

Since your devoted servant has definitively decided after a long and difficult struggle within his soul to accept a professorship in theoretical physics at the University of Leipzig to which he has been called starting from October 1, he requests the high ministry to release him from the position which he currently holds within the University of Vienna beginning October 1, 1900.

He adds to this petition a request to look at his step as not in any way lack of patriotism or ingratitude for the extraordinarily friendly cooperation which he has found from the side of the Government as well as overall in Austria.

---

[1] The Austrian minister of education. See letter #7, footnote 1.

#18

To Wilhelm Ostwald Vienna, April 11, 1900

Just a few minutes ago before noon today in Vienna, I handed in my request to resign. I can therefore be counted upon. Earlier, I had once given my promise [to tell the authorities] about any offer [from another university] before definitively accepting it; thus, against my will, the result has been newspaper reports. I have never concealed that I have a tendency toward nervousness; and so it now happens that once again I have appointment fever. On top of that, I was duty-bound to finish some articles for [Felix] Klein's mathematical encyclopedia, which have required a lot more effort than I had anticipated[1] Instead of being able to recover over Easter, I plunged nervously into the literature. But hopefully, I will pull myself together again soon. Therefore, please excuse me that I have not written you; I have also not written to any Viennese. But since we already spoke some in [the Natural Scientist's Conference at Munich [in September 1899], I don't think any damage has been done. I will probably come to Leipzig soon.

With the heartiest greetings...

---

[1] Hans-Günther-Körber mentions in a footnote in *Aus dem Wissenschaftlichen Briefwechsel Wilhelm Ostwalds, I. Teil,* pp. 29-30, that Boltzmann's articles were on recent statistical contributions to mechanics but that they were part of a larger project which was not completed by the time of Boltzmann's death

#19

From Otto Wiener[1] (carbon copy) Leipzig, April 25, 1900

To my great regret, during your presence in Leipzig, I was traveling. According to what I have heard, your negotiations in Dresden [the capital of Saxony] have been settled to your satisfaction and have confirmed your coming here for the next Winter [i.e. Fall] semester. I am extraordinarily pleased to be allowed to become your colleague, and there are also best wishes from others, for example, Professor Warburg.[2] Hopefully, you will be happy in Leipzig, so that you will not leave us again soon.

Because of my absence from Leipzig, it is unfortunately only now that I can converse extensively with you about what lay close to my heart, especially the planned physical construction. To help orient you, let me clue you in on the plans for the new building as they have been laid before the Saxon Chamber [of Deputies]. I have heard that their approval is very probable. The smaller plans are in the measure of 1:1000; the four larger have the ratio of about 1:400.

These plans originated at a time when it still seemed very probable that colleague Drude[3] would move into the new institute, who not only had a room of his own for his own experimental work, but also had students who were busy with their experiments. Thus, our current institute became too crowded and [it was thought that] he should receive more space in the new one. But I thought that it would be good to ensure that we had sufficient room in case we could attract another theoretical physicist, not merely if he wanted to carry out experiments but also if he wanted to let his students work part-time on experimental problems. I also thought that we could easily avoid what happened with Hertz in his time, who declined to go to Berlin mainly because he would have had no institute. I also believed that under the given conditions we could hardly attract you without the promise of such an institute. From W. Wien[4] I knew exactly that he wouldn't transfer without such. I reported accordingly in the submission to the Ministry that an instiute for theoretical physicists should also be forseen, because scarcely any theorist of significance would come to Leipzig without it.

But during the occasion of a conversation in Munich, you said that you would make no big claim for space, but the question then arises whether or not you would be favorably inclined toward taking over an institute for theoretical physics, particularly as the plans have already been laid down and what wishes for changes you might have. I am also especially interested in learning what you may have discussed with the Ministry about these matters....[5]

---

1 Otto Wiener ( 1862-1927) was professor of experimental physics in Leipzig and is best-known for his work on standing light waves. Much of his later career was concerned with the construction of new buildings in Leipzig and with defending the ether theory. He often typed long letters, but his usual habit of keeping carbon copies has aided scholars substantially, since in many cases it permits us to learn the contents of both sides of a correspondence between scholars. But as Boltzmann suggests in letter #20, it was apparently Wiener's bad handwriting which got him to type more and more of his letters and thus help future scholars. There may be a moral of some kind here.

2 Emil Warburg (1846-1931) became professor of physics in Berlin in 1895. He is an example of a physicist who largely switched from theoretical to experimental work. He was also entrusted with administrative responsibilities, especially running conferences. As Planck's colleague, he often lectured to large audiences of three hundred students or more. Since he is sometimes mentioned as helping to edit the ***Annalen der Physik***, one might suppose that this took place during Wiedemann's last years shortly before his death in 1899.

3 See letter #8, footnote 3.

4 Wilhelm Wien (1864-1928) was professor at Würzburg from 1900 to 1920 when he transferred to Munich. He is best-known for his work on black-body radiation for which he won a Nobel prize in 1911.

[5] The last page of Wiener's carbon copy of his letter is missing.

#20

To Otto Wiener April 30, 1900

I also regret that we were not able to meet in Leipzig. Concerning what I said to you in Munich, I thought of myself more as a mathematician. On the question of the Institute, I did not negotiate with the Saxon Ministry at all about it. Perhaps I made a mistake. About the Institute for Theoretical Physics, at least as represented by the plans which had been sent to me, it seems very practical and pretty.

Unfortunately, I have had some bad experiences about such institutes. In Munich [1890-1894], I always had a couple of people who worked with me, but still the most able [*Tüchtigsten*] went to Lommel[1] where everything was more completely available. But here in Vienna [1894-1900] I have had no luck at all with those working with me; also endless vexation with attendants and assistants. Moreover, because of my bad eyes I cannot do experimental work at length and am also clumsy in setting up and and arranging apparatus. Nevertheless, in terms of my ideas I would gladly let advanced students who don't have to be shown how to do everything work there.[2]

If you believe that I could find enough students of this kind and if the Ministry is ready to approve what is required, then I would be content to take over the institute for theoretical physics which has been proposed by you. But unfortunately, I have some doubts whether I am wholly the right man to manage and lead, since most unfortunately I am once again suffering from my old evil, the most tormenting neurasthenia.

I also wanted to ask you whether Drude in the old Institute had any particular collection of apparatus and especially whether he had access to standing funds.

I intend to teach five hours of analytic mechanics as a foundation during the coming Winter semester in Leipzig. Will that fit in well with the rest of the plans for giving lectures? Please write all of your letters with the machine, since it makes them easier to read.

With the highest of high respect and the most heartfelt greetings...

[1] Eugen Lommel (1837-1899) became professor of Physics at Munich during the middle 1880's and was instrumental in attracting Boltzmann to the university there in 1890. Lommel himself was interested in relating matter to ether in a plausible manner. For more information, see Christa Jungnickel and Russell McCormmach, *Intellectual Mastery of Nature*, Volume 2, University of Chicago: Chicago and London, 1986, pp. 151-153.

[2] In the event, the new institute for theoretical physics in Leipzig while actual work began in May 1901, was not ready for student use until the Winter semester of 1904/1905, that is, well after Boltzmann had come to Leipzig and returned to Vienna.

#21

From Otto Wiener (carbon copy) May 6, 1900

My best thanks for writing in such a friendly way. Because of some questions which your letter brought up, I spoke with Professor Ostwald. It is our foremost wish that you feel comfortable with us, and that your activity here from top to bottom be defined by and adjusted to your own wishes and happiness.

If you want to teach a five hour lecture-course on analytic mechanics next winter semester, nothing stands in the way. Professor Ostwald thinks that it would also be quite enough if you just wanted to take it easy and only teach three hours. But I suppose that you want to put special weight on establishing a good foundation so that you will have sufficient time available for your later courses.

Its not as if you had to follow it, but just for your own information, I would like to share with you what Drude did. He gave a four-semester cycle of four-hours-per-week courses. First came elasticity, acoustics, and capillarity; second, theory of heat; third, theory of electricity and magnetism; and fourth, theory of light. Outside of that, in summer he normally taught another hour on experimental work concerning electric oscillation or electrotechnology. Drude did not teach mechanics at all, since the mathematicians taught a lot of it.

In the Winter Semester of 1898-99 [Carl] Neumann[1] had taught selected topics from analytic mechanics four hours per week. [Christian] Mayer,[2] who has since retired, during the next winter semester taught theory of potentials for two hours per week as a continuation. [Friedrich] Engel[3] at the same time [1899-1900] taught higher analytic mechanics for four hours each week. He is also teaching a one hour continuation during this semester. But it is also good if mechanics is taught every semester.

By the way, the mathematical gentlemen usually have a common meeting for the next semester in order to fix their teaching schedule so that their courses will not collide too strongly with each other. But in any case, no one can object if you announce that you are going to teach mechanics, especially since you need it as a foundation for your later lectures. For your convenience I am sending you a schedule of classes for this and the previous semester.[4]

Drude did not have any special collection of apparatus in the old institute, and I provided everything for him and his laboratory assistants from institute funds or purchased it. But it was intended that in the new institute a special fund should be requested for him. Professor Ostwald thinks that nothing would suffer if you were to have to start from the same situation, since there is no doubt that later you would receive whatever you wanted which you might need.

Drude had an assistant - indeed, so to say, only half of one, since he only received 600 marks yearly - one who assisted when lectures required something experimental and also helped guide more advanced, virtually self-reliant students in their laboratory work. Naturally, at any time and without the slightest need to worry about being turned down you may request a half or [or even a] whole assistant.

It pleases me that you do not fundamentally refuse to take over an institute for theoretical physics. The detailed work of handling laboratory personnel can then be taken off your hands by an appropriate assistant. In short, you need not have the slightest apprehension about such things.

I forgot to write you yet that for this semester Drude had announced that he would teach theory of light, which [Otto] Wiedeberg[5] is handling - and with it he had planned to begin a new cycle of courses [*Kur*] for the next winter semester, a state of affairs which fits in very well with your intentions.

I am definitely persuaded that concerning teaching that you will like it well here. I find on average that the interest and eagerness of the young people is outstanding.

In hopes that your health will no longer give cause for complaint, I greet you with highest respect...

---

[1] Carl Gottfried Neumann (1832-1925) taught mathematics in Leipzig from 1868 to 1911. Since 1869, he was editor of the journal *Mathematische Annalen* which published occasional articles in physics. Though primarily a mathematician, Neumann also contributed to our understanding of crystal optics. He seems to have gotten along well with Boltzmann. Indeed, it appears as if Wiener might have been expressing a common opinion of the time, namely, that theoretical physicists might well feel more comfortable with mathematicians than with "real" i.e. experimental physicists.

[2] Christian Mayer (1839-1908) was a mathematician and son of a wealthy Leipzig family.

[3] Friedrich Engel (1861-1941) taught mathematics in Leipzig from 1889 to 1904, first as an associate professor and from 1899 as a full professor. He is known for his work on transformation groups.

[4] There seem to have been four full profesors of mathematics, an extraordinarius, and two Privatdezenten at that time in Leipzig. Because Sophus Lie was sick, he was apparently replaced by Otto Hölder from Königsberg in 1899. The fact that analytic mechanics and even its higher forms were already being taught at Leipzig by other professors such that Boltzmann might have to compete with these mathematicians for students, and such that many students would already have adopted assumptions about mechanics perhaps incompatible with what Boltzmann wanted to teach, not to mention the very different philosophical assumptions held by many of Ostwald's students may have been factors in discouraging Boltzmann even before he went to Leipzig. In spite of Wiener's optimism and encouragement, the Austrian physicist may have felt he was running into a stacked deck. Nor

was the absence of a laboratory fully welcome. In short, Wiener's letter which was clearly intended to help smooth matters could well have had the opposite effect. See letter #23 plus others written during the Summer of 1900.

[5] The next year (1901), Otto Wiedeburg became professor of physics at the Technical University of Hannover. His Dissertation had been on hydrodiffusion for which he had received his PHD in 1890.

#22

To Joseph Larmor — Vienna, May 20, 1900

Please accept my warmest thanks for sending me a copy of your interesting book on ether and matter.[1] With highest respect...

[1] Joseph Larmor, *Aether and Matter*, Cambridge, 1900. See also letters #3 and #5.

#23

To Otto Wiener — May 27, 1900

Please accept my best thanks for your last two[1] so extremely friendly letters. Please excuse me that I only first answer them now. On the advice of my doctor I had been traveling for a few days, but now I have started giving my lectures again, without however feeling essentially better.

On the other hand, I have written out the announcement of my lectures [to be given in Leipzig]. Hopefully, the form is right. Please take charge of sending it to the right place and correcting any technical errors.

With highest respect and heartiest greetings...

[1] See letter #21. Only carbon copies of the letters from Wiener to Boltzmann and which were kept by Wiener survive. Since no content from the second letter mentioned above appears to be extant, presumably it was either not typed or no carbon copy was made or kept.

#24

Nabl[1] to Meyer — Cambridge, June 12, 1900

. . . . `*A propos* [Lord] Kelvin, have you read his latest insult [*Beschimpfung*] directed against the Maxwell-Boltzmann electro. theory which he declares to be *rubbish* ? I think he's a silly old idiot . . . .

[1] Josef Nabl (1876-1953) like his good friend Stefan Meyer (See letter #1) was a strong supporter of many of Boltzmann's ideas. In 1900, however, Nabl was working in J.J. Thomson's laboratory in Cambridge. His letters to Meyer are full of informative, lively, and sometimes disrespectful comments about Cambridge science and daily routine. In 1905, Meyer apparently helped Nabl become co-author of Boltzmann's last published paper (See letter #108). Nabl was badly wounded in World War I and forced to abandon his work in physics.

#25

Klein[1] to Sommerfeld[2] Göttingen, June 21, 1900

. . . . Boltzmann is sick,[3] which means that he finds himself in a deeply depressed state of mind over the alternatives Vienna versus Leipzig (a problem which is by no means finally resolved). Hopefully, matters will change. It is impossible to make a prediction about his role [*Mitwirkung*] with the Encyclopedia[4] (to which, by the way, he is determined in an almost touching way to contribute). . . .

---

[1] Felix Klein (1849-1925), an outstanding mathematician at Göttingen, was preparing an encyclopedia of mathematics to which Boltzmann was determined to contribute. He had already sent part of a major contribution and the last article he published before his death in 1906 was a co-authored essay with Josef Nabl for the encyclopedia.

[2] Arnold Sommerfeld (1868-1951) who succeeded Boltzmann in Munich in 1894, collaborated with Klein on his encyclopedia of mathematics. He would later (1913) make an important modification of Bohr's theory to allow for elliptical orbits of electrons.

[3] It is not clear how Klein learned about Boltzmann's condition. The last extant letter from Boltzmann to Klein is dated December 9, 1899. (See Karl von Meyenn, "Boltzmann als Kritiker und Rezensent" in *Ausgewahlte Abhandlungen der internationalen Tagung Wien 1981*, (*Ludwig Boltzmann Gesamtausgabe*, Band 8) edited by R. Sexl & J. Blackmore, Graz, 1982, p. 116.

[4] *Encyklopädie der mathematischen Wissenschaften*, 1898-.

#26

Nabl to Meyer Cambridge, July 9, 1900

. . . . Concerning his Lordship's [Kelvin's] invective against the Maxwell-Boltzmann gas theory, it was contained in a speech which he gave in the Royal Institution on April 27 and which was partly cited in the May 4 issue of *Engineer* and where with justification it met severe criticism. You can also find a report on this speech in the *Times* from April 28 or 29. Kelvin takes the field in an aggressive way against [gas] theory, accusing its consequences of contradicting our customary common sense. He comes to the conclusion that it is "absolutely necessary" for the progress of science "to throw it (namely the theory) overboard". It seems that just like so many other ageing physicists (take Mach for example), Kelvin has now been seized by longing to rip apart the structure of science without having put forward even a single building stone for a new edifice. But enough about this . . . .

#27

Franz Josef I to Hartel Ischl, July 14, 1900

I authorize the full professor of theoretical physics at the University of Vienna, Hofrath Dr. Ludwig Boltzmann, to withdraw from his current teaching position with the end of September 1900.

#28

Fuchs[1] to Hartel Seeboden, August 10, 1900

My heartiest thanks for the letter which I received from Your Excellency this morning. I took it at once to Boltzmann, who is staying near here in a small sanatarium because of his neurasthenia, together with some other neurasthenic and anemic patients, who provide exactly the wrong social environment for him.

Since Boltzmann was not there when I came, I took the opportunity to talk alone with Dr. Fasan who is handling his case. He told me that as a result [*in Folge*] of his call to Leipzig that Boltzmann was so tormented by thoughts of suicide that Fasan always took care that Boltzmann was accompanied by someone when he took a walk. Fasan expected, however, that he would soon calm down. Concerning the neurasthenia as a whole he put forward a favorable prognosis, in which the symptoms of a genuine psychosis were not present.

After lunch, Boltzmann came to me and requested that I express to you his deepest thanks and announced that he will request his reappointment in the form of a letter which he will send directly to Vienna. He intends to state that the lack of a laboratory in Leipzig has made him recognize that he will not find all of the necessary conditions there for his scientific work.

For what Your Excellency is prepared to do to help Boltzmann in his so deeply piteous condition, my profoundest thanks again and again.

---

[1] Ernst Fuchs (1851-1930) had been a professor of medicine at the University of Vienna since 1885. He was most interested in the structure, function, and pathology of the human eye.

#29

To Frau Boltzmann [Seeboden, August 1900]

Dearest Mama:[1]

That we are separated so far is terrible for me. I have a guard here . . . and outside of him and some kindhearted nurses no one to talk to. The doctor could hardly care less about me.

Spruceneedle baths, wet compresses, electrical treatment, etc., cost fearfully much and won't do any good anyhow. Everything together will certainly cost thirty marks[2] daily. I sleep badly and am wholly outside of myself with sadness. If someone would fetch me, I would leave immediately. Alone, they won't let me go. Please Mama, Come! Or arrange for somebody else to come. Please have mercy and don't ask anyone else's opinion, but decide yourself. Please forgive me everything! . . .

from Your Lui

One will certainly advise you that the treatment will make me healthy here, but that is absolutely untrue.

---

[1] Henriette Boltzmann (1854-1938), neé Henriette Edle von Aigentler, married Ludwig Boltzmann in 1876. They had five children. She was interested in mathematics and had wanted to attend the University of Graz during the 1870's, but it was difficult for women in those days. She was deeply concerned about her husband's health, as we will see from subsequent letters. He called his wife "Mama". His real mother had died in 1885.

[2] The first sentence of the letter and the use of a term for German money suggests that the sanitarium was not in Austria. It is not certain how many small towns had the name "Seeboden", perhaps several, hence, it may not be easy to make sure where the sanitarium was.

#30

Fuchs to Hartel — Seeboden, August 12, 1900

As postscript to my letter of the day before yesterday, let me share with you that this afternoon I wanted to call on Boltzmann in order to learn more about his health, but discovered instead, that yesterday evening he had suddenly left, even though some hours earlier he had been with me and told me that he wanted to write to Your Excellency. Where he went, no one knows. I hope that he won't take any ill-considered action, as I am fearful that he may, given his disturbed condition.

Let me say now, in order to give prominence to the matter, in case it was not sufficiently apparent, that it was not my freely-chosen idea to send the first letter to Your Excellency, but it was done at the express and urgent request of Boltzmann.

#31

To Wilhelm von Hartel — [Vienna?, August 1900]

Excellency, I earnestly ask your forgiveness once again for the trouble I have caused you. I told Hofrath Fuchs that I had the fixed intention of staying in Vienna, if that could be made possible and requested him to write you to that effect. At that time, I found myself in a sanitarium which I left without having been made healthy again. I am not able to give a cause for changing my decision except my tormenting illness itself. Nor would anything else correspond to the truth. Also, just yesterday, I received a decree from Dresden confirming my appointment [in Leipzig]. This means that I am bound to go there and can no longer strive to regain my position in Vienna.

Pleading once again for your forgiveness, let me sign myself with the expression of my highest respect for Your Excellency . . .

#31A

Jungnickel[1] and McCormmach[2] on Boltzmann [general reference]

Boltzmann's stature and mobility made him repeatedly the first choice of faculties and a participant in an unusual number of institutional developments in theoretical physics. In Vienna, he was the first occupant of an institute for theoretical physics, as he was in Munich and in Leipzig (and he would have been in Berlin, too, if he had accepted the offer that eventually went to Planck). He had done his most important [physical] research before assuming any of these posts, but theorists such as Hasenöhrl,[3] Sommerfeld, and Heisenberg who later occupied these institutes did some of their most important work there.

[1] Christa Jungnickel (1935-1990) received her PHD from Johns Hopkins University in 1978. The rest of her life was very largely spent researching and co-writing the two volume work ***Intellectual Mastery of Nature***, University of Chicago Press: Chicago, 1986, together with Russell McCormmach. The work is mostly about the institutional development of theoretical physics in Central Europe with Ludwig Boltzmann as a major figure.

[2] Russell McCormmach was born in 1933, received his PHD from Case Institute of Technology in Ohio in 1967, became a professor in the history of science at Johns Hopkins University, and from1969 to 1980 was editor of the journal *Historical Studies in the Physical and Biological Sciences.*

[3] Friedrich Hasenöhrl (1874-1915) was a student under Boltzmann in 1894 and recommended by him to an assistantship with Heike Kamerlingh-Onnes in Leyden. Hasenöhrl received his Habilitation in 1899. From 1904 until 1910 he worked on the relation between mass and energy, virtually anticipating Einstein's $E = mc^2$. He became Boltzmann's successor in 1907.

#31B

Kowalewski[1] about Boltzmann [about Fall 1900]

A daughter of Ostwald depicted the evenings which Boltzmann spent playing music in the family circle. With special liking one played Mozart, Schubert, or Beethoven and attempted to work against Boltzmann's depressed mood. The shortsightedness of the pianist Boltzmann had gone so far that to read the notes he needed a second lens and even a third one each behind the other in his glasses.[2]

---

[1] Gerhard Kowalewski (1876-1950) was a *Privatdozent* for mathematics at Leipzig during the first year that Boltzmann was there. He became a professor at Giessen in 1901, Bonn in 1904, and Dresden in 1920, becoming rector at the last university from 1935 to 1937. He did important work in "*Determinanten Theorie* ".

[2] This excerpt from Kowalewski's book, *Bestand und Wandel - Meine Lebens Erinnerungen,* Oldenbourg Verlag: Munich, 1950, pp. 128-129, is included in A. Dick and G. Kerber, *Ludwig Boltzmann - Katalog zur Ausstellung an der Zentralbibliothek für Physik in Wien,* Vienna, 1982, p. 29, and also most of the quotation is in Wolfgang Stiller, *Ludwig Boltzmann,* Verlag Harri Deutsch: Frankfurt am Main, 1989, p. 34.

## B. Opinions about Ostwald and Boltzmann and the Struggle between Energeticism and Atomism from Two Former Students and a Colleague

*Marian von Smoluchowski*: Today, it is no longer easy to think ourselves back into the attitude which reigned [in physics] at the end of the last century. At that time, with few exceptions the leaders in Germany and France were persuaded that the atomistic kinetic theory had already played out its role . . . .

But now, twenty years later [1914], the situation is completely reversed. Atomism is generally recognized as the foundation of current physics; the second law of thermodynamics, contrary to earlier, has become a basic principle of physics, indeed, unshakeable dogma as if valid for all time.[1]

---

[1] Marian V. Smoluchowski had studied under Stefan and Boltzmann. See letter #15. See also his article "Gültigkeitsgrenzen des zweiten Hauptsätzes der Wärmelehre" in P. Debye (ed.), *Vorträge über die kinetesche Theorie der Materie und Elektrizität,* Teubner: Leipzig, 1914 as included in Wolfgang Stiller, *Ludwig Boltzmann,* Verlag Harri Deutsch: Frankfurt am Mein, 1989, pp. 138 & 139.

*George Cecil Jaffé*: Before I . . . continue I must say something about the background to the general situation in the scientific world at that time. Around 1900 there was a hefty fight between two scientific trends in process, which were designated "energeticism" and "atomism". Ostwald was the spiritual leader of the first . . . and on top of that very aggressive. He and his followers made bold claims to represent "presupposition-free science", as they themselves called it. They demanded that physics and chemistry and even a deductive science like mechanics must have a purely descriptive (phenomenological) character which should rest on a very few general principles such as the two main laws of thermodynamics, while only observable data should be brought into relation to these principles. Theory should make no use of detailed models . . . .

On the other side stood the followers of atomism, who in those days fought somewhat on the defensive. They had to confess that their constructions were mere models, but they defended the legitimacy of such models and maintained that a deeper insight into natural laws could not be gained merely by relating observable phenomena . . . The leader of this group was Boltzmann, the only survivor of the three great theoreticians who had created the kinetic theory, Clausius, Maxwell, and Boltzmann.

Even though Ostwald and Boltzmann were rather severe scientific opponents, they preserved the highest respect for each other and on a personal level were good friends. When it was thought to establish a chair for theoretical physics at Leipzig, Ostwald was tolerant enough to try too get the very best man available, and that was Boltzmann. So it came about that students could observe the two leaders of the opposed schools teach in friendly fashion side by side. Naturally the students supported their own mentor, and so it happened as earlier in Troy: While the gods fought in the clouds, the simple people on earth did the same, only with weaker arguments more violently expressed.[2]

[2] George Cecil Jaffé (!880-1965) had studied under Ostwald at Leipzig during the period that Boltzmann was there. This translation is from Wolfgang Stiller's book *Ludwig Boltzmann*, Verlag Harri Deutsch: Frankfurt am Main, 1989, pp. 136 & 137. See also Jaffé's article "Recollections of Three Great Laboratories", *Journal of Chemical Education*, **29** (1952), 230-238.

*Alois Höfler*: The enemies of traditional atomism, who were led by Ernst Mach, liked to call him [Boltzmann] the "last pillar" . . . Some even ascribed his symptoms of melancholia, which went back for years, to the fact that he saw the tottering of that structure but could not prevent it from falling, even with all of his mathematical skill.[3]

[3] A. Höfler, *Süddeutsche Monatsheft*, **3** (1906), 418 as quoted in E. Broda, *Ludwig Boltzmann: Man - Physicist - Philosopher*, Ox Bow Press: Woodbridge Conn., 1983, p.30. Most recent scholars familiar with the situation in Vienna around 1900 agree with the three observations quoted above. For example, Friedrich Stadler, probably the single Austrian scholar most knowledgeable about 20th century Viennese intellectual history, writes: "Thus two physical and philosophical orientations stood opposite each other in Vienna around 1900: the more popular yet in the end defeated "Machists" and the minority tendency under Boltzmann, whose teaching during the course of the 20th century would be confirmed in the most brilliant fashion." See Friedrich Stadler, *Vom Positivismus zur 'Wissenschaftliche Weltauffassung' - Am Beispiel der Wirkungsgeschichte von Ernst Mach in Österreich von 1895 bis 1934*, Löcker Verlag: Vienna & Munich, 1982, p. 68.

## C. Methodology of Science

(From Boltzmann's First lecture at Clark University, Summer 1899)

Let me first return to the reservations that have been urged against the foundations of Newtonian mechanics, or better, to those points where they still seem to need closer elucidation and an analysis of the modes of inference and sifting of concepts. In formulating the laws of motions Newton regards the motion of bodies as being absolute in space. However, absolute space is nowhere accessible to our experience, which only gives us the relative changes of position of bodies. Thus from the outset this goes completely beyond experience, which must surely be questionable in a science that sets itself the task of representing facts of experience. This difficulty did of course not escape the genius of Newton. However, he thought that without the concept of an absolute space he could not reach a simple formulation of a law of inertia which was his first concern, and I believe that he was right in this; for however much this difficulty was elucidated or thought through, there has been hardly any progress. Instead of Newton's absolute space, Neumann introduced a mysterious ideal body of reference, which as with Newton evidently goes beyond experience. Streintz sets himself the task of avoiding such concepts or bodies, by showing how using the motion, relative to a chosen coordinate system, of a gyroscope subject to known forces or none, we can decide whether for this system Newton's laws of motion hold so that it is a useful reference system. However, these considerations of Streintz's seem of little use for the foundations of mechanics, because they already presuppose the laws of motion of a spinning top and a judgment as to the action of forces on it or otherwise, which in turn already requires knowledge of Newton's laws of motion. Lange indeed tries to formulate the law of inertia without any reference system, merely by considering relative motion and he even succeeds, but his account is so complicated and prolix that one will be reluctant to put so unperspicuous a law in place of the simple Newtonian formula. Obviously Mach's suggestion of straight lines determined by the totality of all masses in the world or his suggestion of adopting the luminous aether in place of absolute space both go beyond experience, although of course in different ways. For the former starts once again from purely ideal and transcendental concepts whereas the latter makes an assertion that might possibly be confirmed by experience but certainly is not so at present. The aether would have to obey quite a different mechanics, for it would have to be the cause of inertia but not be subject to it. We meet similar difficulties when we introduce the concept of time. This too is introduced by Newton as absolute, whereas this is never given to us: we always merely experience simultaneity of the course of several processes.

However, the remedy is easier here, for we can start from a process that always repeats itself periodically under exactly similar circumstances. Of course, it is impossible to produce absolutely identical circumstances, but we can make it eminently likely that all circumstances of any essential bearing are the same. We can further confirm this by mutually comparing different kinds of processes that have this property (rotation of the earth, oscillations of a pendulum or of a chronometer spring). The agreement of all these processes in indicating the same time then excludes all doubt as to the usefulness of this method.[1]

A third difficulty concerns the concepts of mass and force. That, Newton's definition of mass as quantity of matter says nothing has long been recognized. However, doubts arise also as to the ratio of force to mass. Does mass alone exist, force being merely a property of it, or conversely, is force alone truly existent, or must we assume a dualism of two separate existents (mass and force), force existing independently of matter and causing its motion? In addition there is the more recent question whether energy too must be accorded existence or whether it alone exists.

It was above all Kirchhoff who on this point objected to the very question itself. Often a problem is half-solved as soon as the right way of asking the question has been found. Kirchhoff rejected the notion that it was the task of science to unravel the true nature of phenomena and to state their first and fundamental metaphysical causes. On the contrary he confined the task of natural science to describing phenomena, a stipulation that he still called a restriction. If however we delve really deeply into the mode of our own thinking, into its mechanism, as I should feel inclined to say, then one should like to deny even that proviso.

All our ideas and concepts are only internal mental pictures, or if spoken, combinations of sounds. The task of our thinking is so to use and combine them that by their means we always most readily hit upon the correct actions and guide others likewise. In this, metaphysics follows the most down-to-earth and practical point of view, so that extremes meet. The conceptual signs that we form thus exist only within us, we cannot measure external phenomena by the standard of our ideas. We can therefore pose such formal questions as whether only matter exists and force is a property of it, or whether force exists independently of matter or conversely whether matter is a product of force; but none of these questions are significant since all these concepts are only mental pictures whose purpose is to represent phenomena correctly. . . .

[Johannes] Classen* has called my account a polemic against [Heinrich] Hertz, presenting the case as if I imagined I had produced something unconditionally superior to Hertz. Nothing could be farther from the mark. I absolutely recognize the advantages of Hertz's picture, but on the principle that it is possible and desirable to set up several pictures for one and the same group of phenomena, I think that alongside Hertz's picture mine still has its significance, in that it has certain advantages lacking in his. The principles of mechanics that Hertz sets up are extraordinarily simple and beautiful. They are naturally not free from an arbitrary streak, but this is kept to a minimum. The picture that Hertz thus constructs independent of experience has a certain inner perfection and obviousness and contains really only few arbitrary elements. As against this, my picture is evidently inferior, for it contains many more features that are marked by an absence of inner necessity, being introduced only in order to facillitate subsequent agreement with experience. Besides it contains a quite arbitrary function, and of the many pictures that result when this function assumes all possible forms, only a very few conform to actual processes. Whereas with Hertz's picture we see at once that if any other pictures are possible at all then certainly only a very few that possess the same simplicity and perfection; my picture immediately provokes the idea that there must be a good many other such that represent phenomena equally well. Nevertheless, there are some points in which my picture is superior to Hertz's. He can indeed explain some phenomena directly from his picture, or perhaps better, represent them by means of it, such as the motion of a material point on a prescribed surface or curve, or the rotation of a rigid body about a fixed point, provided of course that there are no strange external forces. However, difficulties arise as soon as one wants to represent the most ordinary processes of daily experience involving the action of force. . . .

---

[1] Ludwig Boltzmann, "On the Fundamental Principles and Equations of Mechanics", *Theoretical Physics and Philosophical Problems*, edited by Brian McGuinness & translated by Paul Foulkes, D. Reidel: Dordrecht, 1974, pp. 102-104 & 111.

* See "Die Prinzipien der Mechanik bei Boltzmann und Hertz", *Jahresbericht der Hamburg Wiss. Ansichten,* Sonderdruck, 1898, 13 pages, which was produced by the Naturwissenschaftliche Verein in Hamburg.

Johannes W. Classen, born in 1864, was a prolific Hamburg professor who wrote on physics and philosophy of science. See also his book *Vorlesungen über moderne Naturphilosophen,* C. Boysen: Hamburg, 1908, which discusses the methodological and philosophical ideas of Du Bois-Reymond, F.A. Lange, Haeckel, Ostwald, Mach, Helmholtz, Boltzmann, Poincaré, and Kant.

(from Boltzmann's address to the meeting of natural scientists at Munich on September 22, 1899)[2]

I therefore think that Hertz's mechanics is more a programme for the distant future. Should people one day succeed in explaining without artificialty all natural phenomena by means of hidden motions, then the old mechanics would be superseded by the Hertzian. Until then, the former alone can represent all phenomena in a really clear manner without adducing things that are not only hidden but of which we have not the slightest idea how we are to conceive of them.

In his book on mechanics Hertz has given a certain completion not only to Kirchhoff's mathematical-physical ideas but also to Maxwell's epistemological ones. Maxwell had called Weber's hypothesis a real physical theory, by which he means that its author claimed objective truth for it, whereas his own account he called mere pictures of phenomena. Following on from there, Hertz makes physicists properly aware of something philosophers had no doubt long since stated, namely, that no theory can be objective, actually coinciding with nature, but rather that each theory is only a mental picture of phenomena, related to them as sign is to designation.

From this it follows that it cannot be our task to find an absolutely correct theory but rather a picture that is as simple as possible and which represents phenomena as accurately as possible. One might even conceive of two quite different theories both equally simple and equally congruent with phenomena, which therefore, in spite of their difference ,are equally correct. The assertion that a given theory is the only correct one can only express our subjective conviction that there could not be another equally simple and fitting image.

Many questions that used to appear unfathomable thus fall away of themselves. How, it used to be said, can a material point which is only a mutual construct emit a force; how can points come together and furnish extension, and so on? Now we know that both material points and forces are mere mental pictures. The former cannot be identical with something extended, but can approximate as closely as we please to a picture of it. The question whether matter consists of atoms or is continuous reduces to the much clearer one, whether the continuum is able to furnish a better picture of phenomena. . . .

Phenomenology believed that it could represent nature without in any way going beyond experience, but I think that this is an illusion. No equation represents any process with absolute accuracy, but always idealizes them, emphasizing common features and neglecting what is different and going beyond experience. That this is necessary if we are to have any ideas at all that allow us to predict something in the future, follows from the nature of the intellectual process itself, consisting as it does in adding something to experience and creating a mental picture that is not experience and therefore can represent many experiences.

Only half of our experience is ever experience, as Goethe says. The more boldly one goes beyond experience, the more general the overview one can win, the more surprising the facts one can discover but the more easily too one can fall into error. Phenomenology therefore ought not to boast that it does not go beyond experience, but merely warn against doing so to excess.

Even if it imagines that it is not positing a picture for nature, phenomenology is in error. Numbers, their relations and groupings are just as much pictures of processes as the geometrical ideas of mechanics. The former are merely more prosaic and more suitable for quantitative representation but for this very reason less apt to point to essentially new perspectives, they are bad heuristic road signs; Likewise all ideas of general phenomenology show themselves to be pictures of phenomena. Thus the best result will no doubt be achieved if we always use all means of picturing as required, without neglecting to test the pictures at each step against new experience.

---

[2] Ibid., "On the Development of the Methods of Theoretical Physics in Recent Times", pp. 90-91, 96,

(from Boltzmann's article on models, written in early 1900)[3]

Maxwell himself and his followers devised many kinematic models, designed to afford a representation of the mechanical construction of the ether as a whole as well as of the separate mechanisms at work in it: these resemble the old wave-machines, so far as they represent the movements of a purely hypothetical mechanism. But while it was formerly believed that it was allowable to assume with a great show of probability the actual existence of such mechanisms in nature, yet nowadays philosophers postulate no more than a partial resemblance between the phenomena visible in such mechanisms and those which appear in nature. Here again it is perfectly clear that these models of wood, metal, and cardboard are really a continuation and integration of our process of thought; for, according to the view in question, physical theory is merely a mental construction of mechanical models, the working of which we make plain to ourselves by the analogy of mechanisms we hold in our hands, and which have so much in common with natural phenomena as to help our comprehension of the latter.

Although Maxwell gave up the idea of making a precise investigation into the final structure of matter as it actually is, yet in Germany his work, under G.R. Kirchhoff's lead, was carried still further. Kirchhoff defined his own aim as being to describe, not to explain, the world of phenomena; but as he leaves the means of description open, his theory differs little from Maxwell's, as soon as recourse is had to description by means of mechanical models and analogies. . . .

Another phenomenology in the widest sense of the term, maintained by E. Mach, gives less prominence to mathematics, but considers the view that the phenomena of motion are essentially more fundamental than all of the others to have been too hastily taken. It rather emphasizes the prime importance of description in the most general terms of the various spheres of phenomena, and holds that in each sphere its own fundamental laws and the notions derived from this must be employed. Analogies and elucidations of one sphere by another - e.g. heat, electricity, etc. - by mechanical conceptions, this theory regards as mere ephemeral aids to perception, which are necessitated by historical development, but which in course of time either give place to others or entirely vanish from the domain of science. . . .

[But on the other hand], it often happens that a series of natural processes - such as motion in liquids, internal friction of gases, and the conduction of heat and electricity in metals - may be expressed by the same differential equations; and it is frequently possible to follow by means of measurements one of the processes in question - e.g. the conduction of electricity just mentioned. If then there be shown in a model a particular case of electrical conduction in which the same conditions at the boundary hold as in a problem of the internal friction of gases, then we are able by measuring the electrical conduction in the model to determine at once the numerical data which obtain for the analagous case of internal friction, and which could only be ascertained otherwise by intricate calculations. Intricate calculations, moreover, can very often be dispensed with by the aid of mechanical devices, such as the ingenious calculating machines which perform additions and subtractions and very elaborate multiplications and divisions with surprising speed and accuracy, or apparatus for solving the higher equations, for determining the volume or area of geometrical figures, for carrying out integrations, and for developing a function in a Fourier's series by mathematical means.

---

[3] Ibid., "Models", pp. 218-220.

(from Boltzmann's inaugural lecture at the University of Leipzig, which was given during November 1900)[4]

When we propose to introduce new guests into the home that we have long inhabited, we customarily deck out the entrance door with festive ornament. I have been called to this ancient and venerable university in order to introduce you into the vast and imposing edifice of theoretical physics. The entrance gate through which we step into this edifice is analytical mechanics. No wonder then that I wish to show you that science in its most beautiful finery with which it has been adorned not by me but by the choicest spirits over the centuries.

As a true theoretician, I wish to consider the central core rather than all the external detail. The definition of analytical mechanics is very simple; it is the theory of the laws governing the motion of bodies. . . .

The campaign into the territory of heat theory was opened by mechanics with the idea that heat is a motion of the smallest particles of a body; remaining invisible because they are so small, but making itself felt by provoking the sensation of heat when it is imparted to the molecules of our body, or of cold when it is withdrawn. This campaign was victorious because the hypothesis described furnishes a very clear picture of the agency called heat, much more completely so than the earlier view of it as behaving like a substance.

Electricity and magnetism were subsumed under mechanical laws by means of the hypothesis of electric and magnetic fluids whose particles were supposed to be acting on each other according to a law that was merely a variant of Newton's law for the mutual action of celestial bodies and thus rests firmly within purely mechanical territory. Finally, much success was achieved in trying to reduce chemical phenomena and the formation of crystals to a mechanics of attracting and repelling forces and of the mutual motion of heterogenous atoms; indeed, chemical phenomena are closely connected with thermal ones as well as with electric ones. . . .

While then the principle of mechanical explanation has conquered ever wider domains within the whole sphere of science, oddly enough, it has lost some ground in its most central field, namely, in theoretical physics. The cause for this, as often with victorious nations too, lay partly in internal divisions but partly also in external conditions.

While people most successfully aimed at working out applications of mechanics down to the smallest detail, there arose a line of approach that began to shake the basic pillars of the science, and pointed to obscurities in its principles. The fundamental concept of mechanics is that of motion. The concept of pure motion detached from any other change stands out in full clarity only in the study of rigid bodies, where indeed, we have a perfectly immutable structure in which nothing changes save spatial position. Now there are in nature no perfectly rigid bodies but certainly solid ones that are subject to only imperceptible changes of shape during motion. As for changes in shape in liquids and gases, these one tries, without straining the facts, to reduce to the motions of their smallest constituent parts. Indeed, to the eye they resemble changes of shape of a sand heap, which consists of individual perceptible grains. Nevertheless, in the case of actual fluids there is something hypothetical about the assumption that there too each individual particle is identifiable at all times, for experience shows that we are given only the constancy of total mass and weight.

It was now attempted to prove a priori that every change, even if apparently quantitative, must be reducible to a motion of the smallest parts, motion being the only process in which the object moved remains always the same. All such metaphysical reasons seem to me to be insufficient. Of course we cannot avoid forming the concept of motion. If therefore all apparently qualitative changes were representable by the picture of motions or changes of arrangement of smallest parts, this would lead to an especially simple explanation of nature. In that case, nature would appear to be at its most comprehensible, but we cannot compel her to this, we must leave open a possibility that this will not do and that we need in addition other pictures of other changes; understandably, it is precisely the more recent developments in physics that have made it prudent to allow for this possibility.

Mechanist physics had conceived all bodies as aggregates of material points acting directly on each other at a distance. At very small molecular distances it was to be cohesive, adhesive, and chemical forces that were acting, and at larger distances gravitational ones. Besides ponderable matter one assumed that there was a luminous aether conceived as just like a solid body, whereas electro-magnetic phenomena, as mentioned earlier, were explained by means of electric and magnetic fluids whose parts again acted directly on each other at a distance. Only a little over ten years ago did Hertz succeed in giving experimental proof that, as Faraday and Maxwell had suspected already, electric and magnetic forces do not act directly at a distance but are caused by changes of state that are propagated from one volume element to the next at the speed of light. This gave the revered old theory of electric and magnetic fluids a blow from which it soon expired, but another theory too was damaged by Hertz's experiments. For the law of propagation of electro-magnetic waves showed such absolute agreement with those of the motion of light, that no further doubts could be entertained as to the identity of the two phenomena. If this did not as yet refute the theory that light rests on vibratory motion of the smallest particle of luminous aether, nevertheless, it had been established that the aether certainly must have different and much more complex properties than had hitherto been ascribed to it. This gave such preponderance to the theory of electricity and magnetism that from some quarters it was attempted to replace the mechanical hegemony in theoretical physics by an electro-magnetic one, by trying to derive the simplest laws of mechanics from the theory of electromagnetism. . . .

Whereas the most recent views of electro-magnetism merely try to seek salvation in the action of volume elements on their neighbors, some very recently observed phenomena in cathode rays and electrolysis have led to the assumption that even electricity has an atomic constitution, consisting of discrete elements, namely electrons. We therefore see that the old Kantian antinomy opposing infinite divisibility of matter to atomistic constitution still keeps science in suspense, except that today we do not regard the two views as infected by internal logical inconsistencies arising from the laws of thought, but we take each as a mental picture we have constructed and we ask which can be developed more clearly and more easily while most correctly and definitely reproducing the laws of phenomena . . . .

[4] Ibid., "On the Principles of Mechanics I", pp. 129-130, 132, 142-143, & 145-146.

Ludwig Boltzmann in Graz (1887)
Upper row left to right: W. Nernst, F. Streintz, S. Arrhenius, R. Hiecke
Lower row right to left: E. Aulinger, A.v. Ettinghausen, L. Boltzmann, I. Klemenčič, V. Hausmaninger
(people identified by Wolfgang Stiller)
Courtesy of Roman Sexl and Setsuko Tanaka

# Chapter 4

# BACK TO VIENNA

## A. Correspondence

### #32

Einstein to Maric[1] [Milan, September 13? 1900]

. . . . Boltzmann is quite magnificent. I'm almost finished with it.[2] He represents things in a masterly way. I am completely persuaded by the correctness of the principles of the theory, that the question is really about the movement of discrete mass points of gases of a definite finite size which move according to certain conditions. Boltzmann stresses quite correctly that the hypothetical powers between the molecules are not an essential part of the theory, since all of the energy is of a kinetic kind. This is one more step in the dynamic explanation of physical phenomena . . . .

[1] Mileva Maric; also sometimes spelled "Maritsch" and "Marity". For the German version of this letter see *Albert Einstein*, Vol. **I**, edited by John Stachel, Princeton University Press: Princeton, 1987, pp. 260-261.

[2] Einstein was presumably referring to Boltzmann's *Vorlesungen über Gastheorie*, Barth: Leipzig, either Volume **I** published in 1896 or Volume **II** published in 1898. For Einstein's early work on statistical mechanics see Clayton A. Gearhart, "Einstein Before 1905: The Early Papers on Statistical Mechanics", *American Journal of Physics*, **58** (1990) 468-480.

### #33

To Franz Exner[1] Leipzig, December 1, 1900

My best thanks for the friendly postcard, and please extend more thanks to everyone else who has thought of me in a likeminded way.

I also want to thank you heartily as Professor Gegenbauer[2] has written me for taking over the Physical Institute. At the time of my departure from Vienna, I was so disturbed and besieged that I neglected to say goodbye to you and to thank you for the proven collegiality which you gave me in Vienna. I appeal to you not to be upset and to apologize for me to the rest of my Vienna colleagues.

With my best compliments and heartiest greetings to you and all my Vienna colleagues, I am...

[1] Franz S. Exner (1849-1926) became Professor of experimental physics at the University of Vienna in 1891, succeeding Joseph Loschmidt. His father had been an influential philosopher and educator and several close relatives were also scientists. Exner did much work in electrochemistry and in the study of atmospheric phenomena. He also was a very influential teacher,

guiding more students to higher degrees in Vienna than either of his better-known contemporaries, Boltzmann or Mach. See B. Karlik & E. Schmid, *Franz Exner und sein Kreis*, Österreichische Akademie der Wissenschaften Verlag: Vienna, 1982.
2 Leopold Gegenbauer (1849-1903) became Professor of Mathematics at the University of Vienna in 1893.

#34

Maric[1] to Savic[2] Zürich, December 20, 1900

. . . . Albert has written a physical treatise[3] which will probably be published in *Annalen der Physik*. Can you imagine how proud I am of my sweetheart [*Schatzerl*]? And it is by no means just an ordinary work, but very significant. It is about theory of liquids. We have sent it to Boltzmann *privatim*, and would like to know what he thinks of it. Hopefully he will write us.[4] . . .

1 Mileva Maric (1875-1948) seems to have grown up in a Serbian family living in the town of Titel in what was then southern Hungary. She became a fellow student of Einstein in the ETH in Zürich. They had three children, one before and two after marriage. See *The Collected Papers of Albert Einstein*, Vol. I, edited by John Stachel, Princeton University Press: Princeton, 1987, pp. 272-273 & 380-381 and *Albert Einstein Mileva Maric - The Love Letters*, edited by J. Renn and R. Schulmann and translated by Shawn Smith, Princeton University Press: Princeton, 1992.
2 Helene Savic (1871-1943) was born in Vienna and was a history student in Zürich where she met and became friends with Maric and Einstein. She had just married a Serbian engineer, Milivoj Savic, when she received this letter. See Stachel, ibid., p. 386.
3 Albert Einstein, "Folgerungen aus den Capillaritätserscheinungen", *Annalen der Physik*, **4** (1901), 513-523.
4 It is not yet known whether Boltzmann replied to Einstein.

#35

To Wilhelm von Hartel[1] [Leipzig, February? 1901]

My warmest thanks for your extraordinarily amiable letter. Your words which still linger in my ears have given me extraordinary pleasure. But unfortunately, I must confess that returning to my homeland [*Heimat*] which I would have so much desired has become impossible for me at present, and yet, I console myself with the hope which you have raised in me, that in the not-too-distant future it will become possible again.

With the very highest respect I am...

1 It seems that Wilhelm von Hartel, Boltzmann's colleague at the University of Vienna as well as Austrian Minister of Education, felt that Boltzmann's heart remained in Austria and that he could be won back to the University of Vienna.

#36

To Stefan Meyer[1] [Leipzig, 1901]

My best thanks for your friendly letter, and please accept my profuse apologies that I have not already long since answered it. Concerning the Joule-Thomson experiment, I only want to make the following remarks, which perhaps indeed have long been known to you. The gas will be heated at the beginning at *A* by passing through the stopper [*Pfropfen*], and finally at the very end at *B* will cool down almost that much again. The theory presupposes that none of this heat will be dissipated outwards. But if the stopper is elongated as in figure #1, so particularly at *A*, heat could be transferred to the surrounding tubes and less will be absorbed at *B*. I believe, therefore, that the density [*Dicke*] of the stopper must be small with respect to its diameter [*Durchmesser*]. Perhaps its diameter could be made smaller than the diameter of the tube and a very thin, poor heat-conducting layer [*Schicht*] be put between the stopper and the tube.

With heartiest greetings I am...

[1] Meyer, Boltzmann's longtime assistant at Vienna, became his interim successor in 1906, and in 1910 first director of the Radium Institute in Vienna.

#37

To Stefan Meyer [Leipzig, 1901?]

Together with this letter I am sending you the certificate [*Zeugniss*] for Sedlaczek,[1] and I only hope that he won't fare too badly.

Thank you for your friendly letter, and warmest greetings from my side towards yourself as well as all of my Vienna students.

[1] Presumably, Walther Sedlaczek is meant. He was a student at the the University of Vienna from 1894 until 1902. The second part of this sentence possibly refers to the hope that Sedlaczek would pass his coming final examination.

#37A

Jaffé[1] about Boltzmann [for the period 1900-1902]

When I entered Ostwald's laboratory in 1899 it was at the height of its development and fame. The years of militancy when physical chemistry struggled for recognition had passed . . . . As a consequence of this situation the laboratory was as crowded as it could be[2] . . . .

Though Boltzmann was able to wage quite a sharp pen in a written controversy he was not a debater like Ostwald being somewhat handicapped by a rather feeble and high-pitched voice which contrasted with his exterior.

He must have felt that he was on the defensive and was somehow overshadowed by the brilliancy of his friend and opponent; at least so it seemed to us who were in the camp of energetics. As a matter of fact Boltzmann was very unhappy in Leipzig. He was homesick and was longing for his native Austrian mountains. His depressive mood must have reached a first climax for he made an attempt at his own life while he was in Leipzig.[3]

---

[1] George Jaffé, "Recollections of Three Great Laboratories", *Journal of Chemical Education*, **29** (1952), 230-238.
[2] Ibid., p. 231.
[3] Ibid., p. 235.

#38

Müller[1] to Hartel — Vienna, June 19, 1901

. . . . At the same time, the Dean was commissioned to ask Hofrath Professor Boltzmann whether he would accept being reappointed to the Philosophical Faculty of the University of Vienna and whether he agreed with the reorganization of the Institute. Professor Boltzmann's written answer has turned out to be gratifying and the original has been enclosed.

In the [faculty] sitting of June 15, Hofrath von Lang[2] reported in the name of the commission, and the *Collegium* of professors voted by secret ballot for the reappointment of Hofrath Boltzmann with 37 *yes* votes, 6 *no* votes, and one *blank* vote. The reorganization of the Institute was unanimously approved by open vote. . . .

---

[1] David Heinrich Müller (1846-1912), dean of the Philosophy faculty at the University of Vienna,was an Orientalist, becoming professor in the Ring City in 1881.
[2] Victor von Lang (1838-1921) was professor of experimental physics at the University of Vienna and in 1908 Rector. He worked largely in optics and crystallography. Franz Exner had been his assistant. See Karlik & Schmid, op. cit., pp. 37-40. For details of the reorganization of the Institute see Christa Jungnickel and Russell McCormack, ***Intellectual Mastery of Nature***, Volume **II**, University of Chicago press: Chicago, 1986, pp. 185-188.

#38A

Fasol-Boltzmann about Boltzmann — [Concerning Summer 1901]

From the end of July to the beginning of October 1901, Ludwig Boltzmann travelled from Leipzig to the Mediterranean with my father [Arthur Boltzmann]. It was hoped that this would improve his exhausted mental state. However, he suffered terribly from the heat. "Papa sweats and swears all the time", wrote my father on August 8 on board the steamer *Pera* of the German Levant Line. The journey took them from Hamburg via Lisbon, Algiers, Malta, Athens, Constantinople, and Bandirma to Cabakkale.

Owing to an outbreak of the plague, they could not, as planned, continue from Constantinople to Odessa. For a long period the ship lay in quarantine. But both of them enjoyed their daily game of chess. On 4 October they arrived back in Leipzig.[1]

---

[1] Ilse M. Fasol-Boltzmann, "Ludwig Boltzmann and His Family", in Ilse M. Fasol-Boltzmann (ed.), *Ludwig Boltzmann 'Principien der Naturfilosofi - Lectures on Natural Philosophy 1903-1906*, Springer: Berlin, etc., 1990, p. 31.

#39

To Wilhelm von Hartel [*Pera* ?], September 30, 1901

In connection with the words [*Unterredung*] which your excellency vouch-safed [*gewährte*] some days ago, let me put forward in written form the conditions under which I would be ready to return to my teaching post and resume my previous teaching obligations.

What I would most prefer would be to keep the physical institute and my living quarters, but such that the possible separation of some institute rooms would not be excluded. In this case, I would only claim my old pre-pension salary of 7,000 gulden yearly and the previous remuneration of 600 gulden every year for handling seminars.

But should the greater part of the Institute as well as my living quarters be withheld, then naturally the experimental work which was carried out there in earlier times must cease or be reduced to a minimum, and my activity will have to be confined almost completely to the field of theoretical physics. I would also claim 1,500 gulden for the alienation of my living quarters.

In any case, I would give my word of honor never to leave Austria again.

#40

To Franz Mertens[1] [February 1902]

First of all, I would like to express great pleasure that my Vienna colleagues have been thinking of calling me back, and I thank you most honored Dean above all for your worthy communication.

Although I have only experienced love and accomodating treatment here in Leipzig as well, still, I already regret having left Vienna. My attachment to Austria is too great, and I can adjust only with great difficulty to North German conditions.

I would very gladly accept reappointment to the University of Vienna as Professor of Theoretical Physics. As for my activity in the Institute for Theoretical Physics, I have never exercized it on such a scale as Hofrath Stefan[2] did in his time; I would have no objection in principle if there were a change in this respect. Concerning detailed particulars as well as the rest of my duties and responsibilities, they must naturally remain the concern of special negotiations...

[1] Franz Mertens (1840-1927) became Professor of Mathematics at the University of Vienna in 1894. He had replaced Müller as Dean of the Philosophy Faculty.
[2] Boltzmann succeeded Stefan in Vienna in 1894. See letter #15, footnote 2 for more particulars.

#40A

Gerhard Kowalewski about Boltzmann [for 1902]

Why did Ludwig Boltzmann strive so hard to return to Vienna after such a short time in Leipzig? Ernst Mach's heavy illness in 1901/1902 must have already had a decisive effect on him. But Ostwald once revealed that his colleague had a worry which he could not banish. He dreaded that his mind and memory would suddenly fail in the middle of a lecture. This could also be a possible consideration.[2]

[2] A. Dick and G. Kerber, *Ludwig Boltzmann - Katalog zur Ausstellung an der Zentralbibliothek für Physik in Wien,* Vienna, 1982, p. 29.

#41

Einstein to Maric Bern, February [8?], 1902]

. . . . He [*Kleiner*][1] is really excited by my good ideas, and keeps trying to get me to send part of the work to Boltzmann,[2] which has reference to his book.[3] I will also do it.[4] I have almost finished Mach's book[5] and with immense interest, and will finish it tonight . . . .

[1] Alfred Kleiner (1849-1916) was an advisor and supervisor of Einstein at the ETH in Zürich.
[2] It is not clear which paper Einstein is referring to.
[3] Either Volume I (1896) or II (1898) of Boltzmann book on gas theory.
[4] John Stachel gives the following footnotes in *The Collected Papers of Albert Einstein,* Volume I, Princeton University Press: Princeton, 1987, p. 335, notes 7 and 8: " . . . *Kaiser 1930,* p. 69, states that Kleiner rejected Einstein's work on the kinetic theory of gases because it 'sharply criticized' Boltzmann . . . . According to Einstein's sister, he [Einstein] corresponded with Boltzmann until the latter's death in 1906 (See *Winteler-Einstein 1924,* p. 18)."
[5] Einstein presumably meant Ernst Mach's, *Principien der Wärmelehre,* either the 1896 or the 1900 edition.

#42

From Max Wien[1] Aachen, March 1902

Allow me to send you again the reworked sketch for unified physical designations and nomenclature. The wishes and proposals of the Commission members have been taken into account in so far as they can be brought into accord with the system and with each other. The changes have been added with red ink; Where the wishes which you spoke about have not been included, the reasons have been added in red ink inside parentheses.

The view of the majority of members has been that a smaller number of special designations should be introduced at least provisionally. In oppposition to this Professor Boltzmann remarked: "It is precisely when one provisionally fixes only a few designations that one risks the possibility of a summary or digest imposing designations over everything." In order to avoid this, the following working plan would appear to be practical [*zweckmässig*].

1. First of all, general rules upon which the whole system is to be based must be put forward and agreed upon.

2. In connection with this, a trial system has been put forward which includes as many designations as possible. No unity is needed to control the details of this system, but only over the general point of view and the possibility of arranging the most important physical magnitudes and quantities [*Grössen*] in this system.

3. Finally, we will select from the trial system 20 to 30 designations which we recommend as suited above all for general introduction. Naturally, there must be complete unanimity about these designations.

Some members of the Commission do not agree with my sketch of the proposed principle III and have put forward the idea that the same letters are needed to stand for different quantities [*Grössen*]. In fact, this already occurs in practice, but it seems to me that we should not propose it ourselves at all.

I am presenting the sketch of a new Section I, in which proposals are made for important designations for the special list of 20 to 30 definitive cases. Most of these are the same as were included in my first circular and have found no opposition. Some of the same, however, could still create difficulties, for example, mechanical equivalent of heat, temperature, frequency, period of oscillation, and index of refraction. But nevertheless, we must still somehow agree on them, because these quantities [*Grössen*] seem so important at least to me, that they must unconditionally be included in our special list. To prevent the final result from being far too meager, I still believe that for Part II (definitions and nomenclature), we must put forward appropriate, agreed-upon proposals now.

I request the Commission members once again to write their remarks in the margin, and indeed, before all else on the general rules and the special list.

If there is no comment next to the number, it will be considered as accepted by that member. It would be very much appreciated if I could receive the circular back by the end of April,[2] so that I can formulate the definitive resolution and circulate it for ratification by the members...

[1] Max Wien (1866-1938) taught physics from 1899 to 1903 at the Technical University of Aachen and from 1911 at the University of Jena. His work was largely concerned with acoustics, sparks, and electrical systems. He was a cousin of the better-known physicist Willy Wien. It is possible that the commission and its work which this letter describes served as preliminary labor for the 1905 conference in Leipzig which Boltzmann mentions at length near the beginning of the original German version of his article "On the Trip of a German Professor into El Dorado" (1905), which in translation can be found in Chapter 7 of this book.

[2] This circular was apparently kept by Boltzmann and not returned to Max Wien. See the next letter for a possible explanation.

#43

To Svante Arrhenius — Leipzig, April 30, 1902

Please forgive me that I have not until now answered your dear letter of November 7 and your friendly New Year's greeting of January 5. I thank you with all of my best wishes for both of them. How often have I already wanted to answer, but I feel myself now so continuously unwell, am even under treatment by a nerve specialist, and have been in such a contrary mood here [in Leipzig], that I could not even think of writing letters.

Concerning gas molecules which would separate from planetary atmospheres, only the fastest will indeed be able to do it; but by exceeding the attraction of planets, even they will lose velocity, and even if everything were in heat equilibrium they would in my opinion form a gas of the same temperature as they had in the planetary atmosphere. Even the most extensive thinning out of the gas as I see it would still not contradict the Second Law [of Thermodynamics]. But in spite of that, I do not believe in the absolute validity of the Second Law, since I consider it to be a probabilistic law, and what holds merely on average, is not unconditionally valid for everything or everywhere.

It has pleased me very much to hear that you are healthy and satisfied, and I heartily wish that that condition will remain....

#44

Franz Josef I to Hartel — Schönbrunn, June 1, 1902

I name Hofrath Dr. Ludwig Boltzmann, Full Professor at the University of Leipzig, to be once again Full Professor of Theoretical Physics at the University of Vienna under the conditions called for in his contract...

#45

To Stefan Meyer [Leipzig], June 2, 1902

My best thanks for your telegram and friendly letter. I also want to thank all of the other gentlemen who signed the telegram.

Concerning your report on the Institute, I take it into consideration with pleasure. It pleases me greatly that something has been done about the lathe [*Drehbank*]. Otherwise, I have no special wishes, and it would suffice if enough money is available for acquisitiions. But there is still a fatal condition which I must mention now. At the beginning there was a lot of talk about making great reductions in the Institute, which naturally would have also required a reduction in personnel, and since I agreed without thinking about it, this also meant that people who work in the Institute would be let out and I would be left with only enough allocated money [*Dotation*] to pay for one helper [*Aushilfsdiener*].

But now it seems, happily, that the expansion of the Institute will not be reduced. But [Herr] Gänsthaler (who certainly caused me a lot of trouble [*Verdriesslichkeit*] in Vienna) will be assigned to Professor Exner and only one helper will be allotted to me. But that will not be sufficient if the Institute retains its old dimensions.

We will see how things develop further. With heartiest greeetings I am...

#46

To Franz Exner [Leipzig, Summer 1902]

It would be the furthest thing from my mind to want to put in a claim which would hinder you from fully exercizing your teaching activity. I would also never take the position of advocating such an arrangement against [the wishes of] the Ministry or the Collegium of professors, since I myself hold that a merely theoretical education without corresponding experimental foundation, perspective, and exercize for candidates who plan to teach below the university level is completely perverse [*Verkehrtes*].

I therefore request that you tell me all that you need to fully carry out your teaching responsibilities, and thus, what rooms and monies might remain left for the Institute of Theoretical Physics. Concerning apparatus, I gladly alienate most of it, if at least a little usable stuff remains. In its totality, all is merely temporary, since we both still put our hopes in the New Institute.

As is self-evident, I of course have no personal interest in view, but merely wish that the possibility for at least some work in the Institute for Theoretical Physics remains.

Since I have so little certainty of being able to catch you, there is no point in my traveling to Vienna, but would it not be possible on your trip from

Vienna to Göttingen to stop a while in Leipzig[?][1] If you would give me an exact day and hour, we could talk at my place (Leplaystrasse 9 II) or anywhere else that you like [in Leipzig]. Face to face understanding is much easier and more congenial, and I would so gladly avoid every trace of disharmony. With supreme respect and hearty greetings I am...

---

[1] Presumably, Professor Franz Exner was traveling to Göttingen to meet professors Riecke and Wiechert in order to help further their work on atmospheric electricity. For further information on the dispute over the laboratory situation in Vienna, see the next three letters plus Christa Jungnickel and Russell McCormmach, *Intellectual Mastery of Nature*, Volume 2, University of Chicago Press: Chicago, 1986, pp. 184-188.

#47

To Georg Gänsthaler[1] [Leipzig, Summer 1902]

In case Herr Dr. Stefan Meyer comes to Vienna in these days, please tell him before he travels to Carlsbad to lay out correctly for me the inventory accounts which he made for Professor Exner and with the necessary clarifications which will make them fully understandable to me.

I am not going to Carlsbad and will be in Vienna before Herr Dr. Meyer returns from Carlsbad, and I would like to have the inventory accounts in hand as soon as I arrive. With best greetings...

---

[1] Georg Gänsthaler was a laboratory employee in Boltzmann's laboratory in Vienna. For his coming switch to Exner's laboratory and for Boltzmann's opinion of Gänsthaler, see the previous letter. Since neither letter was provided with an exact date or easily determinable temporal clues, their sequence is not certain, though the next letter would suggest that we have probably placed them in the right order.

#48

Gänsthaler to Meyer Vienna, September 19, 1902

This letter from Herrn Hofrath[1] just came. Please write if there is something particular you want.

Otherwise nothing new.

---

[1] Boltzmann.

#49

Exner & Boltzmann to Collegium of Professors [Vienna, undated]

Since the obstacles which have previously stood in the way of building a new Physical Institute for the University of Vienna have fortunately been swept out of the way, [it would be good if] the Collegium of Professors in light of the often depicted inconsolable condition of the present accomodations of the physical professorships [were] to allow itself to direct a devoted request to the High Ministry [of Education] to apply every effort in order to accelerate the building of the new Physical Institute of the University.[1]

[1] The building which currently houses the various institutes of physics in Vienna as well as chemistry was basically finished in 1913. It is a huge building which understandably took a long time to complete. Appropriately, the entrance to the Institute for Theoretical Physics is located on Boltzmanngasse, as is the Radium Institute finished in 1910, which had Stefan Meyer as its first director.

#50

Mach[1] to Wiener November 18, 1902

I thank you heartily for your friendly letter. It especially pleases me that, like your father,[2] you don't follow any religion, nor do you accept the superstition of "Ignorabimus"[3] . . . . Also the considerations of your father about the thing-in-itself were new to me . . . .

Boltzmann visited me a few days ago. I found him in a rare cheerful mood. . . .

If you have occasion to see Ostwald and Hering,[4] please greet them heartily for me. As always, my highest respect . . .

[1] Ernst Mach (1838-1916) moved from a professorship in experimental physics in Prague to "Brentano's Chair" in philosophy in Vienna in 1895, albeit with the title changed to "the history and philosophy of the inductive sciences". But Mach suffered a stroke in 1898 which paralysed the right side of his body. He retired in 1901, thereby apparently planting the idea in Boltzmann's mind that since he too was writing more and more philosophy that he also might be permitted to teach philosophy in Vienna, especially since the Brentano-Mach chair was still vacant. Nevertheless, it is not certain whether his visit to Mach mentioned above was connected with this possibility.

[2] Christián Wiener (1826-1896). He had been a mathematician and philosopher at Karlsruhe, publishing the book *Grundzüge der Weltordnung* in Leipzig in 1863.

[3] Emil du Bois-Reymond (1818-1896), who had studied animal electricity and strongly rejected vitalistic explanations in biology, had became a leading philosopher of science in Central Europe by the 1870's and 1880's. He suggested that there were "seven riddles" which would never be solved in science and to which one must answer "Ignorabimus". Mach thought he was too pessimistic, but it remains the case that the seven problems remain as unsolved as ever over a hundred years later.

[4] Ewald Hering (1834-1918) was a physiologist well-known for his nativism, color theory, and notion of inherited memory. He had been Mach's colleague in Prague for many years before moving to Leipzig.

#51

To Leo Königsberger[1] Vienna, December 7, 1902

My warmest thanks for sending me your work on Helmholtz (Volume I) which deserves thanks. I am sorry that I didn't keep his letters.[2] I suppose it is inconceivable, but at that time I was a young hot-head [*Brausekopf*] who didn't think about the future. They also included some matters which from a scientific point of view were not totally without interest, and they were wholly marked by Helmholtz's very characteristic way of expressing himself.

In one place in your book you mention in passing: "Helmholtz was accustomed to acknowledge the authority of his wife in all things."[3] Alas, if only you had added a small qualification that no one led him against his will, that he was anything but a wet noodle [*Simandl*].

In highest respect and with the heartiest greetings...

[1] Leo Königsberger (1837-1921), a mathematician and longtime friend and colleague of Helmholtz, published his biography, *Hermann von Helmholtz*, Band I, Braunschweig, in 1902.

[2] But Boltzmann's letters to Helmholtz have recently been published. See Herbert Hörz & Andreas Laass, *Ludwig Boltzmanns Wege nach Berlin*, Akademie Verlag: Berlin, 1989, pp. 87-96.

[3] Königsberger, op. cit., p. 375.

#52

Frau to Ida[1] Boltzmann [Wien, December 13, 1902]

Unfortunately, I must reveal to you today something which is very disagreeable [*unangenehm*]. Papa is unwell, and he always loses blood when urinating. But tell *no one* about it. Possibly it won't get worse and it would not be good if it were known. [I am thinking] already, for example, of the Nobel prize which he could still receive. Papa has so little trust in medical doctors and has such a dreadful fear of having these parts of his body examined that he has still not consulted a doctor. Yesterday, I saw Hofrat Lieben[2] . . . He talked to the No. 1 specialist, Professor Hirsch.[3] But Papa refuses to let himself be fetched. He will not be giving a lecture today, and will not start anything before Christmas. It would be dreadful if Papa were to suffer so much that he would need an operation, since that would really make his nerves come apart . . . .

[1] Frau Boltzmann was writing to her daughter Ida who continued to attend a *Gymnasium* in Leipzig after other members of the family returned to Vienna. Ida Boltzmann (1884-1910) studied mathematics and physics and passed the examination for secondary school teachers, but died a few years afterwards.

[2] Adolf Lieben (1836-1914), was a professor of organic chemistry at the University of Vienna. His sister had married the philosopher Franz Brentano.
[3] The medical school at the University of Vienna was often considered the best in Europe. Professor Hirsch has not yet been identified.

#53

To George Cecil Jaffé[1] Vienna, December 23, 1902

I confirm with pleasure that during my two year teaching activity in Leipzig that you were always one of my most eager and gifted students, who in seminars handled the most difficult problems of theoretical physics skillfully and with ease, and that your knowledge of mathematics towers vastly above [*himmelhoch*] what is otherwise found in chemists. I do not doubt for a moment that you will be able to perform very skillfully in the field of physical theory and it pleases me that you intend to devote yourself more to physics now. With high respect...

[1] George Cecil Jaffé (1880-1965) was born in Moscow, studied in Munich and Leipzig, did research with J.J. Thomson in Cambridge and Marie Curie in Paris, taught physics in Leipzig and later Giessen from 1916 to 1933 when he left for America and taught at Louisiana State University until his retirement. See #37A in this book and his article "Recollections of Three Great Laboratories" in *Journal of Chemical Education*, **29** (1952), 230-238.

#54

To Svante Arrhenius Vienna, December 31, 1902

My best thanks for your friendly letter and good wishes for the New Year, which my wife and I heartily return. At the moment things are going rather well, and, apart from slight evils which I will never be completely free from in my life, healthy. Also, my family is healthy.

The little house which we bought in Vienna suits me and my family very well, and I hope that by Spring and Summer when we can sit in the garden that it will please us even more. I have, however, a request. If you come again to the South, please stay a few days with us and live with us; We have our own guestroom. It would make us all happy.

I am now so approximately finished arranging the house and the Institute that I can think again about working for myself. I am working provisionally on Part 2 of my *Mechanik*[1] here and there piling up what is generally interesting, such as what as far as I know is a still unknown generalization of Lagrange's equations, which I lectured on this month in the Vienna Academy [of Science]. I have also begun an experimental ex-amination of the Hall Effect,[2] and my son is also taking part.[3] Unfortunately, everything will again be interrupted if I will begin to write the material [*Referate*] for Klein's strict, detailed, popular, and at the same time highly scientific, short, and yet fully comprehensive way, is such a strain and causes so much headache that apart from my professional duties it prevents me from doing anything else.

I congratulate you on your successes in the important field of serum therapy[5] and am pleased that you will now have more free time for your very successful scientific activity.[6] My wife and I wish from the heart that you will have a happy, fortunate New Year...

---

1 Ludwig Boltzmann, *Vorlesungen über die Prinzipien der Mechanik*, Leipzig, 1897 (Band I) and 1904 (Band II)

2 See Ludwig Boltzmann, 'Zur Theorie des von Hall entdeckten elektromagnetischen Phänomens", *Sitzungsberichte der kaiserlichen Akademie der Wissenschaften* [Vienna], **94** (1886), 644-669.

3 Boltzmann's son Arthur (1881-1952) was still a student. He studied physics in Vienna and Berlin and mechanical and electrical engineering at the Technical University of Vienna. See letter #59, footnote 1, for more information about his adult career.

4 See letter #25 and footnote 1.

5 Svante Arrhenius, "Anwendung der physikalischen Chemie auf das Studium der Toxine und Antitoxine" in *Festschrift v. Inv. af Stat. Serum-Inst.*, Copenhagen, 1902.

6 Arrhenius won a Nobel prize in 1903 for his theory of electrolytic dissociation described in his doctoral dissertation.

## #55

To Otto Wiener Vienna, January 3, 1903

Please excuse me that you have had to go so long without hearing from me, but I was occupied the whole time not only with equipping my Institute anew but also with the modest house I recently bought. I have thought of my erstwhile Leipziger colleagues often enough who were so uncommonly kind and friendly toward me. It has deeply troubled me that in Leipzig where I was treated with such unusual solicitude and where, while I did not have very many students, still, I had some very able ones, that I was unable to feel fully well. Whether it was the swampy climate so different from hilly Vienna or the North German Protestant way of living which is still somewhat strange for me or something else was to blame, I don't even know myself; But now I feel much better again.

I have heard that Des Coudres[1] has become my successor. In any case, he's different [*ein origineller*] and not lacking in bright ideas, even if his head is not specifically mathematical, but all in all, a younger, still promising man. Thus any existing holes left by my departure will once again be filled.

I allow myself to send you a photograph. I belatedly wish you a lot of luck for the New Year and please give my best compliments to your devoted wife, my heartiest greetings, which I also offer my erstwhile colleagues in Leipzig...

[1] Theodor des Coudres (1862-1926) was actually more of an experimentalist than theorist concerning himself at least in the early years with measuring the velocity of cathode rays, and while he would apparently get along with Wiener well, he would publish little after arriving in Leipzig, even in experimental physics. See C. Jungnickel & R. McCormmach, *Intellectual Mastery of Nature*, Vol. 2, Unicersity of Chicago Press: Chicago and London, 1986, pp. 179-183.

#56

Frau to Ida Boltzmann [Vienna, Winter 1902-1903]

Things are going quite passable with Papa; he's even begun iceskating again. Living in a small house [*Cottage-Wohnen*], you know, has its advantages. Had we remained in Türkenstrasse [in the laboratory apartment] he would not be exercizing again so quickly. Why? Because there he was not to be gotten outside, and here he must go out. If things are going better, then he starts going by foot, and movement is best for him[1] . . . .

[1] From photographs and drawings as well as occasional comments it seems that Boltzmann during his last years was very overweight.

#57

To Ludwig Schlesinger[1] Vienna, January 23, 1903

Best thanks for sending the new printing of Bólyai's main work[2] to which you have attached your own attractive-looking treatise...

[1] Ludwig Schlesinger, who was born in 1864, was a mathematician at Klausenburg and later at Giessen. He was largely interested in differential equations and functions.
[2] The suspicion is, given the date of publication, i.e. the beginning of 1903, that Boltzmann received and in his letter was referring to a republication of the brief 1832 appendix on non-Euclidean geometry by Johann Bólyai which had originally been placed at the end of his father's much longer work. A new edition of that longer tome by Wolfgang Bólyai de Bólya, *Tentamen iuventutem studiosem in elementa matheseos . . . introducend, mathematisches Lehrbuch*, was then in the process of being published. The new edition in two volumes came out in 1899 and 1904.

#58

To Otto Wiener Vienna, February 7, 1903

Best thanks for your friendly letter. Also, I have certainly learned much more in Leipzig than I was able to dispense to you and your colleagues. And it was certainly not from any undervaluation of the stirring spiritual life which rules Leipzig or the manysided stimulations which one receives there, even less [from any disregard for] the kindness and consideration with which I was accepted, when I say that I feel wholly better in Vienna.

Furthermore, let me add that I am now living under relatively quite f vorable conditions in a furnished house which is indeed small but has amp place for us with rather healthy air, and it is arranged free from embarrassir luxury but with more actual comfort than one customarily encounters wi rented quarters. On top of that there is salutory peace and quiet, as if one we in the countryside.

I hope that Des Coudres makes out well and will also work his way ev more deeply into theory, where mainly one must finally understar mathematics. Hopefully, the Leipzig Physical Institute will soon be finishe and you yourself will be well satisfied.

With best regards from me and my wife to you and your devoted wi and with heartiest greetings I am...

[1] The Institute was finally finished in 1905.

#59

Frau to Ida Boltzmann [Vienna, Spring 190

Just think dear, Arthur[1] has just flown in a balloon to a height of 5,0( meters with another person and free of charge. They flew in the Prater [par over the Hohewarte and landed on an island in the Danube in t neighborhood of Pressburg.[2] He went with Adjunct Dr. Valentin, of t Meteorological Central Institute [*Centralanstalt*]. Arthur was complete carried away by the beauty of the trip. Papa agreed to let him become member of the *Aeroklub*. Silberer wants Arthur to be trained as someone w could handle it alone. Our garden is not yet in order. The summer hou [*Lusthaus*] hasn't come yet. It is now just before the beginning of Papa lectures and at nights he is very neurasthenic . . . .

Unfortunately, Papa also has Catarrh and asthma at night. When he h strong asthma the last time, without asking him I called Dr. Bräuer. H prescribed something for the catarrh, which Papa won't take. Otherwise I said that it was necessary for him to reduce. Papa is also very depressed by t catarrh . . . .

[1] Boltzmann's second son. He became an assistant lecturer at Vienna University, was artillery observer in a tethered balloon during World War I, and subsequently became Austrian civil servant in the Federal Office for Weights and Measures and later its direct See Ilse M. Fasol-Boltzmann, "Ludwig Boltzmann and His Family" in Ilse M. Fasc Boltzmann (ed.), ***Ludwig Boltzmann 'Principien der Filosofi'***, Springer: Berlin, etc., 199 p. 28. See also Letter #54 above, footnote 3.

[2] The German name for Bratislava, now the capital of the Independent Republic Slovakia.

#60

Frau to Ida Boltzmann [Vienna, Spring 1903]

Unfortunately, Papa becomes more upset every day . . . perhaps things will change when the lectures have begun. But my confidence in the future has become severely shaken. I thought things would turn out better. If only you will become a true medical student or doctor; that would make me happiest . . . .

#61

To Leo Königsberger [undated]

Unfortunately, I failed to keep Helmholtz's letters.[1] I regret that now myself. If I had kept them, then I would have preserved a comical intermezzo. I had carried out an experiment on dialectric constants in order to examine Maxwell's electromagnetic theory of light which has now become so famous. Helmholtz, who did not have the formula in exact terms in his head, told me verbally in passing that the refraction quotient must be the same as the dialectric constant *D*. Since my experiment didn't yield that result, I left Berlin with the firm conviction that I had completely refuted Maxwell, and was already thinking of publication, when I suddenly got the idea to take another look at Maxwell's formula and found there $n = \sqrt{D}$. That was what my letter of November 1, 1872 was about. Since I can well assume that my letter without the corresponding one from Helmholtz cannot be properly printed, let me simply send it back to you.

With high respect and heartiest greetings...

[P.S.] Please excuse the lateness of my reply. I am now somewhat nervous and confused.

---

[1] This undated letter starts out like letter #51 as if Boltzmann had forgotten that he had written the earlier one. See the last line of #61 for an admission of confusion. On the other hand, being undated one cannot be certain that it was in fact written after what we have listed as #51. Nevertheless, #61 is intrinsically interesting and Boltzmann's little story was reproduced by Königsberger in his own autobiography, *Mein Leben*, Heidelberg, 1919, pp. 110f.

#62

Hartel about Boltzmann Vienna, May 5, 1903

Without in any way affecting his existing teaching responsibilities, from the Winter Semester 1903/04, the Professor for Theoretical Physics, Dr. Ludwig Boltzmann, will until further notice be given permission under this teaching contract to give a course of lectures on "Philosophy of Nature and

Methodology of the Natural Sciences" for two hours weekly every semester and for a remuneration of 1,000 Kronen per semester.[1]

---

[1] Technically speaking, while Boltzmann was now allowed to teach philosophy of science as if he were resuming lectures on this topic which had been given by Ernst Mach. Mach, who retired in 1901, had held a chair in "The History and Philosophy of the Inductive Sciences" whereas Boltzmann legally was only given a *Lehrauftrag* or teaching contract for "Philosophy of Nature and Methodology of the Natural Sciences". In short, Boltzmann did not hold two chairs. This editor had failed to understand the distinction in his biography of Ernst Mach published in 1972. He apologizes for the mistake of having believed that Boltzmann held two chairs and especially for having criticized a prominent scholar for holding the correct opinion.

#63

To Leo Königsberger — Vienna, May 8, 1903

Just as the activity of Helmholtz is of unique importance in scientific literature, so your biography has the same importance in biographical literature for the history of natural science. [It is also valuable] for the extent of its instructiveness for the reader. Therefore, I am especially obliged to thank you for sending me parts two and three.[1] I also thank you for the friendly mention of my activity in this work which is so important.[2]

With heartfelt greetings...

---

[1] Königsberger's biography went through Helmholtz's scientific work article by article such that actually it is as much a description of science as it is a narrative story of his life. See also letter #51.

[2] Königsberger included excerpts from Boltzmann's letters to Helmholtz in his biography as well as adding a few comments about their scientific exchanges.

#64

To Eduard Suess[1] — Vienna, May 19, 1903

My best thanks for the friendly invitation to the Academy dinner,[2] which I will not fail to obey. Thus, I am not merely a member in spirit, but as far as my stomach is concerned, actively corporeal. With high respect and hearty greetings...

#64A

Bryan[1] about Boltzmann — [Partly about 1903]

Boltzmann visited the yearly meeting of the British Association for the Advancement of Science in Oxford in 1894 and also took part in one for a short time at Southport in 1903 which was then taking place. In this manner many English scientists came into immediate contact with the Vienna physicist. His visit will long remain in our memory. He was very taken by the friendly

His visit will long remain in our memory. He was very taken by the friendly reception which accompanied him, and when foreign guests were taking their leave he expressed the wish "to often come at England".

---

[1] Eduard Suess (1831-1914) was Emeritus Professor of Geology at the University of Vienna and had been a member of the Austrian Parliament for some thirty years.

[2] The Austrian Academy of Science.

[1] George Hartley Bryan (1864-1928) was born in Cambridge and graduated from his hometown University with a PHD in science in 1895. He was very interested in Boltzmann's work and they collaborated on a paper in 1894 "Über die mechanische Analogie des Wärmegleichgewichts zweier sich berühender Korper", *Sitz. ber. kaiserlich. Akad. Wiss.* [Vienna], **103** (1894), 1125-1134. The above quotation is included in Wolfgang Stiller, *Ludwig Boltzmann*, Verlag Harri Deutsch: Frankfurt am Main, 1989, p. 80.

## B. Student Comments on Boltzmann's Lectures

*Karl Przibram* : "Boltzmann held his [Vienna] lectures in the ramshackle lecture hall of the old Physical Institute at Türkenstrasse 3. He tended to begin every time with a sonorous "Now!" in his high tone of vioice which so ill-suited his heavy stature."[1]

*Franz Skaupy* : In a rather small lecture room, the seats of which descended steeply, were two very long blackboards, one right and one left, and perhaps one more in the middle. In the first quarter hour of the lecture, which he always began in a high-pitched voice with "Now we have, etc. . . . ." Boltzmann usually wrote down all the developments of the previous class in an extremely beautiful and clear way, beginning with the left blackboard, so that even someone who had been absent could cope with what followed. Then the really new lecture began during which the rest of the left and then the right blackboard were slowly and lucidly covered with new equations. If one of the listeners happened to be thinking about something else for, say, ten minutes, he could catch up by reading through what had been written on the board, so clearly was everything arranged. The hall was always full, as obviously not only physics students but also, for example, chemistry students like myself were taking the course. One of the students - I believe his name was [Paul] Ehrenfest - reported on some work or other, and used the opportunity to quote Boltzmann's own work so fully that a smirk grew on Boltzmann's face until he said: "If I only knew my own work so well!"[2]

*Lise Meitner* : "I was able to hear his lectures from 1902 until his death in 1906 . . . . Boltzmann had no inhibitions whatever about showing his enthusiasm while he spoke, and this naturally carried his listeners along. He was also very fond of introducing remarks of an entirely personal character into his lectures - I particularly remember how in describing the kinetic

encountered because he had been convinced of the real existence of atoms, and how he had been attacked from the philosophical side without always understanding what the philosophers had against him.[3]

*Alois Höfler* : "I . . . was one of six hundred listeners of both first lectures of Boltzmann on 'nature philosophy' [October 26 & 27, 1903]. An unchained storm of laughter greeted his opening words: 'How have I come to teach philosophy?' And the laughter re-erupted from sentence to sentence."[4]

*Ludwig Flamm* : "The lectures of Boltzmann on nature philosophy were immensely popular, but his lectures on theoretical physics were also well-attended."[5]

*Karl Przibram* : I also heard Boltzmann's Lectures on Natural Philosophy, which he as the successor of his scientific opponent, Ernst Mach, gave in a large lecture room which was always overfilled.

It must have been in one of these philosophical lectures and not in a physical one that Boltzmann came to speak about curved and multidimensional space - and at that time Einstein's General Theory of Relativity still lay well in the future.[6]

*Abraham Pais* : Boltzmann's qualities as an outstanding lecturer are not reflected in his scientific papers which are sometimes unduly long, occasionally obscure, and often dense. Their main conclusions are sometimes tucked away among lengthy calculations. Also (and especially in regard to the theoretical interpretation of the second law), Boltzmann would change his point of view from one paper to the next without advance warning to the reader. Maxwell said of his writings: "By the study of Boltzmann I have been unable to understand him. He could not understand me on account of my shortness, and his length was and is an equal stumbling block to me." Einstein once said to a student of his: "Boltzmann's work is not easy to read. There are great physicists who have not understood it." . . . . A larger audience acquired some degree of familiarity with Boltzmann's work only after its exegesis by the Ehrenfests, published in 1911.[7]

---

[1] Karl Przibram, "Erinnerungen an Boltzmanns Vorlesungen" in *The Boltzmann Equation - Theory and Applications,* edited by E.G.D. Cohen & W. Thirring, Springer Verlag: Vienna & New York, 1973, p. 641.

[2] Quoted by Engelbert Broda, *Ludwig Boltzmann: Man - Physicist - Philosopher,* Ox Bow Press: Woodbridge, Connecticut, 1983, p. 12. Originally a personal communication to Broda from Franz Skaupy, a metallurgist who had attended Boltzmann's lectures from 1902 to 1904.

[3] Lise Meitner, "Looking Back", *Bulletin of the Atomic Scientists,* **20** (Nov. 1964), 3.

[4] Alois Höfler, "Ludwig Boltzmann als Mensch und als Philosoph", *Süddeutsche Monatshefte,* **3** (October 1906), 18-22. Alois Höfler (1853-1922) was an Austrian philosopher of education. He had been much influenced by Franz Brentano.

[5] Ludwig Flamm, "Die Persönlichkeit Boltzmanns", Wiener Chemiker Zeitung, **47** (1944), p. 30. Ludwig Flamm (1885-1964) became a physical scientist at the Technical University of Vienna and Ludwig Boltzmann's son-in-law.
[6] Przibram, op. cit., p. 642.
[7] Abraham Pais is a well-known physicist and scholar. He became Desley W. Bronk Professor at Rockefeller University and won the J. Robert Oppen'eimer Memorial prize in 1979. The quotation is from his very well received book *'Subtle is the Lord' - The Science and the Life of Albert Einstein,* Oxford University Press: Oxford, 1982, p. 65-66 & 67.

## C. Natural Philosophy

(From Inaugural Lecture on Mechanics, Vienna, October 1902)[1]

It is customary in inaugural lectures to begin with a hymn of praise to one's predecessor. Today I can save myself this sometimes troublesome task, for even if Napoleon never managed to be his own great-grandfather, I, on the otherhand, am currently my own predecessor.

Solipsism is the view that the world is not real, but a mere product of our fantasy, like a dream object. I too once hankered after this whim, which led to my failing to take the right practical action and caused me damage to my immense delight, for this provided me with the desired proof that the external world exists, a proof that can consist only in showing that if we doubt this existence we are less able to act appropriately.

Phenomena of motion are the ones that we observe most often and most directly; all other natural phenomena are more concealed. Besides, we can cope with phenomena of motion by means of the least number of concepts; all we need for describing them are the concepts of position in space and its change in time, whereas for other phenomena many less clear concepts are required, such as temperature, luminous intensity and color, electric tension, and so on.

It is the ubiquitous task of science to explain the more complex in terms of the simpler; or, if preferred, to represent the complex by means of clear pictures borrowed from the sphere of the simpler phenomena. Therefore in physics too the attempt was made to reduce all other phenomena such as sound, light, heat, magnetism, and electricity, to mere motions of the smallest particles of these bodies, and this turns out quite successful for very many phenomena, although of course not for all. Thus the science of motion, that is mechanics, became the root of the other physical disciplines, which seemed increasingly to grow into special chapters of mechanics.

Only in more recent time has there been a reaction to this. The difficulty occasioned by a purely mechanical account of magnetism and electricity raised doubts as to universal mechanical explicability and it was precisely electro-magnetism that became increasingly important not only in practice but also for theory. In the end it became so powerful as to attempt a reversal of roles, explaining mechanics in terms of electro-magnetism. Whereas previously one had to explain magnetism and electricity by rotations or oscillations of the smallest parts of bodies, the present aim was to derive the fundamental laws of motion for bodies themselves from the laws of electro-magnetism.

(from the Inaugural Lecture in Natural Philosophy given on October 26, 1903 with a written version first appearing in the newspaper *Die Zeit* on December 11, 1903)[2]

On one occasion in the assembly hall of the Academy [of Science], I was involved in the liveliest debate with a group of academics, Professor Mach amongst them, on the newly revived controversy about the value of atomistic theories .... In that group of academics during the debate on atomism Mach suddenly said laconically: "I do not believe that atoms exist". This utterance ran round and round in my head".

I here mention in passing that in the assignment that begins with today's lecture I am in a certain sense Mach's successor and I should really have started my talk by paying homage to him. However, I think that to express his special praise would amount to carrying owls to Athens, as far as you are concerned and indeed not only you but any Austrian and even any educated person throughout the world.

It was only with hesitation that I followed the call to meddle with philosophy; philosophers have the more often meddled with natural science. They have now for many years invaded my preserve and I could not even understand what their views were and therefore wanted to improve my knowledge of the fundamental theories of all philosophy.

To go straight to the deepest depths, I went to Hegel; what unclear thoughtless flow of words I was to find there! My unlucky star led me from Hegel to Schopenhauer. In the preface to his first work I found the following passage which I will report verbatim here: "German philosophy stands laden with contempt, derided by other countries, banished from honest science like a ..." I suppress the next bit since there are ladies present. "...The heads of the present generation of learned men are disorganized by Hegelian nonsense. Incapable of thought, uncouth and stupified, they fall prey to shallow materialism that crawled forth from the baslisk's egg". With that I was of course in agreement, except that I found that Schopenhauer, too, really deserved the blows of his own cudgel.

However, Herbart's calculations about psychological phenomena equally appeared like a travesty of analogous calculations in the exact sciences. Even in Kant there were many things that I could grasp so little that given his general acuity of mind I almost suspected that he was pulling the reader's leg or was even an imposter. In this way there developed in me an aversion from philosophy at that time, indeed a hatred of it. In view of these old philosophical systems I should almost say that in my case the goat has been made the gardener. Or was I given precisely the teaching appointment in the way that an old democrat is given the title of Court Councillor [*Hofrat*], in order to turn Saul definitely into Paul? I am afraid that in these lectures I shall be oscillating between goat and councillor, and although I hope that I shall never fall into the style of which I have just read you a sample, I may nevertheless be somewhat rough in places when I apply Mach's method to the perfecting of philosophical systems.

My aversion from philosophy at the time was incidentally shared by most natural scientists. Every metaphysical orientation was pursued with the aim of extirpating it root and branch; but this attitude did not last. Metaphysics seems to cast an irresistible spell on the human mind and all the abortive attempts at lifting its veil have not impaired its power. The drive towards philosophizing seems ineradically innate. Not only Robert Mayer, who was a philosopher through and through, but Maxwell, Helmholtz, Kirchhoff, Ostwald and many others too made willing sacrifice to it and recognized its questions as the very highest, so that today metaphysics figures once again as the queeen of the sciences.

It was clear to me that we unite groups of perceptions into ideas of objects, for example of a table, a dog, a man and so on. Moreover we have memory pictures of these groups of ideas. When we form new groups of ideas which are quite similar to these memory pictures, there is sense in the question whether the corresponding objects have existence or not. We have here as it were an accurate measure for the concept of existence. We know exactly what is meant by the question whether the griffen, the unicorn or a brother of mine exists. However, when we form quite novel ideas, such as those of space, time, atoms, the soul, or even God, does one know, so I asked myself, what is meant by asking whether these things exist? Is not the only correct thing to do here to try to clarify what concept one is linking with the question as to the existence of these things?

If there is something else behind perceptions, how can we even suspect that there is? If there is nothing behind them, would then a landscape on Mars or on a planet of Sirius really not exist if no living being is ever able to perceive them? If all these questions are senseless, why can we not dismiss them or what must we do in order finally to silence them? It is to be the task of my present lectures at least to seek to clarify these questions.

---

[1] Ludwig Boltzmann, "On the Principles of Mechanics II", *Theoretical Physics and Philosophical Problems*, edited by Brian McGuinness and translated by Paul Foulkes, Reidel: Dordrecht, 1974, pp. 146 & 149-151. (This book is largely a translation of Ludwig Boltzmann, *Populäre Schriften*, Leipzig, 1905.)

[2] The English translation is taken from: Ludwig Boltzmann, "Inaugural Lecture on Natural Philosophy", *Theoretical Physics and Philosophical Problems*, edited by Brian McGuinness and translated by Paul Foulkes, Reidel: Dordrecht, 1974, pp. 154-156.

# Chapter 5

# PHILOSOPHY LECTURES

## A. Correspondence

### #65

Frau to Ida Boltzmann [Vienna, November 1903]

Papa has become very well received here. His lectures are cram full (unfortunately, his teaching contract is only for a fixed sum). In the Inaugural Lecture [for philosophy] people were standing in the laboratories, on the stairways, and in front of the building. The blackboard and lecturehall were ornamented with sprigs of fir needles. He was accompanied by thunderous applause. It is reported in all of the Vienna newspapers.[1] He has become very eager once more. *I must write for him again.* He is letting the Inaugural lecture be printed like any other essay. But unfortunately, matters do not stand well with his bodily health. The bronchial catarrh is not good, and added to that, fearful hoarseness. Nevertheless, he continues to lecture. But also his neckwound is not good. It was burned and tomorrow it must be cut again. There is said to be a bean-sized fistula under the skin. But he is ever lively and animated . . . .

---

[1] For three newspaper reports see Ilse M. Fasol-Boltzmann (ed.), *Ludwig Boltzmann 'Principien der Naturfilosofi' - Lectures on Natural Philosophy 1903-1906*, Springer Verlag: Berlin etc., 1990, pp. 147-151.

### #66

Frau to Ida Boltzmann [Vienna, November 1903]

Papa has completely changed. This evening he is in the *Gasthaus* again with his colleagues. He has retired his shiny housecoat and he dresses more respectably at home. He works out in the gym and sometimes comes home by foot. The *Kaiser* [Franz Josef] told him that he was really happy that he had come back again to Vienna, and had already heard how overfilled his lecturehall was. Papa accepted both orders [of merit] and the medal from the *Kaiser*. He is completely changed as through magic; indeed, he is ever wandering about, and suffers from insomnia more than ever. He really likes our home . . . .

### #67

Nernst[1] to Meyer Göttingen, November 9, 1903

Finally, I have enclosed the work for the Boltzmann volume.[2] Just at the last moment of writing down the theory, I considerably expanded what I had done, much more so than I had hoped [would be necessary] at the beginning, therefore the delay of three weeks, but hopefully, it is still not too late.

Since the paper has yielded what is to me a right surprising confirmation of our kinetic views of chemical processes, it is my hope that in fact my contribution will provide a small moment of happiness for our Boltzmann. Does the situation already stand close to being fixed about the kind and manner of the celebration? I would naturally prefer to be present. I would be very grateful for short, timely information about that. With best greetings... [3]

---

[1] Walther Nernst (1864-1941) is perhaps best known for his explanation for the ready ioniozation of compounds in water and for discovering what is often called the Third Law of Thermodynamics, that entropy change approaches zero at a temperature of absolute zero. He had visited Graz during the middle 1880's to learn theoretical physics from Boltzmann and remained a friend of his thereafter.

[2] Stefan Meyer was accepting contributions for a *Festschrift* in honor of Boltzmann's 60th birthday on February 20, 1904. The Meyer *Nachlass* in the Radium Institute in Vienna contains many letters relating to the *Festschrift*. (Also, there are draft letters from Smolukowski to Meyer in the University of Cracow about it.) Unfortunately, for reasons not explained at the time, during the early 1980's when the *Nachlass* in Vienna was examined, Ernst Mach's listed letters to Meyer for 1903-04 were missing. They presumably also had something to do with the *Festschiftt* to which Mach contributed an article. The published book edited by Meyer was simply titled *Ludwig Boltzmann Festschrift*, Barth: Leipzig, 1904

[3] The original of this letter was found By Professor Berta Karlik of the Radium Institute attached to Meyer's copy of Nernst's book *Theoretische Chemie* (1898).

#68

Frau to Ida Boltzmann [Vienna, undated]

Papa has received from Professor Nernst the gift of several of his lamps, which he enjoyed very much. Since they are such a rarity, in the last sitting of the old Physical Society he invited the gentlemen to take a look at them at seven o'clock on Sunday morning. The secretary of the society sent out 55 written invitations. We set up a cold buffet and 50 liters of beer. But only seven people came[1] . . . .

Papa gets up very early every day and has been writing a treatise,[2] which he sent to the Academy [of Science] on Thursday, but always speaks cheerfully. He is a completely different person than earlier . . . .

---

[1] Walther Nernst had invented a lamp which could be heated to incandescence with a comparatively weak current, but it could not compete with Edison's incandescent light, but before concluding that Nernst was just another impractical scientist, he had the last laugh when he sold the patent for a reported million Marks.

[2] The only *scientific* article by Boltzmann listed as published in 1904 had two versions, both very short, one by himself and one co-authored with his son Arthur. The one sent to the Academy was by himself and appeared in the *Wiener Anzeiger*, Volume. **41**, Page 325, November 3, 1904.. Both were titled "Über das Exnersche Elektroskop". See the last article in the last volume (III) of Ludwig Boltzmann, ***Wissenschaftliche Abhandlungen***, edited by Fritz Hasenöhrl, Chelsea Publishing Company, New York, 1968 ($_c$1909), pp. 693-694. Most of Boltzmann's articles by this time (1903-04) were in philosophy.

#69

To Franz Brentano[1] Vienna, November 19, 1903

My best thanks for sending me your book[2] and the friendly letter. It pleases me very much that our judgments on the apriori coincide. I also like your calling it apriori imagination or prejudice. Thus very rightly you drastically illuminate the unjustifiability of belief in the infallibility of everything apriori. Just as [one says] if morality were not grounded all morality would cease, [so one hears] if our behavior lacked causes it would be blind and ruled by chance.

Though what I call apriori judgments are according to Darwin's teaching inherited by particular individuals and their validity is quite probable, since only the well-tested is inherited, it is still not - if you will allow me to use a manner of speaking - apriori certain that they can be false if the fixed belief of the species about what is false was useful.[3]

With hearfelt greetings...

---

[1] Franz Brentano (1838-1917) was a very influential Austro-German philosopher and psychologist. He taught in Vienna from 1874 to 1894. He had been a Roman Catholic Priest but left the church, married Ida Lieben under controversial circumstances, lost his professorship, taught as a *Privatdozent*, and retired to Italy where he wrote voluminously, but only in recent years has most of his work been published. His students included Husserl, Freud, Kraus, Marty, Stumpf, Meinong, and many other influential figures. Only Ernst Mach would come close to conspicuously matching his philosophical impact in Austria, though less conspicuously Boltzmann would also prove rather influential as we shall see. When Boltzmann received a teaching contract to teach philosophy as if he were the successor of Mach and Brentano he also began philosophical correspondence with Brentano. He had known Mach apparently since the early or mid 1860's, but the first surviving letters are from 1892.

[2] It is possible that Boltzmann was referring to a new printing of Brentano's book on the origin of our knowledge of right and wrong titled: *Vom Ursprung sittlicher Erkenntnis* ($_c$1889). If it appeared, the new printing seems not to have been a new edition.

[3] If the last part of this translation of a difficult sentence is correct, then the question arises whether Boltzmann developed this pragmatic notion himself, inferred it from his understanding of Darwin's ideas, or whether he had gotten it from William James himself, whom he had presumably met while teaching at Clark University in the Summer of 1899. Indeed, it is even conceivable that Boltzmann may have helped strengthen James's own philosophy of pragmatism, which was later published in book form in 1907.

#70

To David Hilbert[1] Vienna, November 25, 1903

About your article, "Über die Grundlage der Geometrie",[2] Leipzig, 1899, I know it. Since I am also working on this theme now, I would appreciate it if you would tell me if you have published something on this topic since then. You would also very much deserve my gratitude if it would be possible for you to indicate where I can find other important publications on this matter; naturally only the most important, in so far as you can immediately remember them. You would then spare me a lot of work having to look for them.

My compliments to your devoted wife. With highest respect and heartiest greetings...

---

[1] David Hilbert (1862-1943) was one of the true greats of modern mathematics. He taught at Göttingen from 1895 to 1930. Among his wide range of important contributions, he also did work in the kinetic theory of gases, which was Boltzmann's special field of concentration in theoretical physics. But it seems Boltzmann was also interested in Hilbert's important work on the foundations of geometry and the study of axiomatics.

[2] David Hilbert, "Über die Grundlagen der Geometrie", originally appeared in *Festschrift zur Feier der Enthullung des Gauss-Weber-Denkmals in Göttingen*, im Verlag Teubner, Leipzig, 1899, but the second edition of the article appeared separately in book form under the same title in 1903. An edition titled *The Foundations of Geometry* had been published the previous year by Open Court in the United States.

## #71

Baeyer[1] to the Nobel Prize Committee Munich, December 25, 1903

Proposal for the Nobel Prize of Physics
to be Awarded in 1903[2]

Hofrath L. Boltzmann in Vienna for his work on the Kinetic Theory of Gases.[3]

---

[1] Adolf von Baeyer (1835-1917) was professor of chemistry at the University of Munich, receiving a Nobel Prize himself in 1905 for his work in synthetic organic chemistry. He had become friends with Boltzmann when the latter taught in Munich during the early 1890's.

[2] Originally written in French.

[3] The Nobel Prize in Physics for 1903 went to Henri Antoine Becquerel and Pierre and Marie Curie for their work in discovering spontaneous radioactivity.

## #72

To Leo Königsberger [Vienna, 1903-04]

My best thanks for the friendly words which you have now devoted to my youthful work and with which I naturally very much agree.[1] But please be persuaded that even if you had not written to me first in such a considerate manner I would still not have been one of those people who upset matters by claiming priority in the matter of scientific discovery. I would not even have mentioned the thing at all. Still less can I find fault with you for overlooking a work so little known which had been hidden in the Vienna Academy and which had already passed several people unnoticed (for example Clausius), whose special field of work lay precisely in that area.

With highest respect and hearty New Year's Greetings...

---

[1] See Leo Königsberger, *Hermann von Helmholtz*, Band II, pp. 328ff, and also the former's autobiography, *Mein Leben*, Heidelberg, 1919, pp. 109ff.

#73

To Svante Arrhenius Vienna, January 13, 1904

Best thanks for the friendly letter and the New Year's wishes, which I return in full measure. [I think that] you should be pleased by the bestowal of the Nobel Prize[1] as acknowledgement of your work.[2] I am now busy writing Part 2 of my *Mechanik.*[3] Unfortunately, I was sick during the whole Christmas vacation and was unable to go out, nor was I disposed to work; also my daughter was sick, but we are all healthy again. I wish you maximum success for your future work. I am with best greetings...

---

[1] As already mentioned, Svante August Arrhenius won the Nobel Prize in Chemistry in 1903 for his work on ionic dissociation.

[2] The original German is in the passive voice with both the verb and the acted-upon "subject" at the end. If translated in that word order it would give the unintended impression that Boltzmann was begrudging Arrhenius the Nobel Prize.

[3] It was published as *Vorlesungen über die Prinzipien der Mechanik,* 2. Band, Leipzig, 1904.

#74

Meiner[1] to Meyer Leipzig, January 16, 1904

I am very thankful for your extraordinarily flattering and friendly communication from the 14th of this month. The unrest [*Gährung*] of which you speak, exists not so much within the circle of book publishers and handlers, but rather within the academic community, especially here in Leipzig, who consider themselves either discriminated against or unduly privileged. If this movement, since it is actually without foundation, will soon come to an end, it will still be refreshing, since at least the importance of the work of we publishers will have been been acknowledged by the other side. And it is in this sense, that I thank you once again for your words and am now agreed that the foreword will remain as it is.[2]

Your account has persuaded me that a special biography and assessment of Boltzmann is premature,[3] and thus can be left out. An Ostwald thinks very differently from a Boltzmann.[4]

About the arrangements for the [60th] birthday celebration [for Boltzmann on February 20, 1904], what else do you intend to set up or organize? I am not excluding the possibility that my planned trip to Austria might be worked out so that I could be in Vienna on February 20.

The printing plant has finished a demonstration copy on especially beautiful paper, and I request that you decide whether I should go to the trouble of finding a worthy cover for it here or whether you will have that done in Vienna; if so, I will send you only the printed sheets [*Bogen*].[5]

With best greetings...

---

[1] Arthur Meiner (1865-1952) had acquired Johannes Ambrosius Barth Publishing Company in 1890 and managed it as chief editor.

[2] Stefan Meyer, Boltzmann's assistant, was preparing a *Festschrift* for Boltzmann's 60th birthday. Unfortunately, publishers, like librarians, are rarely enthusiastic about them. Sales tend to be limited to contributors, friends, and former students, and unless they are "misclassified" under the name of the person whose birthday is being celebrated, and not just the editor (whose name is often left out), they can be very difficult to find in libraries, plus the fact that their heterogeneous character usually attracts very few library readers. It is not clear what the "movement" mentioned above was about, but if as seems possible it was connected with reluctance by publishers to put out unsubsidized *Festschrifts* for every professor, whether famous or utterly obscure, their resistance was and still is quite understandable.

[3] Boltzmann's work as mentioned earlier was still opposed by many individuals, groups, and persuasions. Apart from technical objections by many supporters about details of the kinetic theory of gases and those physicists who supported the "electro-magnetic world view", there were also phenomenalists like Mach, energeticists such as Ostwald, plus various other advocates of methodological or philosophical perspectives (such as many Neo-Kantians) who still rejected the reality of atoms or use of the atomic theory.

[4] While it is regrettable for historians that no biography of Boltzmann was published at that time, it might have seemed unduly insensitive for Meiner to have published it in Leipzig, the hometown of Boltzmann's very influential philosophical opponent, Ostwald. One might add that Leipzig was then the center of the German book publishing industry and Barth perhaps the leading publisher of academic books.

[5] Such a copy would presumably have been given to Boltzmann himself.

## #74A

Stiller[1] about Boltzmann [Concerning the 20th of February, 1904]

Austrian physicists under the leadership of Boltzmann's assistant Stefan Meyer (1872-1949) prepared a huge *Festschrift* which Barth Publishing House in Leipzig published punctually. Over a hundred recognized scholars from several continents contributed articles on their physical specialties for this immense book of over a thousand pages.[2]

[1] Doctor of Science Wolfgang Stiller was born in Leipzig in 1940. He has worked as a theoretical physicist and physical chemist at the Academy of Sciences in the former German Democratic Republic. His book ***Ludwig Boltzmann - Altmeister der klassischen Physik Wegbereiter der Quantenphysik und Evolutionstheorie,*** Verlag Harri Deutsch: Frankfurt am Main, 1989 is a well-rounded, well-researched work which is also strong in describing and quoting from Boltzmann's friends and associates.

[2] Ibid., p. 37.

## #74B

Meitner about Boltzmann [Concerning the 20th of February, 1904]

Paul Ehrenfest, who had come from Göttingen to Vienna, asked the audience to remain standing after the professor had come in as he wished to pay tribute to Boltzmann's great accomplishments in a short ceremonial address.[1]

[1] Lise Meitner, "Looking Back", ***Bulletin of the Atomic Scientists***, **20** (Nov. 1964), p. 3.

## #74C

Broda about Boltzmann [Concerning the 20th of February 1904]

Boltzmann himself remarked in jest on his sixtieth birthday that his quick transitions from cheerfulness to sadness stemmed from the fact that he was born in the night between Carnival Tuesday and Ash Wednesday amid the dying noises of the dance. (There was a dancehall in his birthhouse).[1]

[1] Engelbert Broda, *Ludwig Boltzmann - Man - Physicist - Philosopher*, translated by Larry Gay and the author, Oxbow Press: Woodbridge Conn., 1983, p. 30.

#74D

To the Austrian Education Ministry Vienna, April 4, 1904

Your humble servant suffers at present once again from a strong nervous depression, and he fears that he is not able to fulfill all that he has undertaken for this semester without damaging his health . . . .[1]

[1] Boltzmann was asking to be excused from teaching nature philosophy for the spring semester. See letter #84.

#75

From Victor von Lang[1] Vienna, July 31 1904

We are sending you the German translation of a letter which arrived here from the Secretary of the Royal Society, J. Larmor, about an inquiry made by Hofrath Weiss.[2] The original lies ready for inspection in the main office [*Kanzlei*], but we must remark that it is rather difficult to read, which is why we sent you a translation. Our intent in the letter [by Hofrath Weiss] was to make a request to the International Association in order to obtain a a first-hand report, and we politely request that you carry out the job as our man in St. Louis.[3] We have also made known to the Committee of Solar Research that you, most highly respected Herr Colleague, will appear as representative of the [Austrian] Imperial Academy [of Sciences].[4] We thank you in advance for the effort which you will make on our behalf.

The Presidium of the Imperial Academy of Sciences...

[1] Victor von Lang (1838-1821) had held a chair in experimental physics in Vienna since 1866. See letter #38, footnote 2 for a few additional details. There is also a second signature on this rather formal letter but it is unreadable. The tone of the letter does not bespeak a close relation between the authors of it and Boltzmann.

[2] Edmund Weiss (1837-1917) was an Austrian astronomer. He became a professor in Vienna in 1869.

[3] Boltzmann did indeed go to St. Louis, where in connection with the World's Fair there was also a vast scientific program which included over 300 speeches. Boltzmann gave an address on September 24, 1904. For passages from that lecture see Section C. of this chapter. It is not known how well he carried out the Imperial Academy's request.

[4] On returning to Vienna in 1902, Boltzmann was not immediately restored to his regular position in the Austrian Academy of Sciences, but remained a "corresponding member" for some time. He was readmitted as a *Wirkliches Mitglied* only on May 20, 1904.

#76

To Frau Boltzmann [*Belgravia*,[1] August, 1904]

.... The ship is very inferior and has many inconveniences. We [Boltzmann and his son Arthur] are both well. I hope I may remain so on the whole voyage, which will certainly be more strenuous than is good for my age and the state of my nerves ....

---

[1] Boltzmann and his son Arthur left Hamburg on the *Belgravia* on August 21, 1904 for the conference in St. Louis. They returned on the *Deutschland*, arriving back on October 8. See Ilse M. Fasol-Boltzmann (Ed.), *Ludwig Boltzmann Principien der Naturfilosofi - Lectures on Natural Philosophy 1903-1906*, Springer Verlag: Berlin, etc., 1990, p. 31-32.

#77

To Frau Boltzmann [New York? September 1904]

.... I am ill, and if this terrible melancholy which grips me at the moment does not subside in Vienna, I shall be unable to hold my lectures[1] ....

---

[1] Ibid., p. 32.

#78

To Frau Boltzmann Washington D.C., September 7, 1904

We stayed three days in New York and are now in Washington. Even up to the very wharf in New York we have seen nothing new; nor is it gallantry but truth that what is most on my mind is remembrance of thee. Recent days have not been happy ones at all ....

#78A

Millikan[1] about Boltzmann [Concerning Late September 1904]

I was alone working hard on a radioactive problem on which I was to report at St. Louis when a knock came at the door. I thought it was some student and, annoyed a bit, kept on with my work for perhaps five minutes and then went to the door. Imagine my chagrin when I found I had kept the distinguished broadbrowed and full-bearded Ludwig Boltzmann from Vienna waiting all the time. I had not known he was in America. However, he was exceedingly gracious. I showed him what I was doing and spent a most profitable day in his company.... I shall digress a little from this narrative to elaborate upon the final struggle over the atomic theory and the change in thinking that took place at this time. For nothing showed more clearly the change that occurred between say 1904 and 1912. In the first of these two years occurred the great exposition at St. Louis, already referred to, at which was held a scientific congress, one whole day of which was given up to a discussion of the fundamental question as to whether or not an atomic or *discrete*-particle theory was needed for the interpretation of the known facts of observation, or whether a continuous theory of both electricity and matter would work equally well . . . .

Ostwald came over from Leipzig to defend the negative of the proposition, while Van't Hoff came from Berlin to take the other side of the debate. Boltzmann was also there to support Van't Hoff. The amazing thing is that this question could be debated at all at that time, and that outstanding men like Ostwald and Helm, and even the brilliant philosopher Ernst Mach, could at that epoch be proponents of the continuum theory. By 1912 all this had changed . . .[2]

---

[1] Robert Andrews Millikan (1868-1953) received the first PHD offered in physics at Columbia University. After studying in Germany he obtained an assistantship at the University of Chicago in 1896 becoming a full professor there in 1910. His most famous work used oil-drops to help determine the charge on electrons which - contrary to Ehrenhaft in Vienna who thought they could have fractional values - he determined only existed as a whole number of units. This work helped demonstrate the particulate nature of electricity and thereby helped undermine the "electro-magnetic world view" which had supported a continuum approach and opposed atomism.

[2] Robert A. Millikan, *The Autobiography of Robert A. Millikan,* New York, 1950, pp. 84-85.

#79

To Otto Wiener Vienna, November 15, 1904

We have heard nothing from each other for a long time, which is also natural; People have too much to do in order to write to each other regularly, and if one does not accidentally see the other at a natural science conference or some other meeting, then one loses direct contact; also this time it is a request which causes me to write. A catastrophe took place in the Vienna Physical Institute (it destroyed a piece of ceiling [*Plafond*]), which again opens some prospect of getting a new building, and therefore, I am requesting you to send me a copy of the plans of the new physical institute in Leipzig, even if only on a loan basis. I would also be thankful for any advice about matters which we should be careful about. In order to make the exchange somewhat easier, I am sending you a copy of what I would like to see built here just for the Institute of Theoretical Physics.

I thank you and your most honored wife for your proven kindheartedness toward my daughter Ida[1] in Leipzig, and I am with the best feelings and most heartfelt greetings...

[P.S.] Please put your remarks either on the enclosed copy [of my idea for an institute], or write separately on a piece of paper. I don't need the copy anymore, since I have another one.

[1] See letter #54, footnote 1.

#80

To Wilhelm Ostwald Vienna, November 17, 1904

I have just heard that you will be coming to Vienna.[1] You would please my wife and me extraordinarily much if you would come to us one day for a simple dinner at 2 o'clock. The day is up to you, the 25th, 26th, or 27th, or as suits you. Also in the evening if you prefer that, 8 or 9 o'clock as you like. If you want to fill our cup to the top with pleasure, please manage to write us about when you are coming. Also, concerning the [Vienna Philosophical] Society, which you can meet in my company, if you have any wishes [please tell me].

My best compliments and heartiest greetings from us to you and your worthy family...

[1] Wilhelm Ostwald was planning to give a half-serious lecture late in November 1904 "On the Energetic Theory of Happiness", using equations to support his case. Boltzmann would reply in the same half-facetious spirit but for a different conclusion. See Wilhelm Ostwald, "Energetische Theorie des Glücks", *Annalen der Naturphilosophie*, **4** (1905), 457-474 and *Lebenslinien*, vol. 3, pp. 1-6. For Boltzmann's published reply see *Populäre Schriften*, Leipzig, 1905, pp. 364-378 or "Reply to a Lecture on Happiness Given by Prof. Ostwald", in Ludwig Boltzmann, *Theoretical Physics and Philosophical Problems*, edited by B. McGuinness and translated by Paul Foulkes, Reidel: Dordrecht, 1974, pp. 173-184. We have also included a few passages in Section B of this chapter.

#81

From Otto Wiener (carbon copy) Leipzig, November 30, 1904

First of all, I must apologize for answering your friendly letter so late. But I still find myself in the midst of terrible chaos in the not-wholly-finished, but new institute which is already partly in use. My wish: "And let there be light" is only slowly being realized.

I have added some remarks to the [copy of the] building program [which you sent me] and sent it back again to you. I didn't note down anything important, because your program is so rich that it doesn't seem to lack much.[1]

I am also forwarding to you a set of plans from our [overall] plans for the new building. It would please me most if on the next passing opportunity you might take a look at the new institute on location.

I gather that the conception of the enterprise behind your building program must make you feel quite good. I have the same impression about the second volume of your *Mechanik*,[2] which I have already studied with great interest[3] . . . .

[1] See letter #79.

[2] Ludwig Boltzmann, *Vorlesungen über die Prinzipien der Mechanik*, Volume 2, Leipzig, 1904.

[3] Page two of this letter is missing, and partly as a consequence what has not been quoted above from the end of page one and the beginning of page three is fragmentary and not very understandable. The missing page was apparently very critical of something.

#82

To Otto Wiener Vienna, December 4, 1904

Best thanks for the friendly information; I hope that building a new institute will become serious in Vienna now and only wish that I were 30 or 40 years younger so that I could work longer in it; there are also some other reasons.

If you would be interested in having a chance to chat with me again, please do the same as Ostwald recently did and come here over Easter with your revered wife. You will be just as welcome with the Viennese even without making us as happy as Ostwald's four lectures on happiness (He literally clarified how to be happy!). In our house, you will find sufficient but perhaps not brilliant conversation. But there will already be an appropriate retreat from $E^2$-$W^2$ [success squared minus dislike squared].[1]

The best time [for a visit] would [indeed] be the beginning of the Easter holidays, because later I will perhaps travel south.[2] With heartiest greetings from house to house...

[P.S.] If Warburg comes to the *Reichsanstalt* (and I hear that Röntgen has rejected it), who will take over Warburg's position?[3]

[1] Boltzmann is making a humorous but not very understandable reference to one of Ostwald's attempts to put happiness on a mathematical basis. Literally, "E" stands for *Erfolg* and "W" for *Widerwillen* in Ostwald's German. See Section B in this chapter for passages in which Boltzmann replies somewhat more clearly to Ostwald.

[2] See the next letter for a better indication of the direction Boltzmann would travel in Summer 1905.

[3] Boltzmann had been offered a position in Berlin in 1888, but had trouble making up his mind and rejected it, which he later regretted. See H. Hörz & A. Laass, *Ludwig Boltzmanns Wege nach Berlin*, Akademie-Verlag, Berlin, 1989.

Emil Warburg (1846-1931) was a theorist turned experimental physicist and Director of the Berlin Physical Institute during much of the time that Max Planck ran his own Institute for Theoretical Physics. Wilhelm Conrad Röntgen (1845-1923) discovered X-rays in 1895 and suddenly became famous. He taught at the University of Würzburg. For information about Warburg's successor, see the next letter.

#83

Wheeler[1] to Boltzmann (carbon copy) Berkeley, December 14, 1904

I have the honor on behalf of the University of California to invite you to participate in its Summer Session, June twenty-sixth to August fifth, 1905, by giving a course of lectures on such subject as corresponds best to your own interest. The lectures could be given in English or German, as you prefer. You would probably have a somewhat larger audience if you give the lectures in English, but you would still find ample hearing if you choose to give them in German. I feel confident that you would enjoy a summer season at Berkeley, which may fairly be regarded as one of the best summer resorts in the United states. Its climate during the summer is remarkably cool and healthful, as indeed it is during the entire year. It is situated in immediate proximity to San Francisco, which is reached in a half-hour by ferry and electric train. You would find yourself in the midst of an energetic and earnest body of scholars, and this might, after all, be the most interesting consideration. I can assure you of a most cordial welcome to our midst, and I trust that you may be able to arrange your plans so that you can visit us. The honorarium for the work suggested is one thousand dollars. I expect to arrange for special traveling rates between Chicago and San Francisco. With high respect...

---

1 Benjamin Ide Wheeler (1854-1927) was President of the University of California from 1899 to 1919, He had previously studied at Heidelberg and later taught Greek philology at Cornell University. He was known for his admiration of German scholarship and in 1909-1910 was Theodore Roosevelt Professor in Berlin.

#84

To Franz Brentano Vienna, December 26, 1904

To my great dismay, I see that I have still not yet answered either of your two extremely amiable and interesting letters. And I am the one hurt by it! Because by not answering I have delayed this very interesting correspondence from which I have so much to learn. It was not anger at any objections from your side, nor do I wish or even consider it politeness if someone agrees with me in everything; that is quite clear. I only profit from well-grounded contradiction. The cause of my silence was much more the following: First, I fell again into such an unwelcome neurasthenia where precisely philosophy was often the guilty party. For if I reflect on these wholly abstract if not completely abstruse things for too long, entirely too long a time, it attacks my nerves; I am more accustomed to work in concrete fields which are independent from the thinking subject and which are restful in themselves.

Also, my [philosophy] lectures did not have the success which I had hoped for.[1] I included set theory, non-Euclidean geometry, and similar things. That was too mathematical for my audience, and many dropped out. This upset me so much that I dispensed with teaching natural philosophy altogether during the Summer Semester, which troubled me even more. During this Winter Semester I am lecturing again in natural philosophy two hours a week and have made the subject matter much simpler. I teach known philosophers (right now Schopenhauer) and count among the fingers of my hand my own ideas as well as those of other philosophers, wherein I agree with some of their thoughts and disagree with others. But mainly, I no longer consider it important whether I have more or fewer listeners.

But there is still another cause for my delay in writing you. In particular, you mention in your letters synthetic judgments which do not stem from experience but can be understood from the concepts themselves. But I absolutely cannot think of the latter. A concept [*Der Begriff*] is only a word which we have invented [*ersonnen*], that we use as we want, but nothing follows from it. But on the other hand, I couldn't see myself where the related experience lay. And since I was confused, I kept wanting to delay my answer until I could obtain clarity, and thus kept putting it off. But I have now gotten hold of the matter. The judgment "pure red can be the same as not red" certainly does not come from experience in the same sense that one looks at a red object and is persuaded that it cannot be the same as what is not red.

But one has had the experience that parents, teachers, or whoever taught us to speak about certain things as "red", spoke about other things as "not red". Through nothing else but experience has one learned how the words "red" and "not red" are used, and if one subsequently labels concepts which stand in such relationships contradictory, one has certainly not derived from the concepts themselves that they exclude each other, but from experience of what is meant when I say that concepts are contradictory.

If someone doesn't know how to speak yet, I know of no other way to explain to him what is "red" and "not red" than to show him all possible objects and to say about one "red" and all others "not red". One should definitely not suppose that we simply encounter concepts in our soul or intellect. These are not other than experiences, like those acquired when we were taught to speak (and which afterwards we won by our own efforts in finding new words).

Your opinion on this matter would interest me very much. I naturally do not know whether you are now or for the rest of the Christmas holidays in Florence or perhaps even in Vienna; but in order not to delay my letter any longer I'm sending it to you fresh...

---

[1] See letter #74B.

#85

To Svante Arrhenius Vienna, January 2, 1905

It pleases me very much that all of your concerns have turned out so favorably. I also wish you luck with wishes which have correspondingly changed. I had already read in the newspaper that you refused the position in

Berlin and that now Nernst has accepted it; I thought at the same time that you had been promised directorship of the Nobel Institute. Let me also congratulate you for having been offered such an outstandingly fine position in Berlin. The Berliners are beginning to become quite *Nobel.*

I find it quite fitting that someone who has done outstanding scientific work should be called to a special post[1] which suits his individual character and that he doesn't believe that he is doing his maximum best when merely serving as lead ox dragging a cart[ful of students a few inches closer to serious education] by means of conventional lectures.

That you may find it hard to feel at ease in your new home so quickly will not hurt much. Meanwhile you can concern yourself with preparations and do theoretical work. I too have developed a rather strong cough, which had completely disappeared in America, but which has come back again quite strongly in Vienna. I Just received the request to give lectures at Berkeley, which will be stimulating and lively. I am also hoping that the climate will be beneficial for my health. This means I have to learn fluent English now, since I would definitely like to give my lectures in that language. Let me voice my deepest thanks for your role in helping to arrange this agreeable offer.[2]

With heartiest greetings and best New Year's Wishes from me and my family I am...

---

[1]Boltzmann is replying to a missing letter from Arrhenius which presumably announced that he had accepted a position as Director of the Nobel Institute in Sweden while rejecting an offer from the University of Berlin. From 1895 until 1905 Arrhenius had been a professor and for the last few years Rector of the Technical University in Stockholm.

[2] It seems that Arrhenius who had himself been asked to teach in Berkeley interceded for Boltzmann. But it is not clear whether this was on his own or Boltzmann's iniative or indeed came from President Wheeler of the University of California. But in any case, it is simply more evidence that Arrhenius was a good friend and a kindhearted soul.

#86

To Franz Brentano Vienna, January 4, 1905

I must answer your amiable letter right now; otherwise month after month will go by and I will neglect what is truly important and dear, and simply for no reason. Surely, one should carry on such a debate orally; the written way is much too long and cumbersome. I have also neglected something which was very important and useful for myself when I neglected to meet you when you came visiting.

The conflict [*Gegensatz*] between us is a major one. The judgment which you carry out in God's name and which you call apriori is nevertheless just a word. If only one does not draw the conclusion that you are absolutely certain. If we assume that every atom of a fixed body would suddenly divide itself into two original equal parts, then the whole body would occupy almost exactly the same place in space for a second time.

In such a case, it would not be irrational to say that the same body names two different bodies. But just as naming the body at a later time is completely different from naming it at an earlier time, so it is just as practical [*zweckmässig*] to say that one body has two [or more] atoms as to say that a single body has many bodies.

In fact, I am just now occupied with determining the number which plays the same role for time as Loschmidt's [number][1] for matter, the number of time-atoms = discrete moments of time, which make up a second of time. It seems to be on the order of $_{10}\,{}_{10}\,{}^{10}$. Perhaps your son can figure out how often one would go $10$ around the equator if one wanted to write it as 1,000,000.... In a word, this number is much greater than what you can *imagine* [picture to yourself] is a number without end.

Still more. We can say, if all atoms are continuously dividing themselves, as I depicted and the parts do the same in turn, that under those circumstances we could not even fashion [*bilden*] the concept of a body which was the same at different times. But that such a process has never happened anywhere is a pure fact of experience. But if it occured once, how could we apriori forbid it? Objects which at the same time send out red and blue light do exist; they make the same impression on us which spectral violet does. That we cannot distinguish red and blue at the same place on the retina from violet is a fact of experience;[2] But on the piano when two separated keys are played simultaneously, we can distinguish that sound from the sound of the key which is between the two separated keys. This too is a fact of experience. Given such cases, what other conclusion can we draw [by analogy from experience] except that we learn speech from using it? And that is how we distinguish "bird" from "not bird" [or "red" from "not red"].

Let us now turn to the second matter in dispute between us [about the rejection of philosophy in virtually all senses]. That the majority of students don't understand philosophy doesn't bother me. But can any two people understand philosophical [i.e.metaphysical] questions [and agree upon what it is that they understand]? Is there any sense at all in breaking one's head over such questions? Shouldn't the irresistible pressure [*Drang*] to philosophize be compared with the nausea caused by migraine headaches?[3] As if something could still struggle its strangled way out, even though nothing is actually there at all?

My opinion about the high, majestic task of philosophy is to make things clear, in order to finally heal mankind from these terrible migraine headaches. Now, I am one who hopes not to make you angry by my forthrightness, but the first duty of philosophy as love of wisdom is complete frankness. Through my study of Schopenhauer, I am learning Greek ways of thinking again, but piecemeal.

With the heartiest greetings...

[P.S.] On rereading, I see that without thinking I have smeared the letter.

[1] Josef Loschmidt (1821-1895), Boltzmann's friend and colleague during the 1860's in Vienna, made a famous calculation in 1865 that a cubic centimeter of nitrogen at zero degrees centigrade under normal barometric pressure contained about 100 trillion molecules and that the gaseous molecules were spheres with diameters on the order of $10^{-8}$cm. The Loschmodt number as it was called seemed to be proportional to Avogadro's number. Loschmidt's work also appeared to make the reality of atoms and molecules seem very probable to everyone not subject to philosophical or religious prejudice against them.
[2] Boltzmann's extensive use of the word "experience" [*Erfahrung*] without defining it, leaves open the question of whether he always meant conscious experience or allowed for types of experience which were not conscious. Nor did his attempt to dismiss the issue of consciousness make the problem easier to settle.
[3] See Section B of this chapter for more about "migraine headaches".

#87

Mach to Ostwald Vienna, January 5, 1905

I have just read your wonderful Faraday Lecture. In addition to the content, I was also very pleased with the acknowledgment given to Frantisek Wald.[1] I consider it one of my best experiences to have beheld the fall of the atomic theory in chemistry. In physics the thing plays too subordinate and variable a role to cause serious damage. The pressed physicist always says in the end: "Yes, Yes, but chemists can't do without the atomic theory". If the Faraday Lecture exists as a separate publication, please send the details on a card like a friend...

[1] Frantisek Wald (1861-1930) was a chemist and chief metallurgist in Kladno, Bohemia, but like Mach and Ostwald at this time, he opposed the reality of atoms.

#88

Mach to H. Gomperz[1] Vienna, January 9, 1905

The letter to B[2] has been sent. Badwilled is he not, but unbelievably naive and insensitive to his surroundings [*Burschikos*]. More important is that v A[3] is a member of the Commisssion, who from a professional standpoint definitely knows how to handle difficulties.

With best wishes and full of the most respectful greetings...

[1] Heinrich Gomperz (1873-1942) was an original if somewhat sceptically minded Viennese philosopher who revised Auguste Comte's "three stages of progress" into four stages: animism, metaphysics, ideology, and criticism. Like his father the historian of Greek Philosophy, Theodor Gomperz, he was one of Mach's good friends.
[2] The timing of this letter, the fact that it is between two Viennese philosophers, and that Boltzmann later acknowledged that he had been dissuaded from using the original title of his lecture on Schopenhauer before the Viennese Philosophical Society on January 21, 1905 suggests that this letter is *probably* about Boltzmann. The original title of Boltzmann's lecture was as follows: "Beweiss dass Schopenhauer ein geistloser, unwissender, Unsinn schmierender, die Köpfe durch holen Wortkram von Grund aus und auf immer degenerierender Philophaster sei"

Nevertheless, it would be premature to regard the letter as *certainly* about Boltzmann. One scholar, presumably not aware of Boltzmann, thinks that the letter is about Gomperz's academic advancement as *Privatdozent* ; this editor vaguely remembers that Mach used the the letter "B" to describe someone else in a quite different letter; and there is also the possibility that Mach and Gomperz were referring to Theodor Beer who had written a book on Mach's philosophy in 1903 and a short time later was jailed on a morals charge. In general, one of the oldest academic traps is to jump to conclusions as if one were certain about the person a single initial refers to. The world is full of unsuspected possibilities. This and the next letter are probably but not certainly about Boltzmann.

[3] Probably H. von Arnim (1859-1931), who was a philosopher and philologist in Vienna from 1900 to 1914 and again in 1921, but the commission referred to remains unclear, though it was probably connected with the Vienna Philosophical Society.

#89

Mach to H. Gomperz Vienna, January 12, 1905

When the principal resistance disappears,[1] other difficulties should soon follow. I am very happy to hear that everything is running smoothly now.

I thank you heartily for having well-understood my position with respect to B's power play [*Kraftleistung*]. I have no doubt that his will was good, but when left alone he is a man with no feeling for the measure of what is [socially] allowed. Unfortunately, he also lacks this in other things which are more important to *him*. Once again I thank you with heartfelt greetings full of highest respect...

[1] Boltzmann changed the title of his lecture to "Über eine These Schopenhauers". See Section C of this chapter for more details.

B. Some Observations Concerning Mach and Boltzmann

Ernst Mach on Boltzmann:

When Boltzmann made the beautiful discovery that the second law of thermodynamics corresponds to the principle of least action, I was at first no less agreeably surprised than others. Yet there is no reason for being surprised. When once it has been found that quantity of heat behaves like *vis viva,* and thus an analogue of the theorem of *vis viva* is applicable to it, it is not to be wondered at that the remaining mechanical principles (which are not essentially different from this theorem) may also be applied in the theory of heat . . . .

The mechanical conception of the second law of thermodynamics by distinguishing between orderly and disorderly motions and equating increase of entropy with increase in disorderly motions at the expense of orderly ones, seems a very artificial expedient . . . . I agree entirely with F. Wald when he says: "In my opinion, the roots of this [entropy] theorem lie much deeper, and if the molecular hypothesis and entropy theorem be brought into harmony, it is a piece of good luck for the hypothesis but not for the entropy theorem."[1]

But I cannot deny my aversion to hypothetico-fictive physics. Thus I have developed my own particular opinion about Boltzmann's probability investigations concerning the second law as based on the kinetic theory of gases. If Boltzmann discovered that it was very probable that processes act according to the second law and very improbable that they don't, then I cannot accept that it has been proved that nature behaves according to this law.[2]

The simple considerations which Boltzmann advanced in his article "On the Question of the Objective Existence of the Processes in Inanimate Nature" (1897) were long familiar to me. The questions raised there merely concern our taste in words. Nothing is changed by such questions in the matter itself."[3]

I cannot agree with Boltzmann's view concerning the slight use of the *volume element* .... The volume elements with their falls of temperature, behave exactly like finitely extended bodies under similar circumstances; but I have this advantage that I can build up, out of such small volume elements, in any case however complicated, with whatever exactness I desire. I cannot understand therefore, why every differential equation must necessarily be based upon atomistic views.[4]

Philipp Frank, Einstein's biographer, reminisced:

Also, strange as it was, in Vienna the physicists were all followers of Mach *and* Boltzmann. It wasn't the case that people would hold any antipathy against Boltzmann's theory of atoms because of Mach. And I don't even think that Mach had any antipathy. At least it did not play as important a role as is often thought. I was always interested in the problem, but it never occurred to me that because of the theories of Mach one shouldn't pursue the theories of Boltzmann.[5]

Ludwig Flamm,[6] Boltzmann's son-in-law, recalled:

H. Mache [who had collaborated with Boltzmann on two articles in 1899][7] confesses that he suffered at that time and for many years afterwards because he had had two teachers with such different scientific views as Boltzmann and Mach. He thinks that if Mach himself had not taught that the experimental physicist need not trouble himself too much about epistemological matters, that he would have suffered even more from this [intellectual] discord.

Lise Meitner was more direct:

[Anton] Lampa was an excellent experimentalist, but as an enthusiastic follower of Mach, was rather sceptical of the modern development of physics.[8]

But a critic declared:[9]

It was well-known that the whole atomic theory was only a picture, but he [Boltzmann] held on to it as a convenient hypothesis. He thereby found himself in opposition to Ostwald, Ernst Mach, and most other physicists, who for the most part are energeticists, which means that they make no special assumptions with respect to type of matter, but try to understand all physical processes merely with the help of concepts of pure energy. Boltzmann is the last great representative of the atomic theory in the physical world.

Gabriele Rabel on Mach and Boltzmann:

. . . . At that time, Boltzmann in one of his lectures on natural philosophy spoke about psychomonism in an ironic way and named Mach as a representative of it. This caused me to write a letter to Mach . . . . "Most Honored Herr Hofrat! Again and again I am astonished to experience that your unmisunderstandable words have been subject to different interpretations. From the side of a prominent university professor I have heard you repeatedly - and once again today - characterized as a sensualist or even psychomonist, as one for whom the world only exists out of perceptions, as if you thought that only the psychical were real. The men who maintain this stand so high . . . . that I heartily request that you somehow take this occasion once again to declare your point of view, and even more unambiguously than before . . . ."[10]

D.J. Bach was more blunt

It was also very difficult to take him [Boltzmann] completely seriously as a philosopher, and the attempt by Boltzmann to take over a second chair [1903][10a] after Mach's departure [1901] from the University was an experiment which not so much proved its impossibility - since the inner significance of the man covered every gap - as came very close to revealing a ridiculousness which even his deadliest enemy could not have wished . . .

Anders Wedberg[11] on Mach and Boltzmann

Positions close to that of Mach were taken at about the same time by the philosopher Richard Avenarius (1843-1896), and by the physicists Ludwig Boltzmann (1844-1906), Heinrich Hertz (1857-1894), and Pierre Duhem (1861-1916).

Mach's view had its closest historical parallel in the views of Berkeley and Hume, but he himself considered his doctrine to be identical with Kant's when the thing in itself is set aside, thus identifying Kant's "phenomena" with his own "sensations". Several neo-Kantians during the latter half of the nineteenth century also advanced views that pointed in the same direction as Mach's.

William M. Johnston[12] on Mach and Boltzmann:

Mach shared his role as popularizer of empiricist monism with the physicist Ludwig Boltzmann . . . . In 1908 [Hermann] Broch[13] converted to Catholicism, probably to please his wife. Thereafter he bewailed the cleavage of empirical science from speculative philosophy for which he blamed Mach and Boltzmann . . . . [returning] to Vienna as Mach's successor, Boltzmann lectured to 600 students before whom he invited Friedrich Jodl[14] to debate. In these unprecedented sessions, the sceptic Boltzmann demolished the ethical idealism of the grandiloquent Jodl. This event may have deepened the animosity that one student of Boltzmann, Hermann Broch, felt against his teacher.

As a philosopher, Boltzmann asserted the arbitrariness of all hypotheses, a view that startled Newtonian physicists. Rather than retreat into Mach's phenomenalism, he argued that several contradictory theories may be equally correct. He tried to resolve controversies, such as those between atomism and Kirchhoff's deductive use of differential equations, by contending that both approaches involve arbitrary mental constructs. Although all such constructs distort some of the facts, they complement rather than exclude one another. Boltzmann agreed with Mach that no construct can show exactly what a continuum is or how a particle behaves. Similarly, there may co-exist different representations of the world, each accounting for all the facts; such world hypotheses are complementary rather than antagonistic.

Above all, Boltzmann owed his scepticism to virtuosity as a mathematician. Although he commanded a fine literary style, his writings abounded in equations and diagrams. So skilled was he in formulating alternative mathematical constructs, that he came to regard philosophy as analogous to mathematics . . . .

Stephen Brush[15] on Mach and Boltzmann:

In retrospect it seems clear that the criticisms of [Boltzmann's] kinetic theory in this period were motivated not primarily by technical problems, such as specific heats of polyatomic molecules, but rather, by a general philosophical reaction against mechanistic or "materialistic" science and a preference for empirical or phenomenological theories, as opposed to atomic models. The leaders of this reaction, in the physical science, were Ernst

Mach, Wilhelm Ostwald, Pierre Duhem, and Georg Helm. Mach recognized that atomic hypotheses could be useful in science but insisted, even as late as 1912 that atoms must not be considered to have a real existence. Ostwald, Duhem, and Helm, on the other hand, wanted to replace atomic theories by "Energetics" (a generalized thermodynamics); they denied that kinetic theory had any value at all, even as hypotheses.[16]

Christa Jungnickel and Russell McCornmach on Mach and Boltzmann:

In argument Boltzmann liked to magnify the numbers and strength of his adversaries, the anti-mechanists and anti-atomists.[17] This was partly for effect, but from the mid-1890's he seemed truly pessimistic about the fate of his methods, at least in the short run; he published his lectures on gas theory in the hope that when the subject was "again revived, not too much will have to be rediscovered". He interrupted the publication of these lectures to publish his lectures on analytic mechanics, since his gas theory depended on that subject and the published lectures on gas theory could not include it all without becoming too bulky. Boltzmann's published lectures on gas theory and analytic mechanics, each appearing in two parts and all appearing within the ten years from 1895 to 1904, gave his extended audience an ample introduction to his view of the nature of theoretical work in physics . . . .

Throughout the published lectures, Boltzmann treated, implicitly or explicitly the topic of the methods of theoretical physics. He told his readers which parts of Mach's teaching he accepted (for example, Mach's definition of mass), which parts of Kirchhoff's (for example, his characterization of a theory as a description), which parts of Hertz's (for example, his characterization pf theoretical pictures), and so on. He told them as well what he did not accept; for example, he rejected the argument that a mechanics of volume elements instead of a mechanics of material points escapes a commitment to atomism. (In his mechanics lectures, he kept his atomistic assumptions constantly before the reader's eyes by regularly writing the mathematical symbol sigma for a discrete sum instead of the integral sign for a continuous sum, even though he used the calculus throughout.)

Frank on Mach and Boltzmann

Boltzmann once said to me, "You see, it doesn't make any difference to me if I say that the atomic model is only a picture. I don't mind this. I don't require that they have absolute, real existence. I don't say this. 'An economic description', Mach said. Maybe the atoms are an economic description. That doesn't hurt me very much. From the viewpoint of the physicist this doesn't make a difference." Boltzmann had a philosophical viewpoint which did not require that you believe in the real existence of atoms. And there wasn't, I would say, any opposition to Boltzmann's physics from the viewpoint of Mach. This opposition existed only so to speak, in the philosophical realm.[18]

J.L. Heilbron[19] on Mach, Ostwald, Boltzmann, and Planck:

Even before Planck had converted to Boltzmann's approach, he had unmasked Mach as a false prophet, ignorant of thermodynamics. "As far as Mach is concerned, I must say that although I otherwise much appreciate the independence and sharpness of his judgment, I do not think him competent as far as the second law is concerned."[20] In 1896 Planck felt obliged to speak out against Wilhelm Ostwald's energetics, which drew some of its negative arguments from Mach. Energeticists wished to replace all physics, including elementary mechanics, by the equations of energy and its transformations. Planck joined Boltzmann in condemning the program as nonsense. [But for many years as Planck himself admitted: "One could not make headway against the authority of W. Ostwald, Ch. Helm, and E. Mach."[21] -Ed].

---

[1] Ernst Mach, *Principles of the Theory of Heat,* edited by Brian McGuinness and translated by Thomas J. McCormack, Reidel: Dordrecht, 1986, pp. 334 & 335.

[2] Ernst Mach, "The Leading Thoughts of My Scientific Epistemology and its Acceptance by Contemporaries" in *Ernst Mach - A Deeper Look,* edited by John Blackmore, Kluwer: Dordrecht, 1992, p. 136-137.

[3] *Principles of the Theory of Heat,* op. cit., p. 445.

[4] Ibid.

[5] Stephen Brush, "Mach and Atomism", *Synthese,* **18** (1968), 207. See footnote 18 for some qualifying remarks.

[6] Ludwig Flamm, "Zum 50. Todestag von Ludwig Boltzmann", *Physikalische Blätter,* **12** (1956), 410.

[7] "Über die Modifikation der van der Waalschen Zustandsgleichung", *Wiener Anzeiger,* **36** (1899), 87-88 & *Wiedemanns Annalen,* **68** (1899), 350-351; and "Über die Bedeutung der Konstante *b* des van der Waalschen Gesetzes", *Cambridge Philosophical Transactions,* **18** (1899), 91-93.

[8] Lise Meitner, "Looking Back", *Bulletin of the Atomic Scientists,* **20** (Nov., 1964), 3.

[9] Anon., "Der Lebenslauf Boltzmanns", *Die Zeit* (Vienna), No. 1420, September 7, 1906, page 2, column 2. Professor Walter Höflechner has informed this editor that in checking through Wilhelm Ostwald's papers that he discovered that it was Ostwald himself who had written this "anonymous" obituary notice about Boltzmann.

[10] See Gabriele Rabel, "Mach und die Realität der Aussenwelt", *Physikalische Zeitschrift,* **21** (1920), 433-437. For one of Mach's two belated letters in reply, see John Blackmore & Klaus Hentschel, *Ernst Mach als Aussenseiter - Mach's Briefwechsel über Philosophie und relativitätstheorie mit Persönlichkeiten seiner Zeit,* Wilhelm Braumüller: Vienna, 1985, pp. 47-48. Mach apparently did what the young lady wanted by publishing his book *Erkenntnis und Irrtum* in 1905 and republishing it the following year. See Ernst Mach, *Knowledge and Error,* edited by Brian McGuinness with a foreword by Erwin N. Hiebert, Reidel: Dordrecht, 1976.

[10a] D.J. Bach, "Ernst Mach", Frankfurter Zeitung, #49, February 18, 1908, p. 2, col. 1. This article was written to commemorate Mach's 70th birthday. Actually, Boltzmann only received a teaching contract for philosophy in addition to his chair in physics. he did not literally take over Mach's chair. See letter #62.

[11] Anders Wedberg was Professor of Philosophy at Stockholm University from 1949 to 1975. The three volume *Filosofica historia* appeared in Swedish from 1958 to 1966. It was translated into English by Bergljot Wedberg and appeared under the title *A History of Philosophy.* This selection is from Volume 3, p. 8.

[12] See William M. Johnston, *The Austrian Mind - An Intellectual and Social History,* University of California Press: Berkeley & Los Angeles, 1972, pp. 186-188 & 315.

[13] Hermann Broch (1886-1951) continued a family business in Vienna and after 1928 became a prolific writer and novelist. With apparent help from James Joyce he was able to escape Austria for America in 1938. Some of his writings included *Die Schlafwandler - Roman Trilogie*, 1931-1932 and *Die Schuldlosen - Roman in elf Erzählungen*, 1950. Much of his writing was about sleep and death.

[14] Friedrich Jodl (1849-1914) became a professor of philosophy at the University of Vienna in 1896. His epistemological views are normally considered to have been close to those of Mach and Boltzmann and positivism generally, but in ethics and philosophy of life he seems to have been rather more optimistic and idealistic.

[15] Stephen Brush is a physicist-turned-historian-of-science. He is professor at the University of Maryland at Silver Spring, and former president of the History of Science Society. He has written extensively on Boltzmann's work in physics as well as being the translator of Boltzmann's lectures on gas theory. Recently, he has written on the history of cosmology and astronomy.

[16] Stephen G. Brush, *The Kind of Motion We Call Heat*, Volume **I**, North-Holland: Amsterdam, 1976, p. 245.

[17] Christa Jungnickel and Russell McCormmach, *The Intellectual Mastery of Nature*, Volume **2**, University of Chicago Press: Chicago, 1986, pp. 189 & 190-191. Concerning the remark above, this editor is not aware of any exaggeration by Boltzmann about the number and influence of his opponents. Indeed, given the fact that during the 1880's and 1890's and in some cases later he found the following able scientists and mathematicians opposed to his interpretation of the Second Law of Thermodynamics: Wilhelm Ostwald, Georg Helm, Frantisek Wald, Ernst Mach, Max Planck, Ernst Zermelo, Georg Jaffé, Alois Höfler, Hans Cornelius, Gustav Jaumann, Henri Poincaré, Pierre Duhem, Karl Pearson, Lord Kelvin, Michael Pupin, and from his autobiography one even suspects the political historian Henry Adams, he probably underestimated their numbers and influence.

[18] Quoted from a passage in Stephen Brush, "Mach and Atomism", *Synthese*, **18** (1968), p. 207, which in turn was taken from an interview which T.S. Kuhn had with Philipp Frank, deposited in the archives of the Sources for History of Quantum Physics project (July 16, 1962). [Since Frank was born in 1884 and thus almost 80 years old recalling words from over half-a-century earlier and oral interviews rarely allow for the corrections which most scholars would make in published remarks, it is possible that some caution is in order both here and with the earlier quotation from Frank listed above - Ed.]

[19] John L. Heilbron received his PHD in 1964 from the University of California where he has been a Professor in the History of Science for many years. Our quotation is taken from his recent, prizewinning book on Max Planck, *The Dilemmas of an Upright Man*, UC Press: Berkeley, 1986, p. 45.

[20] Planck to Ostwald, July 16, 1891, in Körber, ed., *Briefwechsel Ostwalds*, (1961), Volume **I**, p. 35.

[21] Max Planck, "Persönliche Erinnerungen aus alten Zeiten", *Vom Wesen der Willensfreiheit*, with an Introduction by Armin Hermann, Fischer Taschenbuch Verlag: Frankfurt am Main, 1991, p. 32.

## C. Natural Philosophy II

(Ilse M. Fasol-Boltzmann has reconstructed Boltzmann's lectures on natural philosophy from 1903 to 1906 from shorthand notes. But it must be kept in mind that Boltzmann did not read his lectures; he spoke freely, hence there is no guarantee that his notes accurately represent what he said. What they may do, however, is rather accurately represent his thinking at the different times he wrote them. But even that may not mean that he was satisfied with them, unless the perspective is repeated on later occasions or rather often. Below is an English translation of selected aspects of

Boltzmann's notes for lecture #3 on natural philosophy given on November 3, 1903.[1])

.... Some people have prophesized that the greater part of my audience will fall away; I don't believe that! Just the opposite. I believe that they would absent themselves if I always lecture in a light manner. They did not sign up for this course just to listen to loose talk. They can find that much better at Ronacher's [a variety theater]. If I let a joke slip in here and there, its only for pedagogical purposes. Also, a professor of experimental science must sometimes grab at any, even senseless, means to impress scientific laws [on students]. Sometimes, one can only capture what is most important about beautiful drastic experiments with a joke. Nothing conveys it better!

If someone saves my life [from drowning] - never a bad joke, hence I will repeat it again - and if it is true that everything passes through consciousness during that brief desperate interval, then I will win critical enlightenment and approach a solution to the mysteries of the universe, and above all about the simple concepts of arithmetic, geometry, physics, and chemistry: number, space, time, matter, power, and energy; more in a breath than mankind in a thousand years . . . [but let those of us who are undrowned] start with the proposition, that 2 x 2 = 4 , which far from being merely symbolic [or a joke] is very literally [*buchstäblich*] true.

I am bitterly earnest; I am curious how many lectures will pass until we have understood this. We naturally start in a general manner with an analysis of the concept of number. I once found in a textbook the following definition of number, which I comment on because it is typical and to a certain degree is the crown which brings together previous definitions: "Number [*Zahl*] is the concept for the quantity [*Anzahl*] of a greater or smaller multitude [*Menge*] of real or thought things or other objects." With this archtype, which unites to a certain extent all of the nonsense which is customary with definitions of the concept of number, I will now compare it with definitions which have been given by well-known philosophers. 1) Euclid: "Number is a quantity of things". 2) Kronecker: *Über den Zahlbegriff* page 339. 3) Schmitz-Dumont *Naturphilosophie*, page 100. 4) Helmholtz: *Vorlesungen über theoretische Physik*, Introduction page. 5) Mach: *Wärmelehre* page 67. . . .

I believe that the philosophers are incautious in their use of such words [as "number"]. . . . The question is merely one about which expression is most suitable for the purpose in mind. But how can I decide that if I don't even know what a number is, still less what is apriori as if the truth could depend on what was completely independent of the world? We must ask: Is designating it as such practical [*zweckmässig*]? . . . .

(fashioned from notes for the lecture of November 10, 1903)

At the end of the last lecture, I spoke about a favorite theory of mine: That there are no absolutes, especially no absolute truths. I want to enlarge on that, otherwise I will fall into the very mistake I have just criticized, by presupposing as [absolutely] known what it basically means "Something is absolute".*

(The following passage is from Boltzmann's lecture initially titled "The Relations of Applied Mathematics"[2] given in the Congress of Arts and Sciences in conjunction with the Universal Exposition held in St. Louis. It was given in Hall 7, September 24, 1904, at 3 p.m.)

. . . . My present theory is totally different from the view that certain questions fall outside the boundaries of human cognition. For according to that latter theory this is a defect or imperfection of man's cognitive capacity, whereas I regard the existence of these questions and problems themselves as an illusion. On superficial reflection it may of course be surprising that after recognition of the illusion the drive towards answering these questions does not cease. The mental habit is much too powerful to loosen its hold on us.

It is exactly as with ordinary sense illusions, which continue to exist even after their cause has been recognized. Hence the feeling of insecurity, the lack of satisfaction that grips the scientist when he philosophizes. Only very slowly and gradually will all these illusions recede and I regard it as a central task of philosophy to give a clear account of the inappropriateness of this overshooting of the mark on the part of our thinking habits; and further, in choosing and linking concepts and words, to aim only at the most appropriate expression of the given, irrespective of our inherited habits . . .

---

1 Ilse M. Fasol-Boltzmann (ed.), *Ludwig Boltzmann 'Principien der Naturfilosofi' - Lectures on Natural Philosophy 1903-1906*, Springer-Verlag: Berlin etc., 1990, pp. 83-84. The three types of lecture material which have survived from the Fall 1903 semester were, first, stenographic notes, second, two published lectures, and third lectures written down by an unknown co-worker of Boltzmann. In using and translating the notes, it has often been necesary to make them into coherent sentences, first into German by Frau Dr. Fasol-Boltzmann, and then into English by the editor. The notes not only generally become more and more fragmentary through the years, such that by late 1905 and early 1906 they are often scarcely more than lists of single words or phrases to remind Boltzmann about what he wants to lecture on but their subject matter shifts as well. Mathematical philosophy predominates in 1903-1904, Boltzmann's description of and opinions on Schopenhauer's philosophy in 1904-1905, and epistemology and ontology in late 1905 and early 1906, but by then the notes are so abbreviated that comprehension is very difficult. Selections from 1904 and 1905 seem generally to be most useful, but it is not always clear, especially in lectures for the Fall 1904 semester, whether Boltzmann is expressing his own opinion or merely describing the views of Schopenhauer or someone else.

* Ibid., p. 87.

2 The St. Louis Congress was organized to favor experimental over theoretical science with Boltzmann's lecture listed under "applied mathematics" and the lecture itself labeled as "The Relations of Applied Mathematics". Boltzmann changed the title for*Populäre Schriften* in 1905. Our selection is titled "On Statistical Mechanics" as published in Ludwig Boltzmann, *Theoretical Physics and Philosophical Problems*, edited by B. McGuinness and translated by P. Foulkes, Reidel: Dordrecht, 1974, p. 167. Since Ostwald also came to St. Louis, there was once again a debate over "atomism" versus "energeticism", but Boltzmann did not play quite as prominent a part as in 1895 at Lübeck. He was both depressed and immersed in philosophy. Other leading scientists at the Congress included Paul Langevin, Ernest

Rutherford, Johannes Diderick van der Waals, and Henri Poincaré. The plan for the Congress came largely from the Harvard Philosopher and psychologist Hugo Munsterberg.

(fashioned from Boltzmann's notes for his lecture on natural philosophy given on October 24, 1904)[3]

. . . . Philosophy gets on my nerves. If we analyse the ultimate ground of everything, then everything finally falls into nothingness. But I have decided to resume my lectures again and look the Hydra of doubt straight in the eye, and it can be quite ominous [*verhängnisvoll*] if one values one's life. The title [Principles of Natural Philosophy] doesn't tell us anything coherent . . . .[it is] essentially a joke . . . . I must take care that the lecture is adequate. great difficulties; one doesn't really know what natural philosophy is . . . .

[3] Fasol-Boltzmann, op. cit., p. 107.

(fashioned from notes for the lecture on October 26, 1904)[4]

. . . . One seeks a definition or concept of things for which there is no definition or concept; but a starting point for proof is not proof. Whoever has once clearly seen that the task of philosophy only consists of employing suitable words and not using certain imaginary inborn laws of thought or feeling - and everyone must confess this after laborious reflection - will also see, that if every psychological process really corresponds to a process in the brain, that it would be most suitable [*zweckmässigsten*] for psychologists to refrain from creating any more new words, or at least not to say that there is a qualitative difference when the question is merely one of looking at the matter from a different perspective.

I would like it if everyone could make the prejudice vanish as I have that there is really a problem whether ants are machines, whether my brother is a machine, whether we are in the world, or the world is in us, if perhaps behind the word there is matter, power pushes or not, or if Locke is right that the intellect is between us and things. Or whether we are free or not free . . . .

[4] Ibid., p. 109.

(fashioned from notes for the lecture on November 7, 1904)[5]

. . . . Special forms of overshooting suitable goals are 1) clinging to words like: synthetic, analytic, discursive, intentional, ideal, matter, irritability, sensibility, transcendent, immanent, judicial, empirical, reality . . . 2) forcing everything into categories . . . 3) letting mere feeling determine what is true such as: a) the feeling that man is free b) imagining that self-consciousness is something other than the play of atoms c) that there is an incompleteness in knowledge not to know that things could not be without a knowing subject, that there is particular sense in the question: *whether matter originated before there were thinking creatures* . . . .

Consciousness does not exist, particularly because it is purposeless [*unzweckmässig*] to speak [as if it did]. But it is suitable to say that man has consciousness but the wind not. The question whether ants have consciousness is nonsense. Consciousness is not something special; but one must first accustom oneself to this conception . . . .

One cannot feel the movement of atoms. In fact, atoms are rightly conceived only as thought symbols to be held as pictures They cannot exist independently from thinking. But the model of an atom if it consistently fits the existing rules and is always applied in a suitable way reflects such a world as it existed before there were thinking beings. To think that if I cease, everything ceases, is not consistent with the current rules of thought. but neither is solipsism . . . .

[5] Ibid., p. 111

(from Volume II of Ludwig Boltzmann's book *Lectures on the Principles of Mechanics*, Leipzig, 1904.)[6]

. . . . However, the ray of hope for a non-mechanical explanation of nature came not from energetics or phenomenology, but from an atomic theory that in its fantastic hypotheses surpasses the old atomic theory as much as their elementary structures surpass in smallness those of the old atoms. I need not mention that I mean the modern theory of electrons. This certainly does not aim at explaining the concepts of mass, force, the law of inertia and so on from simpler and more easily understandable concepts and its simplest fundamental concepts and laws will doubtless remain just as inexplicable as those of mechanics for the mechanical picture of the world. However, the advantage of being able to derive all mechanics from other ideas that are in any case necessary for explaning electro-magnetism would be just as great as if conversely electro-magnetic phenomena could be explained mechanically . . . .

We can determine only the distances between different parts of bodies, that is only their relative position. There is no experience in which an absolute space would make itself felt. Nevertheless at the beginning of the book we introduced a definite co-ordinate system which almost plays the role of absolute space. We did this merely because by introducing this system we can state the laws of relative motion of bodies much more simply than by taking other systems, chosen quite arbitrarily . . . .

In conclusion let us return to a fundamental difficulty in the simplest fundamental laws of mechanics, namely the formulation of the law of inertia, if one does not wish to introduce an absolute transcendental space. We have evaded this difficulty in the simplest way by never speaking of anything real or existing, but replacing matter by mere mental pictures, namely material points, without worrying whether this might not also be done equally successfully in other ways (for example on the basis of some other co-ordinate system).

No one can stop us forming mental pictures as we wish, nor therefore including in them a co-ordinate system (the suitable reference frames) over and above the material points. After the event we call these mental pictures true only because they are useful in predicting future phenomena (our future sensations) as completely and effortlessly as possible, that is in adapting our volitional impulses to them.

I can find no inner reason for the assertion that mental pictures containing the reference frame cannot agree exactly with experience. On the contrary, I find that it is precisely this assertion that aims at stating something apriori about experience which the latter did not tell us before. I therefore cannot agree with Mach; If I understand him correctly, he infers from the fact of our thoughts' having to represent only relations between objects that the law of inertia could be determined only by the world of fixed stars. If the picture we have developed here were absolutely correct, then the relations between objects would be such that a mental reference of this kind is required to represent them in the simplest way. it would not be necessary for the firmament to be absolutely free of rotation relative to the reference frame; that in fact it is nearly so, would be explicable from their great distance, at which rotation would require enormous centripetal forces.

For somebody who surveyed the whole world of fixed stars, imagined as finite, it would even be theoretically possible to observe its rotation by means of Foucault's pendulum or a gyroscope as we could do for the Earth if we lacked light and therefore knowledge of other celestial bodies. However, this observation would not be unconditional . . . .

Quite independently of this there is the question whether the mechanical equations here developed and therefore also the law of inertia might perhaps be only approximately correct and whether, by formulating them more correctly, the improbability or rather inhomogeneity of having to adopt into the picture a co-ordinate system as well as material points would disappear of itself.

Here Mach pointed to the possibility of a more correct picture, obtained by assuming that only the acceleration of the change of distance between any two material particles is determined mainly by the neighboring masses, its velocity being determined by a formula in which very distant masses are decisive. This naturally avoids the adopting of any co-ordinate system into the picture, since now it is only a question of distances. Of course, Mach does not avoid introducing other difficulties, for example, that the world is finite, a kind of action at a distance for the greatest distances and so on. These difficulties did not seem to me to be so particularly great as perhaps to some other physicists, but they certainly have the awkward feature that they seem to exclude all empirical test forever . . . .

At all events, I think that such an extension of our vision, by pointing out that what we regard as most certain and obvious may perhaps be only approximately correct, is most valuable. It is in line with the suggestion that the distances of fixed stars may perhaps be constructed only in a non-Euclidean space of very small curvature, which is of course connected with the law of gravity in that a moving body not acted on by forces would then after aeons have to return to its previous position if the curvature in question is positive . . .

---

[6] Ludwig Boltzmann, "Lectures on the Principles of Mechanics", *Theoretical Physics and Philosophical Problems*, edited by B. McGuinness and translated by P. Foulkes, Reidel: Dordrecht, 1974, pp. 259, 261-262, & 264.

(from Boltzmann's published reply to Ostwald's lecture on happiness given before the Vienna Philosophical Society in November 1904) [7]

. . . . That next to the difference E - W [success minus resistance] the sum E + W also contributes to happiness is the persuasion of an action-loving West European. A Buddhist, whose ideal is the annihilation of the will, would perhaps write E - W/ E + W . . . .

Herr Geheimrat Ostwald pays the highest acknowledgement to Mach in all of his writings and certainly with justice, nor do I honor Mach any the less when I do not completely agree with him in everything. But as concerns Ostwald's energeticism, I believe that it simply rests on a misunderstanding of Mach's ideas . . . [Sense perceptions and representations] are therefore the only existent things; the physical concepts are merely added by us in thought. Ostwald only understood half of this sentence, that atoms do not exist; he at once inquired: "But what then does exist?" and gave the answer that energy is real. In my opinion, this answer is entirely contrary to Mach's view, that energy, no less than matter, should be considered merely a symbolic expression of certain relations between perceptions, of certain equations between given psychological appearances . . . .

Why is it that an apparently harmless article like Ostwald's seems to me so dangerous for science? Because it marks a relapse into delight in the merely formal, into the pernicious method of so-called philosophers, the method of constructing theoretical edifices from mere words and phrases and placing emphasis only on their pretty and formal interlacing; and this used to be called a logical foundation, without regard whether this lacework corresponded with reality or was adequately grounded in fact. It is a relapse into the method of submitting to the domination of preconceived opinions, of bending everything to the same principles of classification, of trying to force everything artificially into the same system. It is a refusal to see mathematics for mere formulae, true logic for mere seemingly correctly constructed syllogisms, true philosophy for mere bric-a-brac in philosophical guise, the wood for mere trees. This method, unfortunately, will always please the masses more than does the method of natural science, which leaves less room to fantasy.

[7] Ludwig Boltzmann, *Populäre Schriften*, Vienna, Leipzig, 1925, pp. 368 & 371. The last paragraph is taken from Ludwig Boltzmann,*Theoretical Physics and Philosophical Problems*, edited by B. McGuinness and translated by P. Foulkes, Reidel: Dordrecht, 1974, p 183.

(fashioned from notes for Boltzmann's natural philosophy lecture on November 23, 1904)[8]

. . . . First cause and first change [*Veränderung*] cannot be determined. Laws give the occasion [*Anlass*] for hypothetical judgments. If enough heat enters, then the material or substance burns. A substratum cannot be a cause [*Ursache*] because it cannot change [ändern]. Matter contains inertia apriori. It is just its own bad luck that changes in velocity must have a cause. A thing cannot be the cause of a thing, but merely of change. Also, laws are not causes, since they are eternal and merely apply to appearances [*Erscheinungen*]. it is nonsense to ask for the cause of matter, which is unbounded and remains the same, as it also is to ask for the cause of a natural law. Law = *qualitas occulta.*

An apriori proposition would require that it be its own effect [*Wechselwirkung*], which is nonsense, since the effect must again be the cause of a cause. Every appearance [*Erscheinung*] may therefore have to precede other appearances. Power cannot be a cause; it exists only as *qualitas occulta* . . . .

[8] Fasol-Boltzmann, op. cit., pp. 117-118.

(fashioned from notes for the lecture on December 14, 1904)[9]

. . . . Thus originate general representations of types, i.e. general concepts, *universalia*. Nominalists and realists disagree over their reality. General theories lack consciously immediate [*anschauliche*] content as does the being [*Wesen*] of things. But nothing is to be done about essences [Essenzen] . . . except to throw them away as useless baggage. The causal law has uniform application, but the way things actually work is much more complicated. This is where utility comes in and selections are made, and by tossing out the superfluous one can manage things more easily. Human beings require prudence . . . .

No judgment without sufficient reason, but if this is true, it is only so in *reversed form* [no sufficient reason without judgment]. Truth equals judgment which relates to something other than itself. But naming the existence [*Dasein*] which judgment relates to does not exclude a significant variety of reasons for selecting those names. I would also like to determine - if the truth-seeker does not already know it - which judgment determines this existence or constitutes it, which definition is employed to guess at it. Subsequently, and with enough effort one could certainly find out what was intended, but if one must have a reason for every judgment, then there is

such as tasting it. 3. *Transcendental truth* is about space, time, number, and what must logically exist for us to measure and determine what is material or empirical . . . . 4. There are only four *metalogical truths*: the principles of identity, contradiction, excluded middle, and sufficient reason . .

(from Boltzmann's lecture "On a Thesis of Schopenhauers" given in the Vienna Philosophical Society on January 21, 1905) [10]

. . . . In order to give my talk some local color, I wanted to imitate Schopenhauer's style in my very title. His style distinguishes itself especially by the mode of expression associated in the past with fishwives but which today would be called 'parliamentary'.

In this spirit I had chosen for my talk the following title: "Proof that Schopenhauer is a stupid, ignorant philosopher, scribbling nonsense and dispensing hollow verbiage that fundamentally and forever rots people's brains". These words are quoted verbatim from his book *The Fourfold Root*, except that there they relate to another philosopher. Schopenhauer's style was excused as due to his anger because he had been passed over when posts were filled. Question: Is his anger or mine the holier, given the object of it? Well, this title was denied me, and rightly so . . . .

In the last instance it is not logic, philosophy or metaphysics that decide whether something is true or false, but deeds. "In the beginning was the deed", as Faust says. What leads us to correct deeds is true.

That is why I do not regard technological achievements as unimportant by-products of natural science but as logical proofs. Had we not attained these practical achievements, we should not know how to infer. Only those inferences are correct that lead to practical success.

Of course once a method of inference has been tried and inherited through thousands of years, it seems to us apriori correct, so that we often can go on working with them for long stretches without practical tests, for example if we are confident that a calculation will yield the correct results; however at some time it must have been tested by deeds, and from time to time it must be tried again.

Now it seems to me that just as Schopenhauer's ideas have shown themselves to be untenable, so too are those of all other philosophers in their central core, Kant included, although naturally I have not time enough to prove this here.

The question now arises: was the work of these great minds really in vain? To this I must say no, for those philosophers have put paid to much more naive views still. In so doing they have done useful work in removing defective views, uncovering their mistakes and thus spreading a transition to clearer views.

Similar events often occur in the field of other sciences, for which I cite the example of Wilhelm Weber, who had put forward a theory of electricity and magnetism that is now recognized as incorrect; and yet he is one of those who have most advanced these two sciences. He gave the incentive for many experiments by which the ground was cleared for the newer theories. Although Weber's theory is today untenable, he is one of the greatest theorists of electricity ever.

From this point of view I must express my sincere thanks to those who recommended my appointment to the lecturership of philosophy, since this has given me the opportunity to delve more deeply into its literature. I cannot judge how many so far have really profited from my lectures but I have this consolation that one at least has learnt a great deal, namely myself . . . .

In my view all salvation for philosophy may be expected to come from Darwin's theory. As long as people believe in a special spirit that can cognize objects without mechanical means, or in a special will that likewise is apt to will that which is beneficial to us, the simplest psychological phenomena defy explanation.

Only when one admits that spirit and will are not something over and above the body but rather the complicated actions of material parts whose ability so to act becomes increasingly perfected by development, only when one admits that intuition, will and self-consciousness are merely the highest stages of development of those physico-chemical forces of matter by which primeval protoplasmic bubbles were enabled to seek regions that were more and avoid those that were less favorable for them, only then does everything become clear in psychology . . . .

We are in the habit of assessing everything as to its value; according to whether it helps or hinders the conditions of life, it is valuable or valueless. This becomes so habitual that we imagine we must ask ourselves whether life itself has value. That is one of those questions utterly devoid of sense. Life itself we must accept as that which has value, and whether something else does can only be judged relatively to life, namely whether it is apt to promote life or not. In this we try of course to talk the individual into believing that what has value for him is not what promotes his own individual life but that of his family, tribe, or even mankind as a whole. Since those who believe this (the noble-minded) are in every way furthered and rewarded by the community, they have more oportunities in the struggle for existence and noble-mindedness becomes hereditary, as also unfortunately does self-interest, which for its part offers opportunities in the opposite sense.

However, if we ask whether life itself has value, this means whether life is apt to promote life, a question that has no sense. According to the definition we can ask only how life can be promoted. The valuable is simply what promotes life, the question as to the value of life itself is senseless, although according to

Darwin's theory it is readily explicable why it obtrudes itself. It is another mental habit that overshoots the mark....

The task of philosophy for the future is, in my view, to formulate the fundamental concepts in such a way that in all cases we obtain as precise instructions as possible for appropriate interventions in the world of phenomena. This requires first that if we follow different paths we never reach different rules for further thought and action, that is we never meet internal inconsistencies, such as if one path led us to the conclusion that matter was not infinitely divisible and another that it must be. That sort of event is always a sign that the laws of thought still lack the last finish, that we have placed our words badly. In that case we must alter the laws of thought that lead to such absurd consequences.

A similar procedure is followed in algebra, Operations with negative and fractional numbers are defined in such a way that in applying the rules that hold for positive integers we nowhere encounter inconsistencies.

Secondly, the laws of thought must everywhere lead us to the interventions in actual events in the way that experience shows us to be desirable.

Thirdly, we must, as best we can, work against the irresistable urge to apply the laws of thought even where we overshoot the mark, so that in the end this urge gradually disappears altogether.

That this is not absolutely impossible history seems to indicate. After all, there was a time when it was believed as a result of logical thought that there could be no antipodes. People always saw that the vertical direction is parallel for everybody and if somebody stands in the opposite sense, his head is on the ground and his feet in the air. By constant experience this becomes so much a habit of thought that the antipodes become inconceivable. People even believed it to be impossible that the Earth was turning, because all other rotations make us feel dizzy, but not that of the Earth. In the time of Columbus and Copernicus it was believed that this is a necessity of thought, and even they that had this foisted on them. However, today these mental habits have vanished and any educated man can now barely comprehend how people could be so narrow-minded then.

The prejudice against non-Euclidean geometry and four-dimensional space is likewise in the course of disappearing. Most people still believe that Euclid's geometry alone is possible, that the sum of the angles in a triangle must be 180 degrees, but some have already come to admit that these are mental ideas become habitual, from which we can and must free ourselves.

Thus we must change all laws of thought in such a way that they lead everywhere to the same goal, that they correspond to experience and that overshooting the mark is kept within proper bounds. Even if this ideal will presumably never be completely realized, we can nevertheless come nearer to it, and this would ensure cessation of the disquiet and the embarrassing feeling that it is a riddle that we are here, that the world is at all and is as it is, that it is incomprehensible what is the cause of this regular connection between cause and effect, and so on. Men would be freed from the spiritual migraine that is called metaphysics.

---

[7] Ludwig Boltzmann, *Populäre Schriften*, Vienna, Leipzig, 1925, pp. 368 & 371. The last paragraph is taken from Ludwig Boltzmann,*Theoretical Physics and Philosophical Problems*, edited by B. McGuinness and translated by P. Foulkes, Reidel: Dordrecht, 1974, p 183.

[8] Fasol-Boltzmann, op. cit., pp. 117-118.

[9] Ibid., pp. 121-122.

[10] Ludwig Boltzmann, *Theoretical physics and Philosophical Problems*, edited by B. McGuinness and translated by P. Foulkes, Reidel: Dordrecht, 1974,, pp. 192-193 & 196-198.

# Chapter 6

# BRENTANO AND CALIFORNIA

## A. Correspondence

### #90

Planck[1] to the Nobel Prize Committee Berlin, January 26, 1905

In reply to your worthy letter of September 7, 1904, I have the honor to respectfully submit a proposal for awarding the prize in the field of physics for the year 1906.[2]

I name Herrn Dr. Ludwig Boltzmann, Professor of Theoretical Physics at the University of Vienna. I base my proposal above all on the merit he has acquired through his research in the field of the kinetic theory of gases, which has been included for the most part in his book *Vorlesungen über Gastheorie*, two volumes, Ambrosius Barth: Leipzig, 1896 & 1898.[3]

I consider Boltzmann's main contribution to this field to be the carrying through of mechanical explanation of molecular processes into regions which had proved inaccessible even to the pioneer work of scientists such as Clausius and Maxwell. First and foremost, Boltzmann's work is the foundation which has made possible the conceptual dynamic of today's quickly developing complex system of electrons.

With highest respect...

---

[1] Max Planck (1858-1947), Professor of Theoretical Physics at the University of Berlin and father of quantum mechanics, had resisted Boltzmann's statistical interpretation of the 2nd Law of Thermodynamics for many years, but found himself forced to employ many related ideas from Boltzmann when conducting his own black body investigations which resulted in quantum theory. Planck gracefully admitted his error and became a strong defender of the importance of Boltzmann's work after that.

[2] The date listed on the letter is indeed 1906, but it is reasonable to assume that he meant 1905. See letter #123 for another recommendation by Planck.

[3] For an English translation see: Ludwig Boltzmann, *Lectures on Gas Theory,* translated by Stephen Brush, University of California Press, Berkeley & Los Angeles, 1964.

### #91

To Franz Brentano Vienna, February 17, 1905

I must answer your last letter at once; otherwise, the same thing will happen as with the last one, which I left lying. I will still answer the previous one, but I have not yet fully studied it through. When that has happened, I will ask you for further information on all of my objections.

Concerning my lecture on Schopenhauer,[1] I hit on an idea which like everything radical pleases me. Hopefully, you will not interpret this as arrogance. It was first taken down in shorthand and then beautifully transformed into longhand. I have not yet carefully examined the result, but have already noticed that a lot was incorrectly stenographed in the first place, nevertheless, it does give an approximate idea of the lecture. Would it interest you if I sent you the whole thing? If yes, I ask you to write me, and I will send it by return post. About returning it to me, I do indeed have a request. All of it properly belongs to the Philosophical Society, except the mental input. Should you have enough time and leisure to add marginal notes or comments, then I would be infinitely grateful, whether short or long. You will see there how I exploit your ideas. I know what a rich source they are. I hope you don't mind. Naturally, you can send the lecture back without comment or even reject my impudence. In no case, would I be upset or feel insulted or in any way be offended.

With greatest respect...

[1] which was held on January 21 before the Vienna Philosophical Society.

#92

To Franz Brentano Vienna, February 23, 1905

I am enclosing the manuscript of my lecture, as it was taken down. I don't expect that you will agree with it, since we have diametrically opposed views about so much, but nevertheless your remarks would interest me. Please be utterly frank [*rückhaltloss*]!

I cannot send you [Friedrich] Jodl's reply. My lecture is to be published later.[1] I will find out if Jodl's response was also taken down and if it will also be published. In this case I will ask you to wait until it is published. At the utmost you can write Jodl yourself. I find it hard to ask for his permission to send you his reply (assuming it was taken down at all by the stenographer), since God knows what he would infer from my request.

Jodl won a victory in so far as that while after my talk the applause was louder, after his it was more sustained, as if it would never end. It must have made many friends of the old metaphysics happy that my attacks found [such a vigorous] reply. I cannot say whether it was just Jodl's students, scholars with doctor's degrees, or even *Privatdozenten* who carried on the applause for so long. He scarcely raised any objections against me on matters of fact. He apologized for Schopenhauer's rough terminology [*Grobheit*] with the excuse that he had been passed over in proposals for filling professorships, as if rage about matters of fact were definitely less beautiful than what could come from the deeply personal, as if my criticism were superficial, because it failed to grasp the high value of life, especially since I would maintain that one cannot prove that life has value, but only that something done for the sake of life has value. But after that, Jodl kept speaking for so long that the prescribed time for the end of the session approached, such that my words in reply were cut off.

I would gladly write much more to you, but I must break off, or once again, I won't get the letter sent at all. Please protect your eyes. I also have everything read to me now and others write for me too. My eyes are also weak. Better to lay out a small sum of money, than to become incapable of work. With utmost respect and heartiest greetings...

[1] Ludwig Boltzmann, "Über eine These Schopenhauers", *Populäre Schriften*, Barth: Leipzig, 1905, pp. 385-402.

#93

From Franz Brentano [late February, 1905]

*Nil Temere! Nil Timide!* [1] These are the old proverbial phrases which [I would] apply in an eminent way to the second part of it [your manuscript on Schopenhauer]. But I gladly forgive you for the very many beautiful contradictions which I see there . . . . Nor am I unaware of a heroic force in the field of science . . . .

The second part of your lecture is too short to be able to add a foundation for your theses, hence, it would be ungentlemanly [*unritterlich*] to attack such defenceless naked notions. But I fear that should the lecture really be printed, which I as a friend would not advise, that the larger public would exercize restraint similar to how your listeners reacted.

Also the first part steps very rashly from one point to another. Many of your criticisms show lack of careful judgment. For example, if Schopenhauer is blamed, if he allows a will to dwell in a falling stone, it is assumed to apply to other cases since there are definitely not stronger reasons for them. But you speak here as if Schopenhauer had not in reality taught that the will is the indwelling thing-in-itself with respect to everything else. (Naturally I say this without holding his claim to be justified.)

Also - in order to touch on another case- it seems to me that your criticism of his pessimistic conception of life fails to hit its weakpoint. You allege that nothing can be said about the value or lack of value of life itself, since everything which has value has value *for* life. Should it really be our goal just to live and not rather to live well and happily? Is the life of a scoundrel [*Schurke*] who wants to live long, better? . . . Schiller thought differently when he said:

The highest good is not more life to win.
But the worst evil is indeed guilt and sin.

I close in the hope that the frankness of the frank will not offend. If I could only hold you back from everything which your noble striving will create in the way of obstacles! But still, friendly assistance for those of modest character may not overstep a certain fixed frontier.

I thank you for the words of sympathy for my eyes. You depict yourself as a co-sufferer, but hopefully not to my extent. My hopes for a future life - which you do not seem to share - help me endure all such incidents [loss of eyesight] as something small. And that is definitely not a kind of opportunism, which lets me focus my thoughts on tenets such as the immortality of the Soul and God's ordering of the world. You know yourself that I share such persuasions with first-rank minds of the past and present, also with those who are not blind but who link their views with ours on rational grounds. I can therefore be highly tolerant, since I trust that these high truths of which I speak will eventually shower blessings even on those who deny them.

With sincerest respect...

---

1 "Nothing reckless! Nothing timid!"

#94

To Franz Brentano Vienna, March 6, 1905

Best thanks for the friendly evaluation and return of the manuscript. According to what Jodl tells me, he will not permit his reply to be published, which is the best proof that he fears my rebuttal. I will also not let my lecture be printed in its current state. For example, what is said about the will of the falling stone is not properly expressed, and perhaps it was spoiled even more by the stenographer. But concerning the value of life, I stand my ground. A scoundrel [*Schurke*] is not bad in himself but is often judged so because the [Darwinian?] principle of preservation of life and health must take precedence. He hates the policeman as much as we hate him. Our only superiority is that we are preserving the life and health of the nation which he is undermining.

I interpret Schiller to mean that the individual should not live for himself, but for mankind, that such is man's highest goal. On the other hand, for one's own private advantage, we all voice such an ideal and shout it bombastically in everybody's ear. Still, I intend to publish not just this lecture, but a whole book, where I present my views about the untenability of the so-called metaphysical way of thinking by presenting a series of examples. I know that I position myself in a hail of fire and will be struck by a bullet, and also know that I can fall, but not the view for which I fight.

Since you designate me as a hero, then I intend to do everything to deserve it! And something else! I have suffered since Autumn from nose, ear, lung, and stomach catarrh! I believe that I can escape it only in an air where I don't constantly breathe in the katarrh bacilli of other people. Are most of the people in Florence free of catarrh or does everyone suffer from influenza like in Vienna? Do you believe that I could spend time there completely alone during April in a small house far from street traffic? Could you yourself find something for me outside of town? So much the better if not too far away from you, so that I could visit you now and then. About cottage and cost, anything modest will do; only, I would need someone who will read to me in German for a fee and who could take German dictation.

Will that be possible in Florence? Forgive me that I direct the question to you. If an answer would be awkward, please leave this letter aside. I also have another place in Petto. With fullest respect...

#95

To the Austrian Ministry of Education Vienna, March 9, 1905

Your humble servant has received an invitation to teach for six weeks at a summer school at Berkeley University near St. Francisco in California. The honor which lay in this invitation, which previously only the most renowned sholars of Europe have received, men such as Ramsay in England, Arrhenius in Sweden, and Ostwald in Germany, I would only ungladly turn down. Outside of that, it promises scientific intercourse with scholars who have often acquired their knowledge from a tradition which is essentially different from what we teach in Europe. The visit would also allow closer study of the singular and very large scientific institutions of the Western part of America which would be of great use for my own further education in science. For example, there is nothing west of St. Louis and Chicago to compare with the Lick Observatory.

In order to be able to give this summer school course which already starts before the end of June, I am requesting that summer vacation start next semester from the middle of June. Since the invitation was first proposed some time ago, I have already voluntarily increased my teaching load by lecturing in theoretical physics on most Saturdays from 3 to 4 in the afternoon, lectures which are not mentioned in the catalog. Outside of that, I have also given unscheduled lectures in natural philosophy on Mondays or Wednesdays from 7 to 8 in the evening. With the addition of some extra hours in the coming semester it is certain that the total number of scheduled hours will be surpassed by the actual hours taught before the time when the early summer vacation which I have requested would begin.

#96

To Franz Brentano Vienna, March 20, 1905

The extraordinary friendliness and amiability of your letter places me in an embarrassing predicament, and almost invites the appearance as if I intended to use your friendship only for egoistical purposes. Now, as a philosopher,[1] I am of the persuasion that one cannot act other than egoistically, but in practice I assume that it is better [*schöner*] to act for egoistic ends by means of wanting to please others, so that one does not [conspicuously] act all too egoistically.[2] Under all conditions, I would therefore prefer to cause you as little inconvenience as possible.

I am naturally grateful for the invitation to eat with you, if thereby, I don't create a problem in any way. I am certainly agreeable, but I believe that I can also endure even the plainest Italian meals, since I have already proved to myself that I am able to digest even the most different cuisines from the most diverse countries. I find the least sympathy with cosmopolitan *table d' hote* food.

I pay two *Kronen* per hour for someone here to read to me aloud [in German]. But in Florence I would be glad to pay somewhat more for the short time involved. Five hours a week would be enough, if outside of that I were also able to have three hours of English conversation from another person, since this summer I will be traveling to America and would like to speak as good English as possible. I would also make only very modest requests about my chamber.

I will arrive in Florence on Saturday the First of April at 6:34 in the morning. If you have already sought a room for me or if accidentally there is a room free in your house please send me the address ahead of time. I will then [unload and] carry my luggage myself without having to inconvenience you at such an early hour. I prefer to fearfully burden you only with philosophical conversation. It is precisely in Florence where I intend to write my first work on metaphysics,[3] and there, I will charge [my battery] full of philosophical wisdom. With extraordinarily high respect and heartiest greetings...

---

[1] Boltzmann writes the German word "Philosoph" in this and later correspondence in phonetic style as "Filosof", as he would also do in the "forwort" to his *Populäre Schriften*.

[2] If one thinks in terms of means and ends as Boltzmann appears to be doing here, and helping others is regarded as a means which can only have as an end some way to further one's own interests, then it is natural to consider all human actions to be egoistical. The normal response is that much human behavior is ruled by habit, feeling, duty, or even inherited inclinations where means-end calculation is absent or secondary, such that non-egoistical action is possible. Some interpretations of Darwin's theory also appear to imply egoism, but this was not Darwin's own opinion, since he apparently believed that a measure of non-calculated cooperation between individuals often seems necessary for a species to survive.

[3] Boltzmann often used the words "philosophy" and "metaphysics" interchangably, but on other occasions he regarded it as a special bugaboo which might also be opposed when doing philosophy. In short, it is not clear what he planned to write on. It seems possible but by no means certain that Boltzmann actually began this work, but according to his last extant letter to Brentano (See letter #124 in Chapter #8), he decided instead to write on "so-called a priori knowledge". Neither manuscript, if indeed begun, appears to have survived. Nor does his son Arthur mention it in 1907 when giving an inventory of his father's lecture manuscripts in his letter to Hugo Buchholz (#134).

#97

To Franz Brentano Vienna, March 28, 1905

My best thanks for your friendly letter, which I received during a gathering at Auspitz through your brother-in-law Lieben,[1] who had just come back from Egypt. So I will in fact become wholly your guest! It does trouble me a bit as if actually I had invited myself [to stay with you]. But it was really not so intended. I would have just as gladly stayed in a private place and welcomed an oily Italian repast while exclusively enjoying stirring conversation from you.

But it is certainly better this way. And if two philosophers who have so much to say to each other cannot live together under the most unforced of circumstances, then who else could? I consider it particularly providential that I will be able to live with you this Easter, where I have decided to write my first book on philosophy.[2] Will I also have the capacity, the power of work, and the strength of will? I hope God helps me! Thus you see that I also believe in a God, as symbol,[3] that our own will is not sufficient to achieve our purposes, and that it is useful to remain very mindful of the necessity of having favorable external circumstances.

I fear that you will reject my prosaic way of thinking, but believe me, poetry, spirited response, is absolutely independent of these fortuitous representations which we make about things. I pray to my God just as ardently as a priest does to his. Still, bearing in mind the saying *nulla epistola, vel nulla cogitatio sine philosophia*,[4] I intend to tame my lust for preaching.

I am therefore coming to Florence on Saturday, the First of April, at 6:34 in the morning, and will proceed at once to you at 10 via Bellosguardo. With heartiest greetings...

---

[1] As already mentioned more than once, Adolf Lieben (1838-1914) was a professor at the University of Vienna, a pioneer in Austrian organic chemistry, and his sister Ida (1855-1894) was Brentano's first wife. Brentano married his second wife, Emilie Ruprecht, in 1897.

[2] See footnote 3 in the previous letter.

[3] The original German is in phonetic spelling, i.e. "Simbol".

[4] "Don't write, or rather don't think without philosophy."

## #98

To Benjamin Wheeler[1] (in English) Wien, 26. April 1905

I will leave Bremen the 13th of june with the steamer "Kronprinz Wilhelm" of the North German Lloyd and will arrive the 20th of june or (if the ship has not good weather) at latest the 21th of june in New York. You have me written, that perhaps I could get some reduction of prize or some other preferences or favours from the pacific railway by the journey from New York to Berkeley or at least some care, that I get a good berth in the railway - Waggon. I beg your to send me any legitimation and to write me, what I shall do after my arrival to New York forgetting these preferences, if they are possible at all. I will go from the ship as quickly as possible to the railway.

I would also be glad, if you could me write, how long time lasts the journey from New York to Berkeley and if it is necessary to take a ticket to St. Francisco or if the journey goes directly to Berkeley without passing St. Francisco.

I beg you to give me answer as soon as possible and I am...

---

[1] Boltzmann's English has been preserved in its original form.

#99

To Franz Brentano Venice, April 29, 1905

I feel obliged once more to give my thanks for all of the love and consideration which I have enjoyed at your hands. It will always remain unforgettable, the glorious outdoors, the inexhaustibly interesting art treasures, the magnificent hospitality at your house, but above all particularly the philosophical conversations. That this occupied my attention more than anything else is clear. But you have not persuaded me. Therefore, the conflict endures. I have already become such a *Machianer*,[1] that in fact I am somewhat on tenterhooks whether I will return to normal and find myself actually situated here or whether perhaps I really was dreaming.

In order to fully enjoy the complete success of the "drawer-pulling-experiment"[2] in a maximum manner, I believe that I am still justified in doubting whether the theory that this is not only the best but also the most beautiful of all worlds[3] will succeed again tomorrow in confirming the availability of fresh [fabulous] food. Nor am I ignoring the question whether the collapse of the Campanile[4] [in Venice] constituted an exception, or is compatible [with your interpretation of the experimental results of the drawer-pulling experiment].

I would like to give special thanks to your wife for the self-sacrificing service which she performed at the railway station. I had to wait over an hour and she helped to make things much easier.

I also had some bad luck when I got to Venice. Hotels Cavaletto and Vapore were both filled up. My *fauchino*[5] said it would be the same everywhere. But he knew of a beautiful room in a saloon [*Beisel*]. (He told me in Italian, and that is my translation.) But, as the saying goes: When in need, the Devil gorges on flies. Therefore, I followed him. While walking, I once got stuck in a particularly narrow passage and only became disengaged with some effort. It seemed that if I turned myself around vertically always by 90 degrees I could in unseemly fashion come through. The saloon and its keeper did not inspire confidence. But what choice did I have? I faced a categorical imperative.[6] I needed so much to be cautious., hence, before going to bed I made an agreement with the ruffian, that I would refuse to give him any money if I were murdered that night.

But I must finish the letter. I began in such a serious vein, and then could not resist falling into the most ordinary *Bummelwitz*.[7]

With heartiest greetings and best compliments to you, your fine wife, and Herr Giovanni I am...

---

[1] Ernst Mach (1838-1916) is meant. One of the drawbacks of being controversial and having a short, fairly unusual name is that it allows one to be readily labeled. Mach tried to reduce reality to sensations, and followers of Mach have traditionally been called *Machists*, *Machians*, or in German *Machianer*.

[2] The German expression is *Schublade Experiment*, but it is not self-evident what Boltzmann means. It is possible, and we have dared to presume, that he may well have meant an attempt

to prove that there is an external physical world by putting an object in a drawer, closing it, and opening it again to show that it is still there. But while this experiment may appeal to direct realists, it would not convince either phenomenalists or indirect realists. More evidence would be needed.

3 Leibniz was perhaps the most famous advocate of the position that this is the best of all possible worlds. Voltaire's *Candide* is largely a satire of this claim.

4 Shortly before, it had indeed collapsed. It was quickly rebuilt, however, and still stands as a leading landmark in Venice.

5 presumably a guide or helper, literally "little mouth".

6 a facetious reference to Immanuel Kant's famous moral principle.

7 an idler's low-life joke.

#100

Lampa[1] to Meyer — Salzburg, April 29, 1905

Like a good friend, please reserve a place for me at the Boltzmann lecture.[2] With best greetings...

1 Anton Lampa (1868-1938) was a *Privatdozent* in physics at the University of Vienna, became associate professor there in 1904 and full professor in the German University of Prague in 1909. He supported Mach in philosophy and in 1918 wrote a book about him and his ideas.

2 It is not clear whether this was a special lecture by Boltzmann or was simply an initial lecture starting what Germans and Austrians call the *Sommersemester*, but in any case the fact that a professor wanted to attend and thought he needed a reservation to obtain a seat suggests how popular the lecture was anticipated to be.

#101

To Svante Arrhenius — Vienna, May 2, 1905

It pleases me that you have now found a position which fully satisfies your wishes[1] and also that you have decided to marry again, and I wish you happiness in the heartiest way. I just spent a month in Florence and it has done wonders for my health. But now I will also need a lot of strength, because I still have so much to take care of in Vienna and then I will be going to California.

With best compliments from my wife and heartiest greetings from me...

1 Arrhenius became Director of the newly established Nobel Institute for Physical Chemistry in Stockholm.

#102

To an Unknown Official — Vienna, May 5, 1905

I have allowed myself to send the enclosed photograph for your Imperial Majesty. I have not written my name on it, because nothing was mentioned about doing so.

With highest respect...

#103

From Benjamin Wheeler[1] (carbon copy) Berkeley, May 6, 1905

The privilege of a round trip rate from Chicago of $62.50 has been arranged for the faculty of our summer session. If you will show to any railroad agent the letter from Mr. Horsburgh of the Southern Pacific of which I enclose a copy, the arrangements can be completed...

[1] All of the letters from President Benjamin Ide Wheeler and from the University of California to or about Boltzmann are in English and the texts have been left unchanged with the exception of the name T.J. Rogers, apparently a mistake, which has been changed to F. J. Rogers.

#104

From Benjamin Wheeler (carbon copy) Berkeley, May 15, 1905

If you will show a copy of the enclosed letter from Mr. James Horsburgh, Jr., of the Southern Pacific Company, to the agent of the Southern Pacific Company in New York, he will provide you with a ticket to San francisco at the special rate which it has been arranged for our Summer Session Lecturers. It would be well for you to say to him that you will desire to return by the Santa Fe Railroad. This agent can supply you also with a ticket for your berth to San Francisco. The journey from New York to Berkeley will occupy five days. It would be best for you to purchase a ticket straight to San Francisco and to go upon your arrival to the St. Francis Hotel, telephoning me at Berkeley upon your arrival in order that I may then arrange for someone to call upon you in San Francisco and to see that comfortable accomodations are found for you here in the University town. I may be found by telephone in Berkeley at either North 12 or at Dana 1811. In case you cannot reach me ask for professor L.C. Moore - his telephone number is Polk-2106, San Francisco, or else for Mrs. Henderson at the University at North-10 or Mason-806. Very sincerely yours...

#105

To Otto Wiener Vienna, May 31, 1905

It is extraordinarily kind of you and your wife to invite me to stay in your house during my stop in Leipzig. This allows me to enlarge the study which I have already made of the plans for the new Leipzig Institute [of Physics] in more convenient fashion.[1] I therefore accept the invitation with gratitude. Since I still have a lot to do in Vienna, I will arrive in Leipzig on Friday the 9th of June at 9:50 in the morning, and from the railway station I will proceed at once to the meeting [*Sitzung*].[2] I will fetch the luggage later. I will remain in Leipzig until the 11th or perhaps the 12th.

With heartiest greetings and best compliments to your wife...

[1] See letters #79, #81, & #82 for more information about the new Institute. See also Christa Jungnickel & Russell McCormmach, *The Intellectual Mastery of Nature,* Volume 2, University of Chicago Press: Chicago & London, pp. 175-183.
[2] For an extensive discription of this conference see Boltzmann's account of it in his article "On the Trip of a German Professor into El Dorado" in section A of the next chapter.

#106

To Franz Brentano Vienna, June 3, 1905

As I foresaw, our sympathetic interest in each other has remained undiminished, but our correspondence has ceased. Thus sulphuric acid and caustic potash no longer react with each other when they are spatially separated, but has their physical affinity become any the less? Also, two heavenly bodies do not noticeably influence each other if both primary orbits are sufficiently differentiated, but has the Newtonian attraction between them been reduced? I am not able to give you my hand because I have been invited elsewhere.

Indeed, I am as good as packed and gone, and my bags are full of traveling utensils, notebooks with provisional comments for my lectures, and my spirit is full of the necessary courage and self-confidence, but something is pressing me before departure to write you once more. The mental stimulation which I received from you is still active. Not only do I still think a great deal about the many topics we discussed, but they have made my philosophical lectures during the semester which is just ending much easier. Previously, I had to prepare carefully and memorized a lot beforehand, while now I already speak somewhat more freely.

Philosophy, as lofty as it stands, has its methods and techniques like everything else, and I have been trained by you in a good school. My theory of behavior as the product of pure egoism[1] therefore finds its confirmation again, that is, if one does not perhaps take my own freely confessed egoism [in advocating it], which runs counter to my own interests, as evidence against the theory.

My poor dog had his head so bloodied and smashed that it was hard to get him on his feet again. He was accustomed to bicycles, automobiles, and tramways, but motorbikes were too much. He considered them absolutely evil, and made it his mission in life to eliminate them utterly from Earth. I fear that his striving after this ideal [and disregard of his own selfish interests] will be his undoing. Should human Idealism after Higher Being appear in a similar light?

Please accept my compliments to your fine wife and also tender them to my Florentine acquaintances (Raiti, Herr Zoubow, Miss Stallo,[2] etc.), and I am with the heartiest greetings...

[P.S.] The death of Mustafia must have touched you deeply.

[1] See earlier letters from Boltzmann to Brentano for more on Boltzmann's theory.

[2] John Bernard Stallo (1823-1900), a German-American diplomat and philosopher with views very close to those of Ernst Mach with whom he corresponded, had retired in Florence. Indeed, like Boltzmann, Mach also corresponded with Brentano and was also attracted to Florence, and had seriously thought about retiring there himself. Miss Stallo was presumably a close relative of the recently deceased philosopher of the same last name mentioned above.

#107

To Otto Wiener — Vienna, June 7, 1905

Thank you once again for the new friendly invitation. It would naturally make me very happy to have a pleasant chat with my previous colleagues in Leipzig, but I cannot say definitely whether it will be possible, since I don't know when the trains leave for Bremen where I probably should be on Monday morning. But I rather enjoy traveling at night.

It is absolutely unnecessary to fetch me from the railway station. The way from the Dresden Station to the *Augusteum*, which once upon a time I found so easy, will surely pop back into consciousness.

With best compliments for you and your wife...

#107A

Boltzmann's "fonetic forwort"[1] — Vienna, June 8, 1905

ai left tha nu orthografi in mai last buks aut bekaus ai am tu old tu lern it, but in this forwurd ai wil leev tha laetest vurshun in so that ai wil not be left aut. ai beleev, hauever, that deparchurs from fonetics, sins thae ar not tu be fulli avoided, shud not be kompleetli cawndemd. If wun duz not wahnt tha tael tu wag tha dawg, then tha tael must be kut.

mai publisher haz rekwested that mai popyuler raitings be kolekted tugether intu tha buk wich is befor yu [*Populäre Schriften,* 1905]. Tha kahntent is divers, partli speechus, partli popyuler lekchurs, treetisus with filosofikul kahntent, revyus, etc.[2]

Altho mai vyus hav nachurali experienst modifikaeshuns over tha kors ov taim so that ai wud perhaps not rait in tha saem wae, ai hav left everithing unchaenjd, sins onli that kan and shud giv a pikchur ov mai past vyus . . . .

[1] Boltzmann's Foreword to his *Populäre Schriften* should be understood partly as a humorous reaction to the many phonetic and "scientific language" movements of the period which were led, supported, or popularized by Wilhelm Ostwald in Germany, Louis Couturat in France, and George Bernard Shaw in England. This was the heyday of the Esperanto and Ido movements, even though the number of actual readers or users of the languages has probably increased since then. But much more important than his unusual foreword, Boltzmann included most of his published writings in philosophy in this "popular" book.

[2] But while the "Forwort" may have been written in June, Boltzmann's long paper "Reise eines deutschen Professors ins Eldorado", perhaps the most entertaining contribution in the

entire book, was not added until very late in the year, perhaps November, hence the book as a whole could scarcely have appeared in print before December 1905 or January 1906.

#108

To Stefan Meyer *Kronprinz Wilhelm* Bremen, June 17, 1905

In Leipzig unfortunately, I gave my word that I would have my article for [Klein's] Mathematical Encyclopedia[1] ready by the end of September. I will be back in Vienna on September 5 and then work head over heels [until it is finished].[2] But I will need a co-worker who will hunt through the literature and take down my words. One of the better students would do.[3] If he is really hardworking, he will get the knack of it fast. If he can do the job, he will be named co-author and receive half of the honorarium. Perhaps you are in a position to arrange the matter. But it must be certain. Even better would be to locate more than one person who would be available in Vienna. A lone person could refuse and leave me up the creek when I got back to Vienna. You could leave the addresses with Kornherr; also your own.

Please write me about it in America. My address is Berkeley University[4] near Oakland California U.S.A.

With heartiest greetings to you and all my Vienna friends...

[1] See letter #25, Klein to Sommerfeld.

[2] The article was "Kinetische Theorie der Materie", Volume 5 of Klein's encyclopedia

[3] Stefan Meyer's friend Josef Nabl was chosen. See letters #24 and #26 and their connected footnotes as well as A. Dick and G. Kerber, *Ludwig Boltzmann - Katalog zur Ausstellung an der Zentralbibliothek für Physik in Wien*, Vienna, 1982, p. 35 for more details on Nabl.

[4] University of California at Berkeley.

#109

To Benjamin Wheeler (telegram) New York, June 21, 1905

Will arrive San Pablo train one Overland Limited June 26...

#110

Henderson[1] to Hall[2] (carbon copy) Berkeley, June 21, 1905

Professor Boltzmann telegraphs as follows: "Will arrive San Pablo train one Overland Limited, June twenty-sixth." Very truly yours...

[1] Mrs. V. Henderson, acting secretary for President Benjamin Ide Wheeler.

[2] Dr. Elmer E. Hall of 1825 Arch Street, Berkeley, California. Many years later he became head of the physics department at UC Berkeley.

#111

Hall to Henderson [Berkeley, June 1905]

The men in the Department of Physics at Stanford University who might be interested in Professor Boltzmann's lectures are Prof. F.J. Rogers,[1] J.G. Brown,[2] [and] C.K. Studley.[3] It might also be advisable to write a general letter to Professor Fernando Sanford,[4] Head of the Department extending an invitation to members of the Department. Sincerely...

---

1 Frederick John Rogers (1863-1941) was educated at Cornell, became Assistant Professor of Physics at Stanford in 1899, Asociate Professor in 1908, and Full Professor in 1924. He published papers on coiled springs and the velocity of energy, but his main teaching focus was electricity, but he also taught kinetic theory of gases, electron theory, and even history of physics.

2 Joseph Grant Brown (1869-1967) attended Illinois Normal University in 1898 but came to Stanford for his bachelor's and master's degrees. He rose through the ranks becoming Assistant Professor in 1908, Associate Professor in 1925, and Full Professor in 1933. Several of his publications were on terrestrial magnetism and atmospheric electricity. He also taught history of physics. He had become Stanford's oldest living professor until his death at the age of 98.

3 Clarence Knight Studley was a laboratory assistant for the year 1904-1905.

4 Fernando Sanford (1854-1948) had studied with Helmholtz in Berlin from 1886 to 1888. He taught physics at Stanford from 1891-1919 and was the longtime head of the Physics Department.

#112

To Benjamin Wheeler (telegram) Truckee [Nevada], June 26, 1905

Due Overland Limited Oakland sixteenth street five five this afternoon...

#113

Henderson to Brown[1] (carbon copy) Berkeley, June 30, 1905

Professor Ludwig Boltzmann of the University of Vienna is lecturing here during the summer semester on "Mechanical Analogies of Thermodynamics". Dean Moore[2] joins with me in expressing the hope that it may be possible for you to attend some or many of these lectures. [Concerning] the Summer Session fee, the University would, of course, be glad to waive in the case of yourself[3] ...

---

1 See letter #111.

2 Professor L.C. Moore still awaits proper identification.

3 Perhaps the University of California professor most interested in Boltzmann and his lectures was Jacques Loeb, the German-American physiologist, who was also in written communication with another Viennese professor, Ernst Mach.

#114

Henderson to Hall (carbon copy) Berkeley, June 30, 1905

We have written to Professor Sanford, Professor Brown, and Professor Studley[1] [at Stanford about Boltzmann's lectures]. Very truly yours...

---

[1] He was not a professor at that time. See letter #111, footnote 3.

#115

To Stefan Meyer Berkeley, July 21, 1905

My best thanks for your friendly letter and how well you handled my instructions.[1] Both of these gentlemen[2] will certainly do excellently.

At the moment, my teaching in California is two-thirds finished. My English has been sufficient. It was indeed very interesting, what I have seen and what I still hope to see. But it has also been strenuous and tiring, and so much the more as the California climate is not as agreeable as depicted. This alternation on the one hand between subtropical heat and coldness where one must use a stove or furnace (in July on the same parallel as Palermo) and on the other between a dry-as-dust and a foggy climate where one cannot even see the next house, is hard for us with our European perspective to bear. On top of that, the water which is preserved from the winter rains in giant cisterns, does not agree with my stomach, and wine must be hidden,[3] like a schoolboy's cigars. This is what they call freedom.

I hope that you are having a fine time and are enjoying a good rest during summer vacation, I am...

---

[1] See letter #107.
[2] Josef Nabl worked with Boltzmann. See letter #24. The other person is not known.
[3] There was local prohibition at the time in Berkeley. See "On the Trip of a German Professor into El Dorado" in Section A of the next chapter.

#116

From Benjamin Wheeler (carbon copy) Berkeley, August 9, 1905

I trust that when this letter reaches you you will have completed a very enjoyable trip across the continent, and be in good health for beginning your Atlantic voyage. Let me express to you my very great appreciation for your visit to us, and my thanks for the service you have rendered to those who have attended your lectures as well as to the University community by your presence within it. I send you henceforth the address of the Lick Observatory Expedition in Spain. With best wishes for a pleasant journey, I am...

#117

To Ernst Mach Vienna, August 25, 1905

Best thanks for the interesting book on knowledge and error.[1] I have many greetings to render you from America, especially from Professor Loeb,[2] who is full of touching respect and affection. But more about this when we can talk to each other directly. With heartiest greetings...

---

[1] Ernst Mach, *Erkenntnis und Irrtum. Skizzen zur Psychologie der Forschung*, Leipzig, 1905. It seems that Boltzmann had returned to Vienna ahead of schedule. He mentioned in his letter to Meyer (#108) that he was coming back on September 5. He had intended to see an eclipse in Spain, but felt too exhausted to do so. Also, pressure to finish the promised article for Klein's Encyclopedia may have been a factor in hastening home ahead of time.
[2] Boltzmann spells the name "Löb", which may have been part of his attempt to make German spelling more fonetic. Jacques Loeb (1859-1924) was a professor at UC Berkeley from 1902 to 1910 and became a world-famous physiologist for his partly successful attempt to explain human behavior in mechanistic terms. As mentioned, he had corresponded with Mach for many years. His son, who remembered many stories about Boltzmann's visit, not only became a professor at Berkeley like his father but like Boltzmann would become a professor of physics and would publish on the kinetic theory of gases.

#118

From Benjamin Wheeler Berkeley, September 13, 1905

I am sending you herewith two letters which were returned unclaimed from the New York address which you gave me. I am very sorry they did not reach you before leaving New York...

#119

To Pierre Duhem Vienna, September 29, 1905

Best thanks for your very interesting work on statics.[1] With exceptionally high respect...

[P.S.] Also for all the articles and preprints which I have received from you in such quantity and so often.

---

[1] *Les Origines de la statique*, 2 volumes, Paris, 1905-1906. Pierre Duhem (1861-1916) was a French physicist, historian, and philosopher. Like Mach he was opposed to atoms, statistical mechanics, and Einstein's relativity theory. Unlike Mach, however, he was a strong Roman Catholic and attempted to restrict the scope of science to sensory or traditional "save the appearances" limits not so much to try to make it more certain as Mach had done, but like Cardinal Bellarmine centuries before ultimately to help protect his religion. Duhem's work in history of science, especially on the Medieval period, has also had considerable influence. He often tried to interest Boltzmann in his work but without conspicuous success.

#120

From Arnold Sommerfeld[1] Aachen, October 21, 1905

I have studied your article[2] with great pleasure. I find it, if I may say so, quite first-rate. The leading thoughts are put foward in a graphic [*plastisch*] way and additional material is referred to by means of careful citations. The article will serve as a showpiece for the Encyclopedia. Lorentz's article which is also extremely useful and outstanding does not read quite as well and does not give such an immediate overview of the main aspects of the theory as yours. During my reading some secondary remarks came to mind which I have listed on the accompanying sheets of paper. They are mostly editorial matters concerning agreement with other articles . . . .

Please do not hold up matters by answering my questions. I will figure out the answers later myself from the proofs . . . .

Please give my compliments to your co-author [Josef Nabl]; and to you yourself let me give my fullest thanks for the punctual and extremely successful completion of this work.

---

[1] Arnold Sommerfeld (1868-1951) succeeded Boltzmann in Munich in 1894 and by 1905 was also one of Klein's co-editors for his *Encyklopädie der Mathematischen Wissenschaften*. For more information about Sommerfeld see letter #25 and especially Christa Jungnickel and Russell McCormmach, *Intellectual Mastery of Nature*, Volume **2**, University of Chicago Press: Chicago and London, 1986.

[2] Ludwig Boltzmann & Josef Nabl, "Kinetische Theorie der Materie", *Encyklopädie der Mathematischen Wissenschaften*, Edited by Felix Klein, Volume **V**, Verlag Teubner: Leipzig-Wien, 1907, pp. 493-557.

#120A

Broda on Boltzmann's "Love Lecture" (Concerning October 28, 1905)

Less than a year before his truly tragic death, Ludwig Boltzmann held a lecture with the title "Explanation of Love and the Entropy Law by Means of the Principles of Mathematical Probability". It was given in the yearly main meeting of the Philosophical Society in Vienna. The text was either not written down or has been lost, therefore, I have briefly sketched a reconstruction of the missing lecture . . . .[1]

---

[1] For this quotation plus the reconstruction of the lecture see Engelbert Broda, "Erklärung des Entropiesatzes und der Liebe aus den Prinzipien der Wahrscheinlichkeitsrechnung", *Physikalische Blätter*, **32**, (1976), 337-340. (Had the reconstruction been based on anything surviving from the actual speech or notes for it or from anyone who remembered hearing the speech or even hearing about it, then it would have probably been included in this book. But an invented speech even if based on direct quotations from Boltzmann from other contexts seems unsuitable for inclusion, though many readers, including this one, have found Professor Broda's reconstruction interesting. - Ed.)

#121

Lenard[1] to the Nobel Prize Committee Kiel, January 6, 1906

Ludwig Boltzmann in Vienna for his thermodynamic work ... in gas theory using statistical methods to bring about new insights and successes, placing Maxwell's light theory in fruitful connection with heat theory, and thereby creating in particular means and ways to give an exact foundation for important black body radiation laws.

There can be no doubt, and it is on this which I ground my proposal for the Nobel Prize, that these laws belong to the most magnificent accomplishments of the new physics and with the introduction of improved means of illumination they have already become of practical importance as a rational foundation for the measurement of the highest temperatures.

If one considers the living men most responsible for this development, Boltzmann, Wien,[2] and Planck on the theoretical side and Lummer,[4] Rubens,[5] and Paschen[6] on the experimental, one will find that no individual person among them or even any combination of two or three - except for Boltzmann alone - has been so responsible for success as to deserve the prize.

For the experimentalists could not have succeeded by themselves alone, that is, without laws which are constant, complete, and also theoretical. But among the theorists, Wien and Planck could not have succeeded without Boltzmann and the experimentalists. Boltzmann's work was an essential precondition, but just as important, the results of the experimentalists were needed to correct Wien's work which opened the way for Planck.

Therefore, I consider it fitting that the only thing left to do is to reward this accomplishment with a Nobel Prize and to recommend above all Ludwig Boltzmann as the recipient.

---

[1] Philipp Lenard (1862-1947) had received the Nobel Prize the previous year, hence his letter should have carried weight, but in those days it was rare to give the prize to theorists and though it may have been a close call Boltzmann would die before that mental reservation had been sufficiently overcome. Lenard had been Hertz's assistant and would teach at Heidelberg from 1907 until his retirement in 1931. His Nobel prize was for his discovery of the photoelectric effect in 1902 which Einstein explained in 1905. Very late in life, like many other Germans born outside of the fatherland, he allowed his strong German nationalism to help influence him into supporting Hitler.

[2] Wilhelm Wien (1864-1928) did his most important work on black body radiation during the mid-1890's. He worked out the wavelengths for high frequencies as Rayleigh did for low frequencies. It was Planck in 1899-1900 who developed a theory which seemed to fit both. Wien taught at Wurzburg in 1900 and received a chair in Munich in 1920. He won a Nobel Prize in 1911.

[4] Otto Lummer (1860-1925) worked with Ernst Pringsheim in 1899 to discover problems in Wien's theory of black body radiation. He also stimulated Planck's attempt to understand his own quantum formalism better.

[5] Heinrich Rubens (1865-1922) also taught in Berlin and in 1900 confirmed the results of Lummer and Pringsheim on the shortfall in Wien's black body theory and then with Ferdinand Kurlbaum backed the superiority of Planck's theory.

[6] Friedrich Paschen (1865-1947) an experimentalist at Hannover had independently inferred a black body law equivalent to that of Willy Wien. He also encouraged the idea that it be subjected to rigorous testing, but when the results proved partly negative he seemed more willing than Wien to accept Planck's theory.

#122

Czerny[1] to the Nobel Prize Committee Heidelberg, January 16, 1906

Having been requested to make a proposal for the Nobel Prize in physics, I have had to contact and try to reach agreement with competent people who understand the field, since I myself have not sufficiently mastered the discipline.

According to their judgment the person who is most deserving is above all Ludwig Boltzmann of Vienna.

He should receive the prize for his thermodynamic, gas-theoretical work which has brought new insights and success with statistical methods and connected Maxwell's light theory with heat theory in a fruitful manner and thus created the assumptions needed for an exact foundation of important black body radiation laws.

[1] Vincenz Czerny (1842-1916) became Professor of Medicine at the University of Freiburg in 1871 and Heidelberg in 1877. He was a surgeon at the Cancer Research Institute there and wrote especially on abdominal surgery. Even though like physicists, physicians do treat physical problems, the two professions are not fully identical. Physicians, especially surgeons, tend to operate more on organic matter while physicists generally prefer to experiment on what is inorganic. Czerny may have been surprised to receive a request to suggest a candidate for the Nobel Prize in physics,

#123

Planck to the Nobel Prize Committee Berlin-Grunewald, January 28, 1906

I feel honored that the Nobel Prize Committee for the faculty of physics in their treasured communication of September 1905 has very kindly requested from me a proposal for the distribution of the prize for the year 1906.

As in the previous year,[1] I recommend Ludwig Boltzmann, Professor of Physics at the University of Vienna, and support my proposal in the same way as in my letter of 1905 by pointing out the merit which the above-mentioned has acquired through his investigations in the field of the kinetic theory of gases, and which one can find collected together in his well-known book *Vorlesungen über Gastheorie,* two volumes, Ambrosius Barth: Leipzig, 1896 and 1898.

[1] See letter #90 for Planck's earlier recommendation.

## B. More Philosophy Lecture Notes[1]

(reconstructed from shorthand notes for Boltzmann's lectures in Natural Philosophy, Spring 1905)

. . . . Philosophy = The construction of a world view . . . .

[A.] 1. *completely naive world view.*[2]

2. *naive criticism:*[2] We are able to recognize what and how something is, but only by deduction. [Brentano]

3. *Kant's critique*: We are not able to know what and how things are, but only that they are.

4. *monopsychism*: Things-in-themselves are images; there are only psychic phenomena. The world is a conglomerate of perceptions, memories, pain, desire, perceptions of willing. [Mach]

5. *solipsism.*

6. *solofigne*: Success is the only criterion of truth. opening-the-drawer experiment. [Boltzmann]

[B] 1. What does it really mean to know [*wissen, erkennen*]? Can one know directly? Is that merely mechanism?

2. As a child, I thought I directly saw things. Now I know it is mechanism.

3. Merely aposteriori [knowing] through the experiment of opening the drawer.

4. Perceptions and also all psychic facts are immediately given to us and absolutely certain. They also vary with time and we can only decide by the experiment of opening the drawer whether we have correctly (pragmatically) characterized them; more exactly, they must be compatible with each other.

5. One might think that at least we can characterize something as

existing, [but] we cannot distinguish things existing and not existing. Psychic facts readily harmonize, but it would be best if we don't just introduce them but do so by means of language (spoken language, thought language, and mixed language).

6. We are coming closer to the completely naive way of viewing things, and more significantly: It is good.

7. It is untrue to say that we can merely know [*erkennen*] the qualities of things and not the things themselves, or that we can merely know psychic facts and not their external causes. Really direct knowing doesn't exist at all. We can characterize [*bezeichnen*] one as well as the other.

[Undated, but probably May 1905]

My theory of time:

1. The number of points of time can be made so great, that the probability becomes great, that a very improbable condition can occur in the whole world.

2. Natural law must be so expressed that every condition can be determined by two of the foregoing.

3. This force law must differ at different times, depending upon the different senses in which it is understood at different times.[3]

---

[1] For the original German see Ilse M. Fasol-Boltzmann (ed.), *Ludwig Boltzmann Principien der Naturfilosofi*, Springer Verlag: Berlin, etc., 1990, pp. 143-146.
[2] These labels were almost certainly derived by Boltzmann from reading Branislav Petronijevic's book *Prinzipien der Metaphysik*, Volume I, Heidelberg, 1904, pp. 1 - 4. For a recent account of Petronijevic and his philosophy, see Miodrag Cekic, "The Influence of Mach's Philosophy in Yugoslavia", *Ernst Mach - A Deeper Look*, edited by J. Blackmore, Kluwer Academic Publishers: Dordrecht, 1992, pp 379-382 & 384-390.
[3] Fasol-Boltzmann, op. cit., pp. 282-283.

[undated]

My view: Darwinism does not explain everything, but a lot such as that the deer and caterpillar have developed to be able to survive as a species in their normal or preferred environment. Darwin has also explained how they can do it. He has also shown us that it is possible that everything may be capable of being calculated from the motions of atoms. But God has not shown us as much. If the will is ever completely determined by something [such as a natural law], then it wouldn't exist.[1]

(Lecture Notes from May 24, 1905)

I name Brentano's philosophy "naive criticism", because while he criticizes the conclusions of things, the result is that they exist . . . .[2]

(Lecture Notes from May 31, 1905)

Causal law. Regularities could be merely accidental; a necessary connection might be able to exist, like a conclusion from premises; as not dispensable . . . .[3]

(Lecture Notes from June 5, 1905)

Concepts cannot be further explained. Backwards and forwards in time are different; in a way only forward. Individualization cannot be destroyed. The third dimension is essential; The world cannot other than be complete.[4]

Fall Semester 1905-1906

(Lecture Notes from October 23, 1905)

No one knows what natural philosophy is. Hegel says it is a priori natural science. We connect [*anknüpfen*] science of nature with science of mind; the latter from the mind itself. Psychology [studies everything] which came into existence as the result of the operation of the mind. Mathematics always to natural science. If philosophy is theory of the simplest basic concepts, then natural philosophy could treat the basic concepts of natural science: number, time, space, matter, power, stuff, and energy. So I will introduce it once, very descriptively. Living beings belong to nature; also to the analysis of number, time, space, inner and outer experience; not strict at all; subject and object, the body is the subject.[5]

(Lecture Notes from October 25, 1905)

1. Concept of existence. Matter or energy exists; psychological or physical process. 2. Metaphysical, empirical, psychological. [3.] Materialization [*Verdinglichung*]. 4. three propositions: a) everything inner and outer b) immediate experience is not invented c) Everything object and subject. 5. Relation to other sciences. Introduction to physics and physiology and the reverse. Foundation of the mental sciences. Kant . . . .It is precisely natural science which allows metaphysical judgments; if I think differently, not a priori. Some people imagine it without justification; Also an a priori concept reduces everything to number.[6]

(Lecture Notes from November 20, 1905)

. . . . Contribution to an analysis of basic concepts and fundamental laws of human thought from the point of view of natural science. This analysis was previously and is still today considered to be the task or one of the tasks of philosophy. Philosophy is older than natural science; the latter has developed out of the former. Descartes, Leibniz, Galileo, Kepler. Principles of philosophy; In the beginning the methods were the same; gradual separation; methods of natural science were characterized later and also led to gradual success in specialized investigation. the method was suitable but does not apply to fundamental concepts. One leaves them to philosophy; not the task of natural science. This was a common way of thinking in Kant's time. But this state of affairs cannot be maintained; the tendency to examine basic concepts cannot be fully rooted out, and there is a return to the old methods. Relatively little progress can be detected with respect to detailed questions; nor is [much] understandable. The same evil was also carried over into natural science. It awakens the impression that the problems are much more difficult. I have the feeling that they are not really hard; they are only an artificial way to make problems seem more difficult. With respect to questions in a specialized domain, one knows what one can do and has definite methods to increase understandability. I even believe that these general questions are the easiest of all to solve, but the right method is not yet at hand. I hope to contribute something. We believe that thinking is guided by free will, but we are slaves to our habits of thought, which only slowly become recognized as no longer appropriate and disappear. Other people are also slaves, and even I myself can only struggle to free myself.[7]

(Lecture Notes from January 22, 1906)

In general, we have considered matter, space, and time. But they are always empty, as if they were mere numbers, or only formal [*logischer*] concepts, but which we can rise above by means of specialized treatment which is the job of natural science. Concerning the foundation of philosophy, logic itself is almost entirely philosophical; this is philosophy in the old sense; we discover laws of thought, and we must derive specialized study from what has been so discovered.[8]

(Lecture Notes from January 24, 1906)

A priori and a posteriori methods; *cogito ergo sum* applies proof. Schmitz-Dumont and Petronijevic. Something must be at the bottom; an apriori thing-in-itself. Existence comes before experience. What exists? Matter, power, energy. Question: Is matter possible? Is power possible, or only subjectively? Have we chosen a suitable way of describing the matter? Is experience possible? What is experience? Only designation, in no way freely chosen, but it can be modified. One must confess something: Rules of thought are themselves a posteriori in common with methods, logic, and mathematics. The positive continuum of things, thoughts, and perceptions is validly connected together with the concept of number.

Identity: *a* is *a* remains a logical relation; *a* is *a* all the time. There are objects which do not change; $a = a$, regardless of when we measure it. Contradiction: *a* is not the same as *a*; the concept of negation; causality is also essentially not the same if we don't expect [*erwarten*] that it is the same. Number is something else; again constant with the designation. Concept = *Vorschrift* [specification, instruction, rule], like the art of managing signs. But principles exclude. Something is either red or not red, though something not red may have reddish aspects, that is, at the same time be neither one nor the other. But mere naming is like a finite number; metal and metalloid; exact and inexact numbers; being or non-being. Only psychological acts exist. solipsism . . . . My proof that there is a truth is completely different (not because it must be true that everything is in doubt).[9]

non-being. Only psychological acts exist. solipsism . . . . My proof that there is a truth is completely different (not because it must be true that everything is in doubt).[9]

(Lecture Notes from January 31, 1906)

Connotations. Knowledge through causal law is not at all the same as knowledge through experience. Concerning the source of experience, we emerge [*entstehen*] under its influence. One seeks probability from a priori probability. This only makes sense in equally possible cases. Known causal laws or classification [*Bezeichnungen*] necessarily yield only what is subjective. What is simultaneous or subsequent to an event cannot be its cause.

(After some empty pages and without a date)

Not to understand means that one is not seeking [or not successful] in determining how suitable [*zweckmässigen*] action is caused, or might be caused by something analogous. Continuity cannot be caused by a continuum. If consciousness is inclusive comparison is impossible. What is knowing [*wissen*]? Reason merely selects among inborn drives. Mind does not exist just while we are thinking. Honor is beautiful. Why are we living now instead of Schiller? Is it a priori evident that the order of numbering is indifferent? Is that necessary but not the gravity of it? The concept of number must exist before we can think of inheriting it. Would it be advantageous for ideas to be rewarded in Heaven? Would it be right?[10]

---

[1] Ilse M. Fasol-Boltzmann (ed.), *Ludwig Boltzmann Principien der Naturfilosofi*, Springer: Berlin, 1990, p. 283.
[2] Ibid., p. 143.
[3] Ibid.
[4] Ibid.
[5] Ibid, pp. 143-144.
[6] Ibid. p. 144.
[7] Ibid., pp. 144-145.
[8] Ibid., p. 145.
[9] Ibid.
[10] Ibid.

Drawing by Karl Przibram, one of Boltzmann's students
Courtesy of Roman Sexl and Setsuko Tanaka

# Chapter 7

# EL DORADO & ANECDOTES

## A. On the Trip of a German Professor into El Dorado

by Ludwig Boltzmann[1]

Since I have already been in America more than once [1899 &1904] and was also in Constantinople, Athens, Smyrna, and Algiers [1901],[2] I have not lacked requests to allow my travel experiences to be published. All of this seems unimportant to me, but my last trip which was to California was indeed rather exquisite, so let us take this opportunity to talk about it.

I won't say by any means that one has to travel to California in order to see what is interesting and to enjoy himself. One could also fill the human breast with as much fun and enjoyment on a tour through the beautiful mountains of our Fatherland [Austria]. One can feel as satisfied as a king from the very simplest meal, but a trip to California is Champaigne, *veuve Cliquot* [3] and *Austern*.[4]

The first part of my trip stood under the portent of haste, and with haste the tale will be told. On the 8th of June, I attended as usual the thursday meeting of the Academy of Sciences. On leaving, a colleague remarked that instead of my normal way to Bäckerstrasse, I was turning toward the Stubenring, so he asked me where I was going. I answered laconically: "San Francisco".

In the restaurant of the Northwest Train Station I consumed in what was still pleasant leisure *Jungschweinsbraten* [roast pork] with *Kraut* [vegetables] and *Erdäpfel* [potatoes] and drank in addition several glasses of beer. My memory for meal money owed is tolerably good, but about glasses of beer it is always bad.[5]

---

[1] For the German version see Ludwig Boltzmann, "Reise eines deutschen Professors ins Eldorado", *Populäre Schriften*, 3rd edition, Barth: Leipzig, 1925 [c1905], pp. 403-434. All of the footnotes in this chapter have been added by the translator.

[2] Boltzmann visited these cities from July to October 1901 on a voyage around the Mediterranean. See Ilse M. Fasol-Boltzmann, "Ludwig Boltzmann and His Family" in Ilse M. Fasol Boltzmann (ed.), *Ludwig Boltzmann 'Principien der Naturfilosofi'*, Springer:, Berlin, etc., 1990, p. 31. (There is also a photograph of Boltzmann on the boat on page 21.) Boltzmann had also traveled to Italy, France, Germany, and England many times.

[3] This is a fine French wine. It was also strongly recommended by the Prefect of Police in the famous movie "Casablanca".

[4] oysters

[5] It is common in Austria with each new serving of beer to make a mark with a pencil or pen on the paper-maché *Deckel* underneath the beer mug in order to keep count, but it is not certain whether this practice also existed in Boltzmann's time. Nevertheless, Boltzmann seems to be saying that he often lost track or failed to write the number of beers consumed down.

No fairly frequent traveler will be surprised if I speak about food and drink. Far from being unessential factors, they are the hub of everything. The most important thing on a trip is to preserve the health of the body against various unaccustomed influences, above all to protect one's stomach, particularly the pampered Viennese *digestif.* No Viennese will eat his last meal of *Gollasch mit Nockerl*[6] unmoved, and just as the Switzer's homesickness will focus on memories of *Kuhreihen* [7] and *Herdenglocken*,[8] so ours lingers on *Geseichte mit Knödel.*[9] "Say not that age makes me childish, but that the joys of youth have returned."

On finishing my meal, my wife and children came with my luggage already packed. Once more *adieu* and I was off to the joint meeting of Associations of Learned Bodies [*Kartellsitzungen*] at Leipzig which would start at ten A.M. on the following day. I made myself as clean as possible (though I could have used a washing room like on American trains!), and on leaving the railway car in Leipzig jumped into a *Droschke* [carriage] and arrived punctually at the meeting. At the door of the great hall, I immediately met colleague Credner,[10] who was so kindhearted as to help me put my baggage in the vestibule, since I had found no place to deposit it along the way.

I had not come to this "meeting of associations" without some concern, for should a certain topic be raised, matters could become bitter as wormwood.

Will it bore the reader if I lead him around a work-station of scientific activity for a short time in order to show him the externals and explain something about the mechanism? I hope not. Nowadays, there is scarcely an educated person who has not seen, at least in Baedeker [a travel guide], some kind of iron, leather, or glass factory, but I think that what I will describe will satisfy the curious and be just as informative and educational. Why should I not presuppose some curiosity about the mechanisms which make it *possible* for a factory to function, which I dare say are more important for human culture than even the biggest leather plant, and hopefully without displanting your interest?

Several German academies and learned societies have cooperated with each other in order to hold common meetings yearly and to address subjects of general importance there. This is the "meeting of associations" [*Akademie-Kartell*]. The same body decided several years ago to give material support to a book project, the *Enzyklopädie der mathematischen Wissenschaften.* Mathematics over the past century has enormously increased its comprehensiveness. But with that it seems as if every author uses his own set of symbols and often writes in a manner so hard to understand, that even his closest fellow specialist can only follow it with the greatest difficulty. And

[6] This is a small dumpling much beloved in Austria and Bavaria.
[7] an Alpine dance for cowherds
[8] cowbells
[9] a kind of smoked pork with dumplings.
[10] Hermann Credner (1841-1913) became Professor of Geology at Leipzig in 1895.

yet, among this hard-to-understand, often almost impenetrable mathematical literature which is scattered around the whole world, uncommonly much that is useful, even for practical people in everyday life, indeed, almost indispensable, has been buried.

The task of the encyclopedia which we have been talking about is the well-ordered collection and the easiest possible understanding and representation of all of this material. Everything which has been accomplished in mathematics would be made easily discoverable for the mathematician and at the same time a bridge would be built to practical application in order to make practical mathematics more accessible to the applied mathematician. The need for such an encyclopedic collection of mathematical literature has become so striking that Professor [Felix] Klein in Göttingen has called it a mathematical *Bedürfnisanstalt* [men's room].

Such an undertaking would not be so enormously difficult if it only had to detail the most brilliant contributions without bringing in excessively exact criticism and also included the most important work and naturally what is best-known. But if everything really useful is to be drawn from obscurity into prominence in all fields, everything unessential excluded, references to the literature made as complete as possible, and if one wants to make everything transparent for the use of the reader in the most convenient form, then the difficulties for every [contributor] who has only taken a modest peek into that mathematical literature will be almost terrifying. The already-mentioned Professor Klein has enticed us with the hope that academies will give money for our printing costs, honorariums for authors, and will pay travel expences, while he and his staff will correct errors and refine any rough edges remaining on one's work.

They have scoured all the nations of the planet to locate the most knowledgeable experts for every field. And in fact, Germans, Frenchmen, Russians, and Japanese cooperate together in [amazing] harmony. The selected soul is often a great man with enough money and not enough time, but perhaps too much reluctance to work, yet driven by excessive ambition. First, he must be moved into promising a contribution, then taught and urged by all the arts of persuasion to mold his essay so that it fits the larger project, and last but not least, he must carry out his promise on time.

But hours are used up in discussions about whether an article which it would be better to include later should be brought in now because one already has it in hand while the one which should precede it is still lacking. But Klein himself and his apostles will not shrink from traveling to the farthest corners of the world in order to apply conversational pressure to delinquent contributors. But one writer remains inaccessible, namely, a mathematically-educated Russian officer who is currently being besieged by the Japanese in Port Arthur. I have often taken part in such encyclopedia sittings as this one, and their dramatic ebb and flow is so exciting that German stage poets could well profit from the experience.

Now back to my story. Even at the time when Klein first commissioned me to write an article, I hesitated for a long time. But finally he wrote me:

"If you won't do it, I will give it to Zermelo!" This is the name of a person who holds views which are diametrically opposed to my own.[11] Since that would definitely not set the right tone for the encyclopedia, I answered immediately: "Rather than let that *Pestalutz* do it, I'll do it myself." (All of these quotations, mostly from Schiller which echo the hundredth anniversary celebration of his death [May 9, 1905], are furnished with quotation marks, if one wishes to authenticate them.)

But now was the time when my article fell due. I would have gladly taken a rest in September after the strain of my Summer travels with a stay in the country, but I had given my word, therefore in September [1905] I had to grub through the literature and together with a small cohort of Viennese physicists to grace my "sworn eternal oath".

Professor [Wilhelm] Wirtinger[12] appears to have had similar mixed feelings. He proposed that the the emblem of the Encyclopedia should be a mousetrap - The scent of Bacon beckons and the professor is caught.

But what charm makes the whole project so irresistible? It will not add significantly to anyone's reputation, except to show one's willingness to perform something useful. And the money, ha!. Hardly worth discussing. What has given Klein such psychological insight, enough to stir the envy of philosophers, which allows him to locate everyone's Achilles' heel? To find the exact point where each contributor can be seduced?

He has appealed to our idealism, indeed, to a type which if we open our eyes can be found everywhere, even by the Pacific Ocean. Two thick white towers greet us there, the Lick observatories, the work of an idealist and millionaire, a hundred times over; but more about that later. For a long time I have thought about how remarkable it is, that there exist millionaire idealists or idealist millianaires. Happy the land where millionaires think ideals and idealists become millianaires! Let us honor smoked meat with dumplings; Idealists need a good stomach everywhere.

The idealism of Klein and his co-workers has yielded good fruit. Already after the appearance of the first issue the edition had to be increased. A French translation has started and an English one will begin soon. The academies have climbed on board and the booksellers enjoy good business.

But the Berlin Academy of Sciences unfortunately does not belong to the cartel and is not sharing in the matter at all. They were also not repre-

---

[11] For Ernst Zermelo's objection to part of Boltzmann's statistical interpretation of the second law of thermodynamics, which in German is called the "*Wiederkehreinwand* ", please see the section on physics in the Introduction to this book or the following articles: Ernst Zermelo, "Über einen Satze der Dynamik und die mechanische Wärmetheorie" *Annalen der Physik*, **57** (1896) 485-494; Ludwig Boltzmann, "Entgegnung auf die Wärmetheoretischen Betrachtungen des Hrn. E. Zermelo", **57** (1896), 773-784; Ernst Zermelo, "Über mechanische Erklärungen irreversibler Vorgänge - eine Antwort auf Hrn. Boltzmann's Entgegnung", **59** (1896) 793-801; and Ludwig Boltzmann, "Zu Hrn. Zermelo's Abhandlung über die mechanische Erklärung irreversibler Vorgänge", ser. 2, **60** (1897) 392-398. [These articles have been included in English translation in Stephen Brush, *Kinetic Theory*, Volume **2**, Pergamon Press: New York, 1966.]

[12] Wilhelm Wirtinger (1865-1945) was an Innsbruck mathematician who later taught in Vienna.

sented at the Meteorological Congress at Southport and at the Congress for Solar Research at St. Louis. I fear that this principle of not participating in anything which they themselves did not start will damage more than [international] science; it will hurt Germany itself. It angered me in Southport and St. Louis that everywhere among the foreigners (non-Englishmen) that the French held first place. We Germans should not have been required to stand behind them! But what could I do as an Austrian? If among the meteorologists only Hann[13] had been there, who we all missed! But he is not to be moved into traveling again!

When it comes to chatting, I let my tongue wag with complete freedom. Hence, I couldn't keep quiet even when an American colleague spoke about returning to Berlin. In fact, under Weyerstrass, Kronecker, Kummer, Helmholtz, and Kirchhoff[14] American mathematicians and physicists mostly went to Berlin to study, but now they prefer Cambridge and Paris on the [false] assumption that there is little more to learn from the Germans, and thus America and with it the world declines. This [American] colleague also maintains that it would have been much better if I had not rejected the call to Berlin [in 1888]. And certainly I might have helped matters with my lectures, but concerning creativity and calls to better positions, it may weigh heavily whether a single person [*einziger*] works with Klein's idealism and Klein's inexhaustible energy. Many who are not to be had, would go, if they felt they were wanted in the right way. Even a *klein* wheel[15] which rolls in the right direction can help.

---

If I were to stay in every town with only the number of people as Leipzig, I would not get very far; "but one must not measure the number of souls but their weight", which naturally means their mental significance.

After some highly enjoyable intimate dinners and an official one where I first became personally acquainted with the Saxon Minister of Education, [Paul] Seydewitz, under whom I had served as professor for two years, I continued on to Bremen and then with one of the Hohenzollern princes to New York. This is not to be understood as if I had been worthy enough to accompany His Majesty on a trip to America, but simply that I rode on his back. I went out on the *Kronprinz Wilhelm* and returned on the *Kaiser Wilhelm II.*

---

13 Julius Hann (1839-1921) was a highly respected Austrian meteorologist, who along with Ernst Mach would be nominated by the Austrian geologist-politician, Eduard Suess, for a Nobel Prize in 1910.

14 Leopold Kronecker (1823-1891), Ernst Eduard Kummer (1810-1893), and Karl Theodor Wilhelm Weierstrass (1815-1897) were all famous German mathematicians from an earlier period as Hermann von Helmholtz (1821-1894) and Gustav Robert Kirchhoff (1824-1887) were famous physicists from that time.

15 Literally "small wheel", figuratively "Felix Klein", but the suggestion in English of a second pun: "not a big wheel" i.e. "not an important person" may not have been intended by Boltzmann.

Dear reader! My haste is great, but the sea voyage from Bremen to New York, apart from this banal joke, is simply beyond my capacity to describe. Great Ocean Steamers belong among the worthiest objects of admiration which mankind has created, and one finds that the trip becomes ever more beautiful with every new opportunity. The wonderful, raging sea, every day so different, and every dawn even more astonishing. At noon, white, swirling foam. Afternoon, wild, uncontrolled wind. See that ship afar? Now sunk under a giant wave! No! Already her keel rises victorious, stabbing the angry air.

It is morning and the weather is peaceful. The sea is smooth but fog-gray, as is the mist when one paints melancholy. Then the sun breaks through and yellow and red sparks dance on the waves between the deep, black surface of the cloud-cast shadows. How the golden glint weds the utter blackness! And all at once the whole sky is blue, and the sea moving azure under a white flash, so radiant in its overpowering brilliance that one must close the eyes. Only on rare days does the deep water bejewel herself in such bewitching ultramarine garments, a color dark and yet so luminous with milkwhite spray as if glistening tips might top the heaving beauty of such immense, unhurdled aquininity.

I once laughed to learn how a painter panted night and day for one single true color. But I laugh no more. I weep in wonder at the ocean's hued wiles, a magic lantern. How can mere color cause me to cry? And yet again, tears fall as the glow of wet cheeks mirror the full ivoried-moon on a half-clouded, semi-shrouded night. But this I know well, that only an artist can paint perfection, and never a mere poet at first try.

If any one item can have more value than the beauty of nature, it would be the skill of mankind which since the Phoenicians and well before has struggled with the endless expanse of the sea only to completely triumph over it. How unmercifully the keel cuts through the flood, like a god of the ocean churning the water, propelling it every which way! In truth the highest wonder of nature is art, the genius of the human mind!

If I were asked as Solon was, who the happiest of mortals had been, I would without hesitation answer "Columbus". Not as if there had never been equally important discoveries such as issued from the German Gutenberg, but happiness is co-formed by its visual effect and that of Columbus had no peer! I can never land in The New World without envy joined with rapture which recovers a small part of his joy. Certainly Columbus did not sail with the *Kronprinz Wilhelm*, nor could he see New York with his physical eye, but with mental anticipation he saw so much, perhaps even how it would look in one or two hundred years!

Thus, Christoforo Colon has become the prototype of the discoverer. His "Sail on!" "Sail on!" and ringing cry: "Land Ho!" have become the very spirit of high success and the whole heart of his human persuasion that mere life is not the highest good. "If less than life would weight the scale, then aiming high must surely fail."

Not only the sense of beauty but all others find their satisfaction on a sea trip as well. A fine kitchen tickles the palate with a broad menu. A whole orchestra does wonders for the cultivated ear. One may even have Viennese composers for company, certainly not the very greatest, but Strauss, Ivanovici, Waldteufel, and others.[16] Indeed, on the rising breast of the Atlantic *Danube Waves* [16a] wins worlds of applause, and in fact, if we think of Haydn, Mozart, Schubert, and Beethoven, we can, like Schiller, muse on our famous *Wien* and her flowing friend: "In rippling by, her gentle waves have overheard many an immortal song."

Hence, there is nothing more comfortable than life aboard ship, especially for those whom God has chosen to spare from sea sickness and granted an inner calmness for viewing the many suffering people stretched out on deckchairs. But I must admit that fun may flourish most when a hydrodynamic accident hits the hull and unexpected swells douse the half-slumbering, semi-invalids who leap up screaming.

On entering the harbor of New York, I am always transformed into a mood of intoxication. The towering buildings and the magnificent statue of liberty with a torch in her upraised hand! Around us the ships whistle and sing. One roughly warns and another shrieks in reply; one pipes cheerfully, and her partner moans a melancholy fourth. And then suddenly the inimitable sound of sirens! If I were a musician, I would compose a symphony and simply call it: *Harbor Serenade!*

But I was in a hurry and lacked time for sentiment. In Hoboken I engaged a cab at once, which was first to take me to the office of the Southern Pacific Railroad and then direct to the train station, but at the first edifice I heard that the fast train for which I had paid reduced fare only ran twice a week so that I must stay two days in New York. Facing facts, I engaged a vehicle to take me to Westminister Hotel, which left me some time to explore the city.

But it was not dull. Even a simple trolley ride offered rich rewards in conversation and observation! Tickets were not handed out or punched, and there were no limits to overloading and no different prices. With eyes like a hawk the conductor spotted every new sardine. I squeezed five cents into his hand, a cord was pulled, and the payment was acknowledged in a register above our heads. At the same time everyone on board heard a sound like a clock striking. If one finds a seat near the driver, one can only admire his *Führergabe* [driving ability, leadership], which not even a Napoleon or Moltke can match. How fast they race when they find an open lane, and how quickly they stop in front of an automobile (which simply goes around coaches and trolleys in a half-circle). Everything which one could see in the wild [*Natur*] and a lot more can be found on the streets of New York, and it is worth seeing.

---

[16] Alas, not fully so by 1905. Johann Strauss and Iosif Ivanovici were dead and Emile Waldteufel was 67 years old.

[16a] Iosif Ivanovici (1845?-1902), a Romanian orchestra leader and composer was the author of *Danube Waves* (1880), possibly his most popular work.

On the third day, I picked up speed again. From New York it took four days and four nights to reach San Francisco. One is simply hurled, to a certain extent, shot forward. The paths through the very long trains which one must traverse in order to reach the dining cars are not entirely agreeable. Furthermore, the observation cars are completely open in the rear and if one leans against the guardrail which fences them in while gazing at the countryside, one has to keep in mind that any sudden jolt could cause the dreaming poet to fall over and kill himself.

For the most part the scenery was monotonous, nevertheless, directly experiencing how fast we were going was interesting. If one looks back from the observation car, the rails themselves look like an endless ribbon furiously unreeling from the bottom of the train itself. Also, the trip on top of the giant wooden dike which runs straight across the Great Salt Lake was interesting together with the vast fields of salt crystals like an ocean of snow. And toward the end of the trip it was wonderful to pass over the Sierra Nevada Mountains. This section reminded me of our own winding way up and over the Semmering in Austria. The Sierras were, to be sure, not as pcturesque, but more magnificent in length and in the height of the mountains.

Thanks to my delay in New York, I failed to reach Berkeley on time. Summer School had already started during the morning of July 26, when I arrived that evening. Nevertheless, since it is the habit to fritter the first day with introductory speeches and filling out forms, even if I had arrived on schedule and begun class at 9 A.M. nothing in fact would have been lost. But I must add as I acknowledged to myself that I would almost certainly have not been in the right frame of mind. The effect of the four days of shaking and jolting hit me hard. I was unable to take a single step on the still ground without fear of falling, and at night in bed I woke time and again scared because I was not actually woozy when I still dreamed I was.

But I must confess that I always have a little stagefright before giving the first lecture, and especially in this case where I had planned to do it in English. Regrettably, on the train there had been less occasion to engage in English conversation than I had hoped. The German passengers who could speak English lapsed after a few words back into German, and genuine Englishmen didn't say anything at all.

My English conversation followed this pattern: I: "When lunch will be served?" He: "ieöö". I: "I beg you, could you say me, at what hour lunch will be served?" His gurgled answer now turned a fifth deeper: "aoouu". I see the mistake in my plan of attack and cry dispairingly: "Lunch", "lanch", "lonch", "launch" and so forth. I utter vowels which cannot be found in Gutenberg's vast collection of minted metal printing blocks. Suddenly, a face lights up with dawning comprehension: "Ah! Loanch!" Now, a bridge of communication appears. I: "When? At what hour? When o'clock?" He: "Half past one!" We have understood each other. And now I should give thirty lectures in this language?

I passed up Tuesday the 27th, but started that Wednesday. I was a bit tentative in the first lecture but was already more natural in the second, and as I heard, moreover, that the students understood me well, and even found my representation clear and lucid, I soon began to feel more easy and at home.

I cannot help but express my thanks to Miss May O'Callaghan for this success [back in Vienna]. Without her untiring efforts to tame my disobedient tongue, I could never have succeeded. How proud I became when I could easily pronounce tonguetwisters like "blackboard" and "chalk" with obvious confidence when I had to acquire sufficient blackboard space and chalk which could actually write! How smooth and fluent flowed prickly pears such as "algebra", "differential calculus", "chemistry", "natural philosophy", and so forth!

Even a magnificent *Hummersalat* is due to my studying English with such diligence. Looking at a menu in English my eyes almost passed over "lobster salad" when I suddenly remembered my lesson where I could hardly believe that *Hummer* meant lobster. Therefore, "Let's have the lobster!" and it was wonderful.

---

The University of Berkeley [i.e. California][17] where I taught is the most beautiful place one can imagine. A square kilometer of park with trees which have already seen a hundred years, or is it a thousand? Such an eon of racing time robs me of words! Beautiful, modern, well-equipped buildings were nestled within the scenic atmosphere, though admittedly, perhaps already a little small; but construction sites are there, one makes new buildings, and the money exists.

But there is a certain philosophical cloud hanging over the whole thing. The name "Berkeley" is that of a highly esteemed English philosopher who has the posthumous glory of being the discoverer of the greatest folly [*Narrheit*] which has ever been hatched by a human brain, philosophical idealism, which denies the existence of the physical world, therefore, idealism in another sense than I am accustomed to using it.[18]

The university philosophy department even has its own building for teaching purposes, but it is not made out of mere phrases and phantoms [or Berkeley's "spirits", "ideas", and "notions"]. But pardon me; what I wanted to say was that the physical structure the philosophers use is not literally fabricated from rational concepts and logical hypotheses, but from stone and wood, where research into the human psyche is carried out with physical

[17] Traditionally, German and Austrian universities were named in Latinized form after their noble founders or patrons, but in practice almost everyone called them by the name of the city in which they were located. Boltzmann simply continued this practice when talking about American universities, where the custom seems odd and is vitually unknown, especially since many of the larger and even smaller cities often have more than one institution of higher learning.

[18] Boltzmann probably means that he himself is an ideal-ist but not an idea-list like Berkeley, who tried to reduce all physical reality to ideas, sensations, or conscious experience.

instruments such as tuning forks, colorwheels, kymographs, and other scientific measuring devices.

But another building proved more important to me. A speculative proprieter had read in an encyclopedia that Berkeley had been an English Bishop whose residence had been called "Cloyne Court" which apparently inspired him to build a faculty lodging with that name, and it is here that I lived. But he made little effort to make the architecture similar since his impressionable mind had also noticed that the front street was called "Euclid Avenue". Thus, the building became an exact parallelepiped without any trace of anything non-Euclidean. But the inside was comfortable. I had a small bedroom, a somewhat larger study room, and a bathroom, with everything electrically-lighted. Hot water circulated through thick pipes and had the effect of moderately warming the rooms, which was often welcome even in July given the occasional icy winds from the Pacific on a latitude below that of Palermo. But on the other hand, the winter temperature is only slightly cooler than that in summer; but it is rich in rain which is wholly lacking during "the hot season".

The food was good. As a rule, at least elements of one type of food placed in front of me could be eaten and kept down. There was no printed menu. The items were voiced by a waitress wearing glasses who muffled her words in a softly musical monotone.

"Yet, our Earthly lot lacks life's unmixed joy", but it is also mixed or lacking even on a lark to Berkeley. First, let my stomach have its say. Previously, I had neither drunk naked nor bottled water, neither river nor carbonated, and so my stomach inspite of the unaccustomed nourishment remained healthy. But Berkeley is abstinent; It is strictly forbidden to drink or sell either beer or wine. But refusing to die of thirst, I tried the water but without ice; perhaps it would be healthier in Berkeley than in New York or St, Louis. Unfortunately not! My stomach revolted, and after a night in which I had to keep my clothes on because it was too much trouble to keep putting them on and taking them off while going back and forth to the place I had to go, I took courage in my hands and asked a colleague about the location of a wine shop.

The effect on him reminded me of a scene in the smoking car on the train between Sacramento and Oakland. Someone was talking to a man from India, who being wholly naive because he was Indian, asked him where the Indian dancers' house [*Bajaderenhaus*] was in San Francisco. Most of the men in the smoking car were from San Francisco, and young girls, "ladies" with the line "Give me money, I give you honey", certainly exist in San Francisco, but all of the men put on shocked, embarrassed faces.

My colleague managed exactly the same stunned look when I asked him about a wine shop. He glanced anxiously around to make sure that no one overheard him. Measuring me carefully to make sure that trust was well placed, he finally took me to an excellent wine shop in Oakland. I succeeded in smuggling a right large battery of wine bottles back to Berkeley, and since then, I often found my way there.

My belly said amen and recovered amazingly fast, even though my board remained the same. But when I took my glass of wine after meals I had to drink in secret, so that I almost got the feeling I was depravity's slave. Temperance is therefore on the best way to becoming a hypocritical pose, of which there are already enough types without adding new species.

Scarcely having pacified my stomach, other evils popped up. Previously, my irksome athsma would always disappear as soon as I set foot on the steamer and would remain gone until I stepped on European soil again, and so it was once more until I reached California. But what makes this climate famous, the humid coolness, at least in my case revived that old uninvited guest that sought as always to narrow the air passages in my thirsty throat.

Then I developed a boil under one of my arms (I believe as a result of having bought and worn a new shirt without having had it washed before wearing it). I had it cut out in Roosevelt hospital. It was highly interesting to become acquainted with an American hospital from the ground floor up, and its elegance far surpassed that of Kaiser Wilhelm II (I naturally mean the ship), but it cost thirty-five dollars. This was the most expensive luxury of the whole trip and robbed me of a few mixed pleasures.

Tuesday the Fourth of July was Independence day, the greatest American holiday, and since I gave no lectures on Saturday and Sunday, I only needed to miss Monday or catch it up later in order to have a four-day holiday in order to sneak in a quick visit to Yosemite Valley. But the fates were against it. Therefore, on Sunday, July 2, I heard the "half-hour of music" with the orchestra doing its best which was given free every weekend in the Greek Theater. This theater is a true copy of the Sophoclean Amphitheater in Athens, only, as it seemed to me, somewhat enlarged. Since it never rains in Berkeley in the summer and because of the frequent fog the sun rarely shines, this completely uncovered theater is quite suitable. Only the music in this space which was surrounded by architecturally wonderful rings of Eucalyptus and Live Oaks was infinitely thin. If [Gustav] Mahler had been heard there with the [Vienna] Philharmonic playing the Third Symphony,[19] so that the trees would have quivered with delight and the Pacific become even more pacific, pacified in its attentive listening, most of the audience might still not have noticed the large, let us say, Continental difference.

On Tuesday I saw the magnificent fireworks from the roof of Cloyne Court, which are part of the annual celebration of Independence day. Cloyne Court is located on the upper part of a gradual rise and preserves a view of San Francisco Bay, The Golden Gate, Mount Tamalpais, etc., as could

[19] Given Mahler's current high reputation as a composer, it would be natural for us to imagine that Boltzmann was referring to Mahler's Third Symphony, but given Mahler's greater reputation in Vienna as a conductor at that time and Boltzmann's love of Beethoven's *Eroica*, it seems very possible that he was referring to the latter. Additional research will be needed to settle this matter.

scarcely have looked more beautiful from the old English Bishop's residence of the same title.

Even Dear God himself seemed to take pleasure from the celebration; for he gave us a sunset worthy of his own greatness and that of his creatures. Once again as so often before on that trip to America, I had the wish to be able to paint what I was seeing so vividly!

As the last rays of the lingering glow fell dark and the lights of San Francisco illuminated the scene, the lesser powers of man loosed her own fire works but whose own size was by no means nothing. Down below skirting the shore a brightly colored ray beamed upward and soon rose above the horizon as a flaming star. But how can one describe it? It flashed above Berkeley and San Francisco, but in Oakland it had an even more beautiful effect! And I have almost neglected to mention, since it was scarcely visible to us, that it was probably even more beautiful above Alameda. It was so overwhelming that I promise henceforth to celebrate American Independence Day in my own Garden in Vienna by setting off fireworks. George Washington's struggle and that of his loyal band was not just a product of local patriotism but has world-historical significance.

Schiller once said: "If there were only a few more thousand like me, the German Nation would become a vibrant, vigorous Republic, against which Rome and Sparta would pale in comparison like a faint-hearted cloister of nuns." Naturally it never happened. A few thousand like Schiller? The world hasn't even produced one. But ideas don't die. The Republic, which in comparision makes Rome and Sparta appear feeble still exists, though admittedly far across the Ocean, and how colossal it is, and how it grows! "Freedom gives birth to Giants!"

---

During the next period I was invited somewhere every Saturday and Sunday. The first time was with Mrs. Hearst in her splendid home near Livermore. Who is Mrs. Hearst?[20] It is not so easy to make this clear to a

[20] Phoebe Apperson Hearst (1842-1919) was the wife of George Herbert Hearst (1820-1891). He was a brilliant and fortunate prospector who opened or controlled many famous mines in the Western states including the Ophir mine in Nevada, the Ontario mine in Utau, the Homestead mine in South Dakota, and the Anaconda mine in Montana. After their marriage in 1862, she became a philanthropist and helped to establish or fund numerous kindergartens and libraries. She was herself a teacher and after her husband's death she also provided the money for a number of buildings on the campus of the University of California at Berkeley. She became a regent of the University in 1897. Their only son, William Randolph Hearst (1863-1951), became famous for his newspaper empire, starting with the *San Francisco Examiner* in 1887 and soon spreading his acquisitions across the country. He had immense influence in his day, was partly responsible for the American declaration of war against Spain in 1898, and ran several times for public office, coming close to winning the mayorship of New York in 1905. At the time of Boltzmann's visit he was almost certainly better known to the general public than his mother. His most lasting legacy outside of his family's surviving newspapers and his own vigorous newspaper tactics would probably be

European. One would perhaps come closest to the truth by saying that she *is* the University of Berkeley. In Europe, one's *alma mater* [literally: "nourishing mother"] is an Ideal from the Classical Age, but in America she is a real person, and what is most important, every year she gives part of her real millions for the improvement and extension of the University; indeed, the cost of my trip to America came from her largess. The President of the University, who we would call the *Rektor,* except that he holds the position for life [and is not appointed by the faculty], is only the executive officer of the Trustees, at whose head Mrs. Hearst stands. In order to reserve a modicum of freedom of action, give himself a little genuine authority, the current President had to make his acceptance of the position contingent on being permitted certain allowances free from interference by the Trustees.

Even worse - yet how can one say that when one is so indebted to the friendly hospitality and beautiful hours with Mrs. Hearst? - How can one maintain that such an *alma mater* conceivably could be bad? Especially when conditions are even more difficult [more demeaning for the faculty and administrators] over at Leland Stanford Junior University across the Bay at Palo Alto, which I happened to confirm on a one-day visit there.

Mr. Leland Stanford Senior had managed the construction of the first railway to the Pacific, that is, the first one which offered continuous rail connection between the Atlantic and the Pacific Oceans. He knew how to put forward in Congress the importance of the undertaking in the right light, and he was influential there. He persuaded them to underwrite half the costs which he had a certain prerogative to manage, but which for the most part were to be left in the hands of the railway contractor. Mr. Stanford then established a second company with a completely different name, whose head he was which delivered all of the materials and labor for the railway company. Since he was the boss of both, he found it easy for the construction company to pay doubled prices to the resource and labor-providing company, such that the government which was contracted to pay half the expenses in effect ended up paying for the whole project.[21]

When he had become enormously rich, sudden misfortune carried off his only son [Leland Junior] for whom he had collected everything. He and

---

"Hearst Castle" (not to be confused with his mother's hacienda which Boltzmann describes) which is still a major tourist attraction in California.

[21] Leland Stanford Senior (1824-1893) combined a successful business career with high public office, being governor of California during the Civil War when he helped keep the state in the Union and later becoming a senator in Washington D.C.. He was also fortunate in having brothers who were also highly successful in business in California who helped him at important times with money and in organizing the collateral supply companies needed to construct the Central Pacific and Southern Pacific railroads. Though he and three equally famous colleagues obtained what seemed like unreasonably lucrative fortunes in the process, the fact is that his companies helped build the railroads over difficult terrain during the 1860's and later much more quickly than was generally imagined to be possible at the time, thereby substantially aiding the settlement of the Western part of the United States. He married Jane Lathrop in 1850 and she like Mrs. Hearst became an active philanthropist, helping to establish libraries and after the death of her husband making sure that Stanford University obtained the financial support it needed.

especially Mrs. Stanford fell into a kind of religious madness [*Wahnsinn*]. In Europe if an old lady loses her mental balance or goes dotty she usually buys a dozen cats or a parrot, but in this case in this country she called in a building contractor of the first rank (What can money not buy?) and had him build a university which will definitely be the family seal for the future.

While Berkeley University is built according to the pavilion system [i.e. separated buildings and colleges], Stanford University has been laid out according to already fashioned plans as architecturally unified, which certainly strikes me as unsuitable for teaching [since one cannot predict future needs and enrollment that exactly]. But architects are dreamers everywhere. But especially magnificent is the university church which is richly decorated with statues, colored glass, and paintings on the walls and ceilings. The organ, which was played in my presence, sounded so wonderful that I gladly became piously attentive.

After the death of her husband, Mrs. Stanford became the nourishing mother of the University for a long time. But she has also died now, but not without generously providing for the University in her testament.

It is said that once a professor of economics at Stanford University had openly spoken in his lectures against the founder's business fraud. The President believed that he could take a page out of Mrs. Stanford's book by firing the professor at once, but Mrs. Stanford herself was so magnanimous that she recalled the professor and chided the overeager President.

At such universities it is perhaps self-evident that males and females possess equal rights both as students and among the teaching corps, but I would like to bring forth drastic evidence for the far-reaching rule of the "weaker sex". One of my teaching colleagues, and I remember her name: "Miss Lilian Seraphine Hyde", a not unattractive lady, announced her teaching schedule just as I had done mine, two hours of lectures per week. In what subject one may ask? "The Preparation of Salads and Desserts". [Would any "masculine" person be allowed to teach the same at a highly respected university?] If you want proof the course was offered, I can still show you the listing in the catalog of lectures.

Every open space in the University swarms with women, whose number stands little behind their male counterparts. It is particularly striking that in every room there is a lady's hat. In the faculty chamber there is a lady's hat, in the washroom a lady's hat, in the telephone booth a lady's hat, and serving some purpose there is even a lady's hat in the photographer's dark room. In fact, as I got up to go after the aforementioned operation for a boil and still somewhat weak and confused I nearly put a woman's hat on my head by mistake!

But now back to Mrs. Hearst, *alma mater berkeleyensis*. As already mentioned, she had invited me along with a series of other summer school teachers to her country seat in the vicinity of Livermore, which is a jewel of a place, as fine as luxury, wealth, and good taste with the help of lavishly lush nature could make it. The coach was waiting for us at the train station and whisked us off through an entry gate into a park of fabulous content, a rich array of trees framed with beautiful flowers. The florid realm is surrounded with water and wherever possible there emerges a flood of blossoms which bloom the whole year, summer and winter. For a long time, but much too short for me, we traversed the park, which also offered a good view of Mount Diablo and Mount Hamilton. Finally we reached the main residence. It was built in a Mexican-Portuguese style with a group of buildings around an enclosed courtyard sealed with a heavy iron gate; evidently it was a kind of fortress. There was an antique marble well in the middle of the courtyard, which the owner had bought herself in Verona and had shipped to its current location near the Pacific Ocean. Indeed, the whole country villa had been named for the well: "Hazienda[22] del Pozzo di Verona".

While riding in the vehicle, my neighbor explained to me that the owner had employed a German architect with the name of Schweinfurt who had studied many old Spanish and Portuguese buildings in Mexico before attempting this one. I remarked that he must have had good taste, but received the response "Yes, that's what killed him." I asked: "How did that happen?" Whereupon he said: "He loved good taste so much that he kept drinking California wine until he died." These Californians have a dreadfully droll idea about their admittedly rather strong wine. But at the end I got to rather like it. I will also die one day and have to stop drinking, so that I too may "guzzle 'til gone".

The inside of the hacienda is a small treasure chest full of the most magnificent artistic works and rarities which the lady who possesses them has bought from all areas of the Old and New World. It was the most original [*originellste*] mixture of Greek, Roman, Medieval, Mexican, Chinese, Japanese, and Indian curiosities [*Seltenheiten*].

As the only European present, I sat on the right of Mrs. Hearst at the dinner table. The first course was blackberries. "[No] thank you." Then came a melon which the lady of the house made more appetizing by sprinkling it with table salt. Again "[No] thank you".[23] Then came an indescribable paste made from oat-meal, which in Vienna might be used to fatten geese, but I doubt that true Viennese birds would touch the stuff. While the melon was being removed I had already noticed a somewhat uncharitable look from *alma mater*, and let me assure you that an *alma mater* takes special pride in her cooking. I therefore turned my face away, gagged on the food [*würgte*], and thanked God that no other human weakness fell from my flesh. That is the most disagreeable thing about invitations in America. In restaurants, one doesn't have to eat everything; one can leave food on the plate, but what can one do with a housewife who is proud of

[22] The current Italian spelling of "hacienda" seems to be "azienda".

[23] One of the differences between English and German is that when food is offered Americans normally say "Thank you" if they want it, whereas Germans say "Bitte". In this context, when a German says "Danke schön" or "Danke", he normally means *No thank you*. When Boltzmann said "Thank you" it probably confused the hostess, since he actually meant "No thank you".

American cooking and particularly her own? Fortunately, then came chicken, preserves, and other food with good taste which helped my shocked tongue recover.

After dinner we went into the music room, a space if I reckon rightly almost as large as the *Bösendorfer Saal* [for pianos] in Vienna, but what fantastic Baroque ornamentation! I know of none of the smaller Viennese concerthalls with which to compare its beauty. Favorable reports of my miserable piano playing had penetrated far into the hacienda, hence I was requested to open the concert. After some reluctance, I sat down before the grand piano, a Steinway at the very highest price range. Without any inkling or suspicion I touched the keys and pressed them. I had perhaps heard concerts before with such magnificent sound, but I had never used my own fingers to produce it. If I had sometimes rued the rigors of traveling to California, from now on no more.

I played a sonata by Schubert. At the start, the mechanics were somewhat strange, but how easily one accustoms oneself to what is truly good! But already by the second part of the first movement things went well, and in the second movement, an andante, I forgot myself as if it was not I who played the melody but the melody which directed my fingers. I had to forcibly restrain myself from attempting the Allegro, which was good, because there my technique would have fallen short.

After me, a female student of [Heinrich] Barth[24] in Berlin played with as much skill as musical understanding. Among those present too was a professor of music from Milwaukee with a very manly, martial exterior, who would surely have made a magnificent grizzly hunter. He had been solidly exposed to fine sounds, and like the lady above even pursued his studies on the piano with Barth, but without catching this particular bear. He knew that Beethoven had written nine symphonies and that the last was the Ninth. He even paid me the undeserved compliment during a debate over whether music could occasionally be humorous by requesting me to play the Scherzo [literally joke] from that work. Should I admit that I couldn't do it [because it doesn't exist] to a professor from Milwaukee? Choosing my words, I responded humorously to his comical suggestion by saying: "Yes, of course, if you will join me on the kettle drums. It will add substance."

That night I slept in the hacienda in a wonderful chamber with a bathroom attached, and there was even a Moor who served me and also polished boots. There was a picture of a protecting angel hanging exactly over my bed. I especially love art when it has a sharp point. Ah, what I could learn from a beautiful picture of The Battle of Abukir[25] as a soldier in a barracks

---

[24] Heinrich Barth was a noted turn-of-the-century piano teacher in Berlin. For an interesting account of his character and teaching methods as seen through the eyes of perhaps his greatest student, read: Arthur Rubenstein, *My Young years,* Popular Library: New York, 1973. Rubenstein also discusses how Barth normally treated budding American female pianists.

[25] Of the three battles of Abukir, Boltzmann is probably referring to Napoleon's defeat of a Turkish army on July 25, 1799 which had just landed from ships offshore. For the other two battles (1798 & 1801), which were English victories over the French, but not over Na-

[looking at needle-sharp bayonets]? But a protecting angel over one's head in a resting place certainly voices the wish of the housewife that I enjoy a good sleep. Yes, I am superstitious. It was precisely at that time that I suffered most severely from asthma and was already thinking about leaving the hacienda, but my protecting angel consoled me and it was she who broke the viciousness of the attack that night.

During the next few days there was no end to the remarkable things we examined in house and patio, field and wood. Among other things, we were shown giant Live Oaks with immense, wide-spreading boughs. A whole hut had been built in these branches with a ladder leading up to it. On the top of it was a hunting turret. I left that afternoon and arrived back in Berkeley by nightfall in order to be able to appear in my lecturehall on time the next day.

The following Saturday and Sunday was devoted to the Lick Observatory. Already on Friday I had gone part of the way by traveling to the friendly little city of San Jose where some streets were palm-bordered boulevards. Not only could one walk under palms but squeeze them on bicycles, automobiles, and tramcars. At seven o'clock the next morning in a somewhat defective post coach I set out for Mount Hamilton, a high hill about the same elevation as The Semmering, but it seemed taller since the starting point of the trip was only slightly above sea level. The road was very good and wound slowly upwards in measured serpentine fashion past fruit and wine gardens as well as forests and meadows. But the latter at this time of year only offered sun-bleached hay. Cows ate fresh grass in winter and hay in summer.

My coachman was a course, grumpy old man, who was also an official of the postal service. As soon as we left the inn he sorted the letters into different sacks which lay at our feet while mumbling and muttering the whole time. Soon we were outside the city. In front of the door of a large, well-fenced property, a beautiful dog greeted us with cheerful barking. My coachman put a couple of letters inside a rolled newspaper and skillfully threw it into the mongrel's jaws who at once secured it under his teeth and was off. This kind of mail service was repeated at several addresses. At others a big stick is stuck into the ground with a nail hanging out. Without stopping, my coachman lifts the outgoing bag and leaves the incoming bag hanging. Only twice were people waiting, and this was for baskets of food or other big packages. Both servants were maids but my knowledge of anthropology was too limited to guess their race. We changed horses twice and lunched once (but one shouldn't ask me how).

At about one-thirty we arrived at the observatory. Only the younger astronomers led by Dr. [Robert] Tucker were present, since Director [William Wallace] Campbell[26] with the others was already in Spain preparing themselves for the total eclipse of the Sun. Nevertheless, since at

---

poleon, see the 1963 edition of the *Encyclopedia Britannica* under "Abu Qir" (which strangely does not mention the Napoleonic or French victory at all).

26 William Wallace Campbell (1862-1938) directed Lick observatory from 1891 to 1930 and wrote several books on astronomy including *Elements of Practical Astronomy* (1899).

that time I also still intended to go there to see it for myself I asked Dr. Tucker where the eclipse would be observed. He answered: "In Daroca-Ateca-Almazan". Somewhat startled, I involuntarily replied that this must be a Spanish village [i.e. Greek to me]. But he casually countered, it was indeed a Spanish village northeast of Madrid. But that was little geographical help and the name of the tiny place also sounded rather odd. Spain was already becoming too *spanisch* [troublesome] for me.

I was then shown all of the equipment of the magnificently organized observatory, which because of its favorable location could be used in the most fruitful ways. The best feature is the giant telescope itself which was polished by Alvan Clark[27] with its twenty-eight [i.e. thirty-six] inch lens (which is simply called "the big glass"). It was used for one of the most interesting discoveries of recent times, namely, the two moons of Mars. Inside the giant pillar which carries the telescope, Mister [James] Lick is buried, who had the giant observatory constructed from his own private means.[28] Could anything be more idealistic? I think I know his mind. He certainly knew that it was all the same to him where his bones rested; but the world wanted visible testimony of the final goal of a millionaire. Verily, he had bought eternal rest with lucre.

If I were a poet, I would depict under the title "Two Idealists" the meeting of Schiller and the philanthropist in Heaven. Schiller would have Wisdom say to Wealth: "I don't need you!" And Lick would demonstrate the opposite. Certainly, the emotional bonus which money can buy is only 2nd rate, and love for money not even 3rd, but one may also receive a Steinway grand piano for money, an Amati fiddle,[29] a Böcklin [landscape],[30] and now even immortal fame.

At this point I would like to tell a true story which links Idealism with money grubbing. The American physicist Rowland[31] had given a speech that

---

[27] Alvan Clark (1832-1897) was a well-known telescope maker who had also directed construction of the Yerkes Observatory with its 40 inch telescope which had been able to discover that Sirius had a companion.

[28] James Lick (1796-1876) was a strange wanderer, living for many years in various South American countries but showing up in San Francisco during the Gold Rush with enough money to buy up important city land. This soon made him truly wealthy. "Lick House" was long a landmark in the city. Though changing his will several times, a large part of it went to build the observatory which meant so much to him.

[29] The Amati family is famous among musicians for their work on the violin during the 16th and 17th centuries in Cremona in Italy. As improved by Stradivari (1644-1737) who also worked in Cremona, their violin has often been considered the best in the world.

[30] Arnold Böcklin (1827-1901) was a German artist who liked to paint landscapes with mythical figures from Ancient Greece but often in a rather brooding, unclassical way. He was quite popular in some circles for a time. Boltzmann may have become acquainted with his work in Munich during his stay in the early 1890's.

[31] Henry Rowland (1848-1901), who had studied under Helmholtz, taught at Johns Hopkins from 1875 until his death, and for a time was probably the best-known American physicist. Realizing the importance of accurate refraction gratings for spectrum analysis, he increased the number of gratings even up as high as 20, 000 per inch. His gratings became used in laboratories around the world and helped make the modern study of spectroscopy possible.

teachers and scholars should not strive for the acquisition of money. A year later he fell sick, had a doctor examine him, and learned that at most he had three years to live. He had a wife and four vulnerable children. In the conflict of duties, family love triumphed. He invented a telegraphic typewriter [*Typendrucktelegraph*] which he patented, and which on his death which soon followed provided his wife and family with a fortune of $200, 000, that is, about a million [Austrian] Kronen. But by doing that he exposed the main point of his speech as a lie. But dear reader, do you know what I most admire about Rowland? That he was able to develop such a profitable invention so quickly as if he had had it ready in his mind all along. When he reaches Heaven, may he shake Schiller's hand with honor!

The [typical] American businessman is certainly realistic in how he acquires money. For example, when I discussed the purpose of my trip with a very prudent businessman, he absolutely could not conceive why I would visit San Francisco if the money made available to me only covered travel expenses to Berkeley.

But back to higher things. I stood lost in thought before Lick's column and the giant Lick telescope on top of it, which carried in its belly the bones of citizen Lick. I then wandered through all the rooms of the observatory. The observation results have been well-organized and put to good use here, [licked into shape as it were]. Every region of the sky has its own chest and every star its own drawer so that observations can be found again at once and made ready for future use. The content of the shelves is growing rapidly. It is no wonder that with so much work in this mountain loneliness that time passes quickly; on top of that it goes without saying that pretty lady astronomers are also present.

After having seen Mars that evening through The Big Glass as large and well-lit, almost like the face of the moon, I departed in the night for the valley. Remarkably, we could observe the contours of the ground fog. The starry heaven was above and a flat ocean of mist below. With a sudden movement the waggon entered the fog, the stars disappeared, and the forward lantern could only penetrate a few feet in front.

I used the next day (Sunday) to look around San Jose a bit more, but returned to Berkeley in good time that afternoon in order to have abundant opportunity for a long sleep and preparation for my lecture on Monday.

My excursion on the following Sunday to the beaches of Monterrey, Pacific Grove, and Santa Cruz was just as interesting. I had already been to San Francisco and traveled far enough to abe able to see and enjoy the Great Pacific Ocean [by Seal Rocks?], but now I had an opportunity to admire a broader expanse of rocky coast as well as the play of the waves. But a small house in Pacific Grove where Professor Löb [Jacques Loeb][32] had his laboratory was far more interesting to me than all of the above.

---

[32] Jacques Loeb (1859-1924) as already mentioned was a world famous physiologist who taught at the University of California from 1902 until 1910. He is perhaps best-known today for his work on tropism as if it applied to animals as well as plants, but in Boltzmann's time his fame was largely based on his discovery that environmental changes could fertilize frog

How great the difference still is between the giant works of industry and the workshops of science! How imposing and colossal the ocean steamers are! But if one travels often enough, it becomes clear that ships' officers, machinists, and sailors always carry out the same duties. In the areas reserved for passengers, one always hears the same people say the same things lounging on the same deckchairs or on the same upper deck shuffling the same boards toward the same painted triangle [in the same pallid amusement unoriginally called "shuffleboard"]. Massive muted matter but no new notions!

In science, however, while much is accomplished by mass means (as we saw at the Lick Observatory) what is truly great will always be produced by modestly-sized methods (though certainly our Minister of Education *darf das nicht hören* ).

It must be great to help mobilize millions of the best men of a great people, win battles at the head of hundreds of thousands. But it seems to this observer still greater to discover truths in modest rooms with modest means which will remain fundaments of our knowledge when the memories of those battles are only historically preserved with difficulty. What survives today from the whole Greek and Roman people which is still fresh as life, richer, and more powerful than ever? The fighters at Marathon have been outdone at Vionville[33] and Lioyang.[34] The people who read Homer or Sophocles for pleasure are dying out.[35] But the Pythagorean theorem and the principle of Archimedes are truly immortal.

These are my general views. How far they apply to Pacific Grove only the future can determine. The [scientific] discovery made there [by Loeb] had already caused me great embarrasment years before while it was still quite new. Full of enthusiastic fire I discussed the matter in a society of persons, not suspecting that something so purely factual where there was no intent of stimulating sexual lust, indeed, it could not, could be obscene. My first hint that something was wrong came with the rather conspicuous

---

eggs withhout input from a male frog, which suggested that maybe human males were superfluous. See Boltzmann's account of this shortly. It is possible that Boltzmann on this visit also met Loeb's son Leonard who was about fourteen at the time. The son would become interested in Boltzmann's work and become a noted physicist in his own right.

33 This battle between the Prussians and the French was fought on August 16, 1870. While it was a Prussian victory, most military commentators believe that the French almost certainly would have won had Marshal Bazaine brought up his available reinforcements promptly.

34 The city of Liaoyang is a Manchurian rail center between Mukden and Port Arthur with another rail line leading to Korea. The Japanese fought the Russians in a battle for this strategically located city which started on August 28, 1904. Though the Japanese suffered the heavier losses and came very close to disaster, the subsequent Russian decision to withdraw from the field toward Mukden gave the impression to many observers that it was a Japanese victory. In wartime, however, perception is often more important than reality, and it is often only after the war is over that the actual facts about what happened can be determined.

35 At the end of the 20th century, almost a hundred years after Boltzmann's death, it seems probable that more people still read Homer for pleasure (at least in translation) than study physics with or without pleasure, hence Boltzmann was very likely wrong in the "short" run. But outside of Europe and America or in the future Boltzmann could be correct.

departure of the lady neighboring me at the table. Afterwards, the same delicate creature sang a very risqué song by Aletter,[36] which made it hard for me to avoid wondering how her ballad could be pure and my theme impure. "We don't understand your theme" said the lady, to which I automatically replied: "But you do Aletter". In short, that is the old hypocrisy while the temperance crusaders would like to add a new one. I will now try to present Loeb's investigation in such a way as to make it clear without stirring any such objections.

For a long time it was widely believed that all chemical connections which are characteristic of living organisms, so-called organic connections, could only be produced with the help of a special power, a vital force. Today, we know that numerous organic connections are formed by everyday chemical reactions without any trace of any trans-material force. But many people still believe that life itself is something very special, completely different from accompanying chemical processes, and that special characteristics which constitute life can never be brought into existence by something non-living. This view is still far from having been refuted by the investigations of Löb [Loeb], but it does add weight to the negative side of the scale.

There are known species of animals where the egg of the female can develop into a living organism without any aid from her masculine counterpart (parthenogensis). But working with sea urchins and sea stars, animal species where such never naturally occurs, he has shown [starting in1899] that an unfertilized egg with the help of carbonic acid, butyric acid, or propylessig acid can develop into what appears to be a normal living organism.

If one grasps how important this discovery is, that a process which had previously been seen as an exclusive result of a special vital force, can be produced by purely chemical activity, and if the same also holds for higher forms of life even for man himself, then what social upheavals could well be in store! Such as an emancipation of women which current advocates can only dream! Man simply becomes superfluous. A small bottle with skillfully mixed chemicals replaces him entirely. Also, rational inheritance can be carried out much more easily than now, where so many accidental factors intervene. Nor will it be long before we learn which mixtures produce boys and which girls, and since the first will be completely superfluous, only a few exemplars will be produced, namely for zoological gardens. This could also make drinking wine unnecessary.

From Pacific Grove I then traveled to the seaside resort of Santa Cruz, which is worthy of mention because of the large number of bathers who appear to be living not in houses but in linen tents with linen windows which they rent. Other people live in small wooden dwellings which are built on

[36] It has proved difficult to identify any composer having this name. It is possible that the actual spelling is somewhat different.

ships and can soon be transported by them to shallow bays and to the mouths of rivers which run into the sea.

In general, it is surprising how small the houses are everywhere as well as the fact that they are made almost exclusively out of wood. In Berkeley there are many houses which remind one of the wood-whittling farmers who live around Graz [where Boltzmann taught for many years] who sculpture and make wooden boards to decorate their small homes.

After that, I saw the so-called "big trees", whose age one counts in the thousands. I was shown one which was no longer living but whose stump had rings which one can count exactly year by year. But I lost count and forgot its full age.

The rest of the week outside of Saturday and Sunday were devoted to work but in no sense without entertainment. There was lots of socializing, including a couple of celebrations. My colleague for one of these gatherings who had come to fetch me had already warned me with genuine English solicitude that I would have to put on formal wear. As he entered my room, I called out to him: "Am I not beautiful?" But no, how dreadful! I had forgotten to polish my boots. But my colleague had the answer. He led me into a subterranean room, put coat, vest, and collar aside, found the required instruments in a cabinet and polished the boots himself with virtuosity. On top of that, he drank out of the glass which he had earlier used to put water on the polish. How American it seemed!

Nor were ladies lacking in social circles. Wives of Berkeley professors gathered around the newcomer. But more, there was the hostess with her charming daughters, one of whom bewitched us with song, and various friends. Given such fair attention, I often fall into a form of unfathomable poetic illness called inspiritus [*Dichteritus*]. As with other ailments, I longed to conceive its severity, and here is the product of my search:

To My Wife

Should I ne'r discourse with ladies in foreign lands
When you are loved most dear, most near, all the year?
Must I ne'r kiss thy photóed face nor hold it in my hands
When thy image, so like thee, becomes thee, is thee?
Yes, I am theory-bound, and toe to brain, theory-crowned,
But kissing can be theory too, so trust me, I am kissing you!

Therefore apologies from all sides and also because of kissing. It was nothing more than poetic obligation.[37] I would like to see more than mere music, conversation, tennisplaying, and walks together. Let us create fine-finished verse!

---

[37] Aesthetic kissing was apparently the rage in Vienna in those days. One need only recall the paintings of Gustav Klimt and other *Jugendstil* artists.

By the way, the women of California are remarkably large and strongly built, and often there is also a rather unwanted wisp of mustache as if there were the beginnings of a beard. I also had to honestly agree with a colleague when he asked: "Don't you find that American women have something masculine about them?" But he was reluctant to accept my answer "and the men somewhat feminine?" But by the last I only meant their unbearded chins. In terms of courage, strength of will, adventurous spirit, and constancy of character they hold their jaws at manly height.

Among the things which helped divert the pattern of working days was the visit of [William Howard Taft] the American Secretary of War who was passing through San Francisco to the [recently acquired and pacified] Philippines accompanied it is said by Miss [Alice] Roosevelt [the beautiful daughter of then President Teddy Roosevelt], but I never got to see her. The War Minister [who would become the next president of the United States] had been invited to a large gathering of people in the large Live-Oak Forest on the grounds of the University of Berkeley. You should have heard the naive bluntness, the bold talk, and the enthusiastic nature of the speeches! Only one example. After a few opening words, the Mayor of Berkeley introduced the Secretary of War as follows: "This is Mr. Taft! A good War Minister, a good citizen, and in every respect a jolly good fellow." But it sounds far better in English.

Yes, America will accomplish great things in the future. I believe in these people, even after seeing them integrating and differentiating in my seminar in theoretical physics where they were clearly not at their best. They were about as prepared [in this subject] as I was to jump over ditches or climb the many hills which one passes by or over at Berkeley University.

Finally the evening came to decipher the monotone of my bespectacled waitress for the last time. When I had cut my last omelette, my colleague stared at the number of pieces with falcon eyes and said: "You have exactly half-a-minute to eat each one." After hurrying to catch the train, I was rushed straight north (two nights on the train) with the first stop at Portland. Though attracted by an exhibition there, I left at once for Livingstone (again two nights on board). The trip was wonderfully beautiful, that is, if it had only been daytime the whole way! The most magnificent sight was Mount Shasta with its tall snowcovered peak and the sub-tropical vegetation at its base. Some lakes which I drove by were mountain-ringed and forest-fringed and made the Gmunden and Atter lakes [near Salzburg] seem insignificant. One sees here no house on the bank and I don't even know if they all have names. I won't say anything about Yellowstone Park. It is a wonder, which it would be hard to find anywhere else in the world. One rereads it in Baedeker or contemplates good pictures or best of all one sees it in its natural surroundings if one has sufficient time, money, and good humor. But none of these are enough for me. If I could, I would go at the beginning of June when the heat is not so great and devote 14 days to it or better a month, so that one can see everything in a leisurely way and pass from astonishment to enjoyment.

As it turned out, I had sent too much of my luggage on ahead. I would have to spend four more nights on the train with my capacity for enjoyment exhausted; nor did I have any more fresh clothes. I was both mentally and physically, so to say, cleaned out. On top of that there was the terrible [August] heat. I always had a dry towel in my hand, which fortunately, one receives without limit on American trains in order to wipe away the sweat. I now understand what the expression "sweatshirt" means. Furthermore, the Americans love to hermetically seal their train cars, not to keep out draughts which they ignore, but soot [from the soft coal used to fire the engines]. On the stretch of track I was taking there were none of the beautiful observation cars which being so far back have less of that problem. For a long time I left my window open in my section of the forward part of the train, but [my face] became so black that I would not wonder if in future centuries some scholar might not raise the hypothesis that the reason negroes were black was because they had been so extensively used as railway personnel.

Now my stomach catarrh began acting up again. One indeed receives wine in the dining car, but only unwillingly after the meal, provided most of the guests, especially the ladies, had already departed. The first thing one gets at food time is a glass of icewater and a scrap of paper on which one writes what one wants to eat. Then an eternity passes until someone picks it up and takes it to where he thinks it belongs while one sits with parched palate staring at the icewater. Dazed and dumb with drought (which would have killed an angel) I finally drank the poison.

Indeed, one day it even came to pass that I was given no wine at all. The "explanation" was that whole state of North dakota was abstinent which meant that while the train passed through it the passengers would remain as dry as the wheat fields around us. I protested. What did I care about North Dakota? I only wanted to see lush Vienna. And here for all I could tell I was traveling through nowhere *wo der Pfeffer wächst*. "Oh pepper, a lot is raised in this region" someone volunteered. Oh the devil take it! Even my deadliest curse, who could understand it? To be sure, I finally got my wine by tipping the porter and adding a lot more under the table, nor was it allowed to be listed on the "bill of fare".

The American railway officials customarily add a new waggon only when the others are full, and in spite of that the trains are of enormous length. The individual cars have names like ships, since otherwise they would be almost impossible to locate. I passed through the series of waggons named *Jesabel, Pembina,* and *Vernedel,* and they were crawling with people of all nationalities and races who in the great heat were often half-dressed. One baby was completely naked while sitting on a velvet chair and reminded me of a picture of the Christ Child, who certainly had certainly never rested on a Pullman car. I wanted to say something flattering to the lady, but how ordinary everything sounded in English: "Äs de tscheild Tschises Christ!"

It was fortunate that I was not born English. I would never have taken a bride home. One can imagine that I felt somewhat embarrassed during my declaration of love. "Look as resigned as when you enter the lecture hall" said Mephistopheles [in Goethe's *Faust*]. Indeed, if it were only as easy as entering the room where I gave instruction and confess my feelings to Lady Physics and Metaphysics! But before me stands a beautiful demure maiden, and how difficult my speech becomes. Yet, as I reach the *punktum saliens* the wisdom and good taste of our ancesters helps me for they found the word with both the highest feeling and the most moving tone, the word *Liebe*. As with the Steinway grand piano about music, so it is not I who voice the words, but the words which lead me on until I win full success. But if I would have to say in terms of the genuine English language: "Ei lowff ju", my chosen [bride] would have run away like the chickens before the heavy breathing Styrian farmer who had tried to attract them with a goiter on his throat.

The whole colorful crowd of passengers must be packed into bed in the evening. The organization of the American sleeping car is as follows: There is a rather narrow pathway down the middle of each waggon. There are upholstered benches on both sides where two people sit. Each passenger receives such a seat by assignment. In the evening the space above the two-seat benches which are always facing each other on the same side of the aisle are transformed into two beds which are placed on top of each other with curtains facing the aisle. The direction of the beds is parallel to that of the movement of the train. One can indeed reserve both the upper and lower bunks to oneself - a whole compartment - but it costs twice as much. Since there are no drawers one must put the night wash, slippers, and what else one needs in a bag which the Moor places in the bed. Then one disrobes behind the curtain in the bed itself but the clothes and bag are also brought into the bed and then one tries to sleep without suffocating.

Every space has a small opening covered with a fine screen, which leads outside and becomes with hot weather so smoky and steamy that I preferred to sleep like the little baby in his birthday suit, which also saved me the trouble of always packing and unpacking clothes. One night I left the whole window open, but the next morning as I got out of bed the Moor addressed me as *Herr Kollege*.

In order not to lose my watch, wallet, eyeglasses, and other things I put them in my hat which I placed in a wide-meshed net which was the only place to put things in the bed area. But while the bed was being made the Moor always hung the hat way up high in an unreachable place and it was amusing how perplexed he was that I needed the hat in bed.

The critical period is when the benches are transformed into beds. One can no longer sit on the benches and the bed is not yet ready. I fled to the washroom, but there one passenger was brushing clouds of dust from his clothes and another was squirtung wash water in all directions. Then I tried to reach the salon car, which is lacking on many trains, but that meant I had to pass through seven or eight cars all with closed curtains on both sides of the aisle, but out of their seclusion soon an arm and later a leg and then even later a softer object would poke me in the side as I marched along. Furthermore, I kept stumbling over pieces of luggage left in the aisle.

Finally, I found a place in a compartment which still had untransformed benches. But on the other side there was already a bed whose curtain wiggled in constant motion. There was something about the eternal feminine. In my lecturehall in Vienna, when checking to determine which students have had their inscription books signed for classes, I always recognize at once the hands of every lady. Hence I felt sure that this particular curtain concealed a female limb, and soon it exposed itself by an incautious movement during undressing, and I saw that I had been correct.

I escaped many inconveniences with my custom of getting up early in the morning. Thus I was alone when first dressing and washing which reminded me of the words from [Prince Otto von] Bismarck which he in turn had quoted from Mozart's *Die Entführung aus dem Serail*: "To keep me back, try Dawn's crack."

Because of the heat and soot, stomach catarrh and thirst [in the Middle West in August] I had become so tired of the trip that I not only gave up my intention to see the eclipse of the sun [in Spain], but even making connections with the *Kaiser Wilhelm II* [in New York] if there was a faster way to take me home. But on top of everything else a telegrapher's strike broke out which put us six hours behind schedule. I fell into a rage [*Wut*]; but one should see the patience [*Phlegma*] of Americans! They look at the man who utters insults and oaths almost with commiseration as if they would give voice [in English] to what the good man was thinking [in German], which helped somewhat. In my case, the conductor simply said: "We don't want to risk an accident."

In Chicago, I only had twelve minutes to get from the Canal Street Union Station to the Nickelplate platform. Loaded with luggage I hurried aimless here and there. Two people I asked for information did not answer. A young lady noticed the situation and courteously asked me what I wanted. She herself did not provide the information; perhaps she did not even understand my question, but she did direct my attention to a policeman, which in my distress, which was so great, I had not even noticed. When I called after her not from flattery but from the depths of my heart "You are an angel", I first saw that trait for trait [*Zug für Zug*] she was the guardian angel of Pozzo di Verona. Was belief in my guardian angel accepting a fairy tale? How well did angelic behavior fit the Halls of the Canal Street Station in Chicago? From guardian angel to police guardian I quickly obtained the vital information, so that I was able to reach the other train station in time.

Another surprise awaited me in New York. The beautiful pier which had led from the railway to the ferry had burned down, and it became necessary while heavily laden with baggage to stumble over the burned-out pieces of wood.

But in spite of all obstacles I reached the *Kaiser Wilhelm II* on time. When I finally got all of my large and small pieces of luggage on board, oh how my heart beat with happiness!

Raise thy voice in cheerful song;
Remember thy family hearth.
Thy battered boat is still afloat
And returns from distant parts.

The trip back was favored by the most wonderful weather. The good ship's fare completely calmed my stomach. I drank no more drops of water, also not much beer, but so much the more noble *Rüdesheimer* wine. It really suits a sea voyage, since tottering on deck is blamed on the roll of the vessel.

Now only the brief traintrip from Bremen to Vienna, a dashing drive in a Viennese *Fiaker*, and I'm back! Yes, such a lengthy voyage has much that is interesting and magnificent. California is beautiful, Mount Shasta is grand, and Yellowstone wonderful, but by far the most beautiful on the whole trip is reentering the door of my home.

## B. Some Stories about Boltzmann

(Dieter Flamm is a grandson of Ludwig Boltzmann and like his grandfather is a professor of physics at the University of Vienna. He has offered the following information.)

> Boltzmann ... constructed an electric sewing machine for his wife, who made dresses for the whole family. Boltzmann was fond of his children. Once his youngest daughter, my mother, desired to have a small monkey. But his wife disliked animals in the house. As a compromise Boltzmann bought rabbits for his daughter and installed a cage for them in his own library. The room of his youngest daughter was next to his study and in the evening he used to tap at her door as a sign of friendship.[1]
>
> He used to rise early in the morning. In his later years he often worked already at five o'clock in the morning. Nevertheless, he enjoyed parties until late at night. [Indeed,] there were regular parties in Boltzmann's house to which he also invited his graduate students. He liked social life. At parties he was a favorite guest since he used to entertain the people with his splendid humor. At such occasions he even wrote funny poems. One of these is "Beethoven in Heaven".[2]

---

[1] Dieter Flamm, "Life and Personality of Ludwig Boltzmann" in *The Boltzmann Equation - Theory and Applications*, edited by E.G.D. Cohen and W. Thirring, Springer Verlag: Vienna & New York, 1973, p. 15.

[2] Ibid., p. 16. For the poem see Appendix B.

(Engelbert Broda [1910-1983], for many years Professor of Physical Chemistry in the University of Vienna, wrote the first major biography of Boltzmann in 1955 and has collected a number of personal accounts from colleagues and students of the great physicist-philosopher)

> *Fritz Hasenöhrl*: Characteristic of the good teacher - at the elementary as well as at the university level - is ability to understand those who are learning, interest in their development, goodwill, and sympathy. In a word, the good teacher is characterized by a kind heart. Those were the personal traits that made Boltzmann a brilliant teacher and that assure him of the everlasting gratitude of his many students.[3]

> *Lise Meitner*: His relationship with the students was accented by its humanity. He considered not only the knowledge in physics of candidates who were, for instance, examined at the end of the semester but he also tried to appreciate their general characteristics. External forms meant nothing to him, and he didn't shrink from using words laden with sentiment. From time to time he invited home the few students who took part in his advanced seminar. At those times he played for us - he was a very good pianist - and told us about various personal experiences.[4]

> *Franz Streintz*: The relationship between the master and the research students [in Graz] was exceedingly lively and cordial. Boltzmann took the warmest interest in everyone's work. He could give good advice in every difficult situation; and in this connection never took it amiss when he was interrupted in his own work at home by someone seeking guidance out of some labyrinth in which he had suddenly fallen. The unpretentious great man made himself available to all for hours on end, always equipped with unending patience and delightful humor.[5]

(Here are some observations by a well-known mathematician who was a *Privatdozent* when Boltzmann was in Leipzig. The following was written down in his own later book *Bestand und Wandel - Meine Lebenserinnerungen,* Munich, 1950, p. 128 & 129.)

---

[3] See Engelbert Broda, *Ludwig Boltzmann*, Ox Bow Press: Woodbridge, Conn., 1983, p. 19. The original German quotation can be found in Fritz Hasenöhrl, *Bericht über der Errichtung eines Denkmals für L Boltzmann*, Vienna, 1913.

[4] Broda, ibid., p. 17. (Personal communication from L. Meitner to E. Broda.)

[5] Ibid., p. 19. See also F. Streintz, *Grazer Tagblatt*, February 20, 1904.

*Gerhard Kowalewski*: In the room for teachers in the *Augusteum* [in Leipzig] where the professors gathered during the pauses between classes, Boltzmann was always quite talkative. He made no distinction between full, associate, and assistant professors [*Ordinarien, Extraordinarien*, and *Privatdozenten*]. He especially enjoyed chatting with young teachers like us. A main thread of his being was unbounded human friendliness. Everytime he would have me tell him what I was lecturing on, and was so eager to comprehend the subject better that he sometimes took out pencil and paper. The most beautiful experience for him was to understand something. These conversations with Boltzmann have remained unforgettable.[6]

(Martin J. Klein[7] in his biography of Paul Ehrenfest has also related a number of stories about or connected with Boltzmann.)

Ehrenfest looked up Lise Meitner shortly after his return to Vienna from Göttingen, when he heard that she had a detailed set of notes on all of Boltzmann's lectures. He proposed that they study Boltzmann's ideas together, as well as some of the background work by C.G. Jacobi[8] and others in analytic dynamics, a suggestion she quickly accepted since his gift for clarifying the ideas of theoretical physics was already quite evident. Their joint study was a great stimulus to her scientific development.[9]

Another fellow student of those years was Felix Ehrenhaft, a Viennese contemporary of Paul Ehrenfest's. Apart from the similarity of their names the two Viennese physicists had little in common. Felix Ehrenhaft became professor of experimental physics in Vienna, and is best known for his long and eventually bitter controversy with Robert A. Millikan over the existence of a natural unit of electric charge. Millikan showed conclusively that Ehrenhaft's 'sub-electron' did not exist.[10]

---

[6] Included in A. Dick and G. Kerber, *Ludwig Boltzmann - Katalog zur Ausstellung an der Zentralbibliothek für Physik in Wien*, Vienna, 1982, p. 29.

[7] Martin J. Klein received his PHD in theoretical physics from MIT in 1948 and has been Professor of the History of Physics at Yale University for many years. He has contributed extensively to recent scholarship on Boltzmann, Planck, and Einstein.

[8] Carl Gustav Jacobi (1804-1851) received his PHD in Berlin in 1825 and in 1827 became a professor at the University of Königsberg. He is known as an outstanding mathematician, doing work on partial fractions, elliptic functions, differential equations, theory of the last multiplier, theory of determinants, and Abelian transcendents,

[9] Martin J. Klein, *Paul Ehrenfest*, Volume I, North-Holland: Amsterdam, 1970, p. 49.

[10] Ibid.

> Not surprisingly, the names of Ehrenhaft and Ehrenfest were often confused: even Boltzmann is reported to have asked Ehrenfest, after the latter had made some clever observation, to use his cleverness to devise a way for Boltzmann to tell Ehrenfest and Ehrenhaft apart.[11]

(A. Dick and G. Kerber in their very useful work, *Ludwig Boltzmann - Katalog der Zentralbibliothek für Physik in Wien,* Vienna, 1982, have given us further stories, starting of course with still another tale about Ehrenhaft and Ehrenfest)

> *E. Hlawka*: Ludwig Boltzmann was a decided opponent of vector analysis. He always said that mathematical calculation did not have to be elegant, that elegance should be left to tailors and shoemakers. Nowadays, vector analysis plays a big role in the schools, with scalar and vector products. But Boltzmann told us in his lectures: "Before I learn to distinguish between scalar and vector products I will tackle an even tougher problem first: How to distinguish Ehrenhaft from Ehrenfest."[12]

> *Karl Przibram:* He still remembers after sixty years a tavern sheet [*Kneipzeitung*] with a distich presumably inspired by one of the very many listeners to Boltzmann's lectures on natural philosophy, especially about curved space:

> When normal people step on worms
> Bending is how they come to terms,
> But Boltzmann when he ties his lace
> Stands and eyes bending space.[13]

> *Hans Thirring*: Albert Einstein on the occasion of the marriage between one of Boltzmann's daughters and a former student of the bearded physicist said: "Now one knows which physicist loves Boltzmann's works best."[14]

---

[11] Ibid.

[12] E. Hlawka, (unpublished) *Autobiographie,* p.9 in A. Dick and G. Kerber, *Ludwig Boltzmann - Katalog zur Aussstellung an der Zentralbibliothek für Physik in Wien,* Vienna, 1982, p. 10A. That Boltzmann may really have had difficulty is indicated by the following comments from Martin Klein, op. cit., p. 49, footnote 33: "I might add that the typed copy of Boltzmann's opinion of Ehren*fest's* thesis which was sent to me from Vienna actually refers in one place to Paul Ehren*haft*. I am not sure whether the slip was Boltzmann's in 1904 or the typist's in 1963, but in either case it shows how easily the names are confused."

[13] Ibid., p. 38A.

[14] Ibid., p.32A (See also H. Thirring, *Almanach der Österr. Akad. d. Wiss. in Wien,* **115** (1965), p. 394.) The bride was Elsa Boltzmann (1891-1965) and the groom Ludwig Flamm (1885-1964).

(Since it would impoverish history to fully suppress either Boltzmann's personal life or his jokes, either those by or about him, let us continue. Perhaps the most unrestrained description and humor - apart from Boltzmann's own jokes - has come from his former student and unconditional admirer, Dr. Eleonore (Lily) Minor, later wife of the mathematician and member of the Vienna Circle, Hans Hahn)

> Though born in 1885 she still remembers (1967) her first visit to a Boltzmann lecture (1904). Out of curiosity she wandered into the classroom, but couldn't follow his mathematical reasoning until Boltzmann, himself, stood back, puzzled. Then it came to him. He discovered a variable with the power of "3" instead of the desired "2". He reacted humorously, whetted his finger, and went back and wiped away each "3" all across all three blackboards until he discovered the place of his mistake. Fräulein Minor, amused and informed, soon attended his lectures on a regular basis.[15]
>
> He had three daughters, gave special dances and parties for them, played musical instruments with them, and had box seats for his whole family at the Vienna Opera House and at the Burgtheater. Boltzmann had studied piano under Anton Bruckner and continued to take lessons from a tutor until his death.[16]
>
> At parties he loved to dance but because of his bulging physique would have to hold his partner at arms length. When it was lady's choice, Lily Minor liked to choose him.[17]
>
> He was a member of a bowling group, but tended to follow the ball down the alley. A student once commented on his bowling, and Boltzmann replied, "Careful now, you may talk about my lectures all you want but not about my bowling".[18]
>
> One Day Fräulein Minor was telling other students about some silly questions which came up in the adult school where she was teaching mathematics. She mentioned that one student even went so far as to ask her a question as absurd as whether one should sleep with the window open or closed. Boltzmann,

---

[15] John T. Blackmore, *Ernst Mach - His Life, Work, and Influence,* vol. **2**, (University of California Doctoral Dissertation), Los Angeles, 1970, p. 766-767. (based on private conversation with Dr. Hahn.)

[16] Ibid., p. 785.

[17] Ibid.

[18] Ibid.

who was standing nearby, overheard and remarked with evident interest, 'I have always wondered about that myself.'[19]

Boltzmann once wore a pair of shoes so long that his wife nagged him to buy a new pair, but he refused. Then one day someone told him it was strange that when a container develops a leak the water flows *out*, but when a shoe has a hole or tear the water seeps *in*. The old physicist reflected on this and the next day bought a new pair of shoes.[20]

Ludwig Boltzmann decided to attend a conference at the St. Louis World's Fair in the summer of 1904. On arriving in America the stout, short-sighted professor sent his baggage on ahead, but somehow it was delayed. As a result he had no fresh clothes to change into in St. Louis and looked even shabbier and more bedraggled than usual; so much so, that an official mistook him and asked him to pick up the luggage for the famous professor Boltzmann. From deep within his throat a voice burst out: 'But I am Boltzmann!'[21]

One day, Professor Boltzmann decided to buy a house. He was shown a houseplan, approved it, and then asked, 'How much?' On being given a figure he reached into his pocket and paid the man on the spot in cash. he was given the key. He then returned to where he and his wife were currently staying to tell her the good news. Boltzmann, face beaming with joy, entered the room and said: 'We have bought a new house!' His wife thought a moment and then innocently asked 'Where is it?' He slowly responded, 'Oh, I forgot to ask about that!'.[22]

Boltzmann's best remembered response to student riots - a constant problem at the University of Vienna - was to tell a story about farm animals. He told Fräulein Minor that when he was young he had been on a farm and noticed that the tails of some pigs curled to the left and others to the right, but it was never clear to him if the pigs with the left-curled tails grouped together against those with the right-curled ones or not.[23]

---

[19] Ibid., pp. 785-786.
[20] Ibid., p. 786.
[21] Ibid., p. 779.
[22] Ibid., vol. I, p. 325.
[23] Ibid., p. 786.

(Ferdinand G. Smekel recollects two Boltzmann stories in approximately the following terms.)[24]

> It is said that Boltzmann while concentrating on a problem once handed the letter he was carrying to the waitress behind the lunch counter, as if it were a post office.
>
> Few ceremonies are more memorable than receiving a doctor's diploma, but giving them out may also have its moments. The story is related that Boltzmann who was in charge of handing out the diplomas gradually became lost in thought, but when the ceremony was finished and the Rector of the University came up to congratulate him for a job well-done Boltzmann simply put a diploma in the Rector's outstretched hand as if he were just another student.

(A few stories which Jacques Loeb, then a UC Berkeley professor at Pacific Grove, told to his son L.B. Loeb* about Boltzmann's trip to California have also survived)

> On crossing the bay to San Francisco, Boltzmann insisted on taking an umbrella with him. He was told by several longtime residents that 'It never rained in San Francisco in summertime', but on this day it did, and he crowed: 'I told you so!'[25]
>
> He was served excellent food as a rule, but he could not adjust to oatmeal, which in Austria was only for farm animals. On the other hand, his American hosts were dismayed, though they tried to overlook it, when the well-rounded physicist spilled food from his mouth.[26]
>
> The Austrian professor prepared his lectures in German, but addressed his audience in English, unless he failed to find the right English expression, in which case he went ahead anyhow with sentences such as: "Dass ist the truth!"[27]

---

[24] Ferdinand G. Smekal, *Alma Universitas*, Verlag Eine Welt der Vereinten Nationen, 1967, page 254 as included in A. Dick and G. Kerber, *Ludwig Boltzmann - Katalog zur Ausstellung an der Zentralbibliothek Fur Physik in Wien*, Vienna, 1982, p. 18A.

* It is possible that Boltzmann's trip to Berkeley may have helped L.B. Loeb develop an interest in physics and kinetic theory. Loeb published a book on Boltzmann's favorite field titled *Kinetic Theory of Gases* in New York in 1922 and over thirty year's later *Basic Processes of Gaseous Electronics* (1955).

[25] Based on a private communication from L.B. Loeb to John Blackmore circa 1967.

[26] Ibid.

[27] Ibid. This story was first told by Elmer E. Hall to L.B. Loeb who then related it to John Blackmore (1967). For further mention of Hall see letters #110 and #111.

Ludwig Boltzmann's gravestone,
Zentral-Friedhof, Vienna
Courtesy of Setsuko Tanaka

# Chapter 8

# FINAL MONTHS & AFTERMATH

## #124

To Franz Brentano Vienna, January 16, 1906

Unfortunately, I am not very well at all at the moment, but precisely because of that I am in a hurry to answer your letter immediately, since knowing myself, if I don't do it now it will just wait unanswered for a long time.

Over the Christmas holidays I had a painful and very troublesome illness which kept me chained to my room for almost fourteen days. I have recovered from it, but a great mental depression has remained. How I envy you and your constant cheerfulness and satisfaction. You are truly a genuine philosopher. I am already sixty-two years old, but have won no peace of mind. In such times of melancholy I am downright pessimistic concerning my own philosophy, indeed, even about the very possibility of philosophical knowledge.

I have long had the intention to work out in written form not indeed my last views on metaphysics which I reported to you in Florence, but my thoughts on so-called knowledge [*Erkenntnis*] a priori. That all knowledge is to be equably chosen [*Dass alle Erkenntnis gleichmässig zu wählen*].[1] Only I now find that in carrying out my plan I have run into insuperable difficulties. So momentarily I have reached an unbelievably low level of mental energy.

With the heartiest greetings from me and my wife to you and your worthy family members...

[P.S.] Thank you for your New Year's
wishes, which I heartily reciprocate.

---

[1] Boltzmann last known lecture notes two weeks later (January 31, 1906) appear to rephrase this obscure point as follows: "One seeks probability from a priori probability. This only makes sense in equally possible cases." (See page 169.)

## #125

To Svabte Arrhenius Vienna, January 22 [and later], 1906

I am very pleased that you are in good health and that your new institute is making such satisfying progress. I would like to add to that my heartiest wishes for your happy marriage. Unfortunately, things are not going especially well with me. I am suffering fearfully from my old evil, neurasthenia. I definitely overexerted myself in California.[1] That was also the cause for not traveling to Spain to see the eclipse of the Sun, which I had previously decided to do.

I am very pleased that you have taken the trouble to work out the vicious circle by means of which Ostwald has derived the law of multiple proportions for chemical connections, so to say, from nothing. I didn't take the time to go into the matter in detail, but it was clear at once from reading Ostwald's essay that there had to be a logical circle in it somewhere, since the premises from which his conclusion would honestly follow were just not there.

Oh my dear colleague! To my horror I see that my letter which I began in January is still lying here. That is the most disagreeable consequence of my neurasthenia, that I have such a reluctance to finish anything or to send it off, even such a simple thing which is so dear to me as sending a letter to you. I must therefore send it to you at once, so that it is not again delayed.

At the same time, I am sending you a copy of my popular writings[2] which have now been collected together and published, and where also the paper on [Ostwald's] theory of happiness is included.

With the heartiest wishes for your happiness and the compliments from me and my wife to you and your dear wife, be she ever so unknown to me...

---

[1] Boltzmann may well have overexerted himself *after* his return from California too, given his hurried work on a long, technical essay for Klein's Encyclopedia, writing his article about El Dorado, and preparing *Populäre Schriften* for publication, not even counting his Fall lectures in physics plus those in philosophy plus his special "love and entropy" lecture. Indeed, it was Ernst Mach's opinion that overwork was largely responsible for his final depression and suicide. See *Die Zeit*, #1420, Abendblatt, Dec. 7, 1906 & *Neue Freie Presse*, #15,732, , Morgenblatt, Sept. 8, 1906. (Both are Viennese newspapers.).

[2] *Populäre Schriften*, Ambrosius Barth: Leipzig, 1905.

#126

To Stefan Meyer — Vienna, [undated][1]

Unfortunately, my asthma has become stronger again, so that I cannot lecture today or tomorrow. I am sending over all of my class attendance credits [*Belegscheine*][2] which I have. Perhaps it will be enough to send everything to the right official [*Questur*]. Otherwise, I will have to ask you to send out [or hand out] those which are most necessary yourself...

---

[1] A, Dick and G. Kerber in their *Ludwig Boltzmann - Katalog zur Ausstellung an der Zentralbibliothek für Physik in Wien*, Vienna, 1982, p. 48 list this letter as having been written in 1906, so we have done the same.

[2] It was common to give "Belegscheine" to students as proof of regular attendance and passing the course [i.e. with or without having to take and pass an examination for that particular class]. By collecting enough of them plus proof of legal enrollment to be found in their personal *Belegbuch*, the students became eligible for major examinations which would determine if they would receive a state certificate or a PHD. German and Austrian universities apparently did not normally offer bachelor's degrees and only rarely master's degrees at that time. But while obtaining the PHD degree was normally easier than it would normally become in the United States, one had to take a further examination to become

eligible to teach in a university on a permanent basis. This was called a *Habilitation*. There have, however, been some changes since Boltzmann's day.

#126A

Ludwig Flamm about Boltzmann [Concerning Early1906]

I, myself, as a student was able to hear the last lecture which Boltzmann held on theoretical physics; it was in the winter semester 1905/1906. A nervous complaint prevented him from continuing his teaching activity. Together with another student I took and passed my oral examination in his Villa in Währing. On leaving after the examination was over we heard from the front hall his heartrending groans.[1]

[1] L. Flamm, "Die Persönlichkeit Boltzmanns", *Wiener Chemiker Zeitung*, **47** (1944), 30.

#127

To Hugo Buchholz [Vienna], March 1906[1]

Please excuse me for answering both of your letters so late. I was away from Vienna for a long time, and am so unwell now that I cannot even hold classes. There was also considerable delay in receiving your letters.

I have absolutely nothing against you making whatever use you want of my lectures on theory of potentials for your book, if you provide brief mention of the source. Just the opposite. It pleases me very much that seeds I scattered at that time [lectures given in Munich and Leipzig] have finally borne fruit.

Unfortunately, I feel very sick and unhappy...

[1] Hugo Buchholz (1866-1921) included this letter and added the date to it on page vi of the foreword to *Das Mechanische Potential nach Vorlesungen von Ludwig Boltzmann bearbeitet und die Theorie der Figur der Erde*, Leipzig, 1908. This letter can also be found in A. Dick and G. Kerber, *Ludwig Boltzmann - Katalog zur Ausstellung an der Zentralbibliothek für Physik in Wien*, Vienna, 1982, p. 39-42 plus additional biographical material.

#127A

Mach about Boltzmann [Concerning Spring 1906]

Boltzmann had announced lectures for the spring semester, but had to cancel them, because of his nervous condition. In informed circles one knew that Boltzmann would most probably never be able to exercize his professorship again. One spoke of how necessary it was to keep him under medical surveillance, for he had already made earlier attempts at suicide.[1]

[1] Ernst Mach, "Ludwig Boltzmann", *Die Zeit*, #1420, Abendblatt, September 7, 1906, p. 1, column 3.

#127B

Höfler about Boltzmann [Concerning the Easter holidays 1906]

When I visited him during the Easter Holidays for the last time, he expressed his physical and mental suffering thus: 'I would never have believed that such an end was possible.'[1]

[1] Alois Höfler, "Ludwig Boltzmann als Mensch und als Philosoph", *Süddeutsche Monatshefte,* **3** (October 1906), 418.

#127C

Dick & Kerber about Boltzmann [Concerning April 23, 1906]

[Franz] Exner and [Ludwig] Boltzmann were responsible for judging the doctoral dissertation of Richard Dobner, but when the examination actually took place on April 23, 1906, [Viktor] von Lang substituted for Boltzmann.[1]

[1] Ibid., Dick & Kerber, p. 45. For a list of dissertations taken under Boltzmann see Dick & Kerber pp. 42-45 or Appendix A in this book.

#128

Pernter[1] to the Ministry of Education Vienna, May 5, 1906

Herr Hofrat Professor Doctor Ludwig Boltzmann is suffering from severe neurasthenia and according to the doctor in charge must stop every scientific activity . . . . The Institute for Theoretical Physics as well as the lectures are being taken care of by the Assistant and Privatdozent Dr. [Stefan] Meyer. Only the lectures on Natural Philosophy will be left without a teacher, but no pressing pedagogical need will be damaged by that.

[1] Josef Maria Pernter (1848-1908) was Dean of the Philosophy Faculty at the University of Vienna at that time. His main areas of research were geophysics and meteorology.

#128A

*Neue Freie Presse* on Boltzmann [about Events shortly after Easter]

On the advice of the doctor, Hofrat Boltzmann was brought to a clinic for nervous diseases in Munich shortly after Easter, but he would not suffer being there for long. After a few days he left and returned home.[1]

[1] Anon., "In der Wohnung des Hofrates Boltzmann", *Neue Freie Presse* (Vienna), #15102, September 7, 1906, page 3, column 3.

#128B

*Neue Freie Presse* on Boltzmann [about events in July and August 1906]

In spite of all the arts of loving persuasion by his family members and doctor, at first Boltzmann refused to even think about leaving home. Only with great difficulty was he able to decide to take a summer trip. The departure was repeatedly delayed. The family left Vienna around the middle of July.[1]

The three daughters and their mother accompanied Boltzmann. The two older daughters have devoted themselves to science. The eldest, Henriette, studied natural science and has received her doctor's degree. The second daughter is studying medicine. The third is only twelve years old.

The family first went to Schwannberg in Styria where they stayed for several weeks. It was only in the middle of August that they decided to travel south [toward Grabo on the Adriatic].[2]

---

[1] For easier reading and comprehension, the last sentence has been transposed from the beginning of the paragraph.

[2] Anon., *Neue freie Presse*, op. cit., page 3, column 3.

#128C

*Die Zeit* on Boltzmann [Concerning September 5, 1906]

He had brought his family to Duino, a small, beautiful seaside resort near Gradicka [on the Adriatic] to go swimming, yet he was also upset and nervous, because he was anxious to return to Vienna. His condition otherwise seemed better. [But] on the day of his death Boltzmann showed himself to be particularly disturbed [*erregt*]. While his wife and daughter went swimming, he carried out the deed.[1]

---

[1] Anonymous, "Weitere Nachrichten die letzten Tage", *Die Zeit*, #1420, Abendblatt, September 7, 1906, p. 2, col. 2.

#128D

Lily Hahn about Boltzmann [Concerning September 5 1906]

He was very melancholy for a long time and did not want to send his suit to be cleaned because it would mean further delay in returning to Vienna. After his wife left, taking the suit with her, he hanged himself.[1]

---

[1] Private communication from Frau Dr. Lily Hahn (Vienna, 1967).

#128E

*Die Zeit* about Boltzmann [Concerning September 5, 1906]

He hanged himself with a short cord from the crossbars of a window casement. His daughter was the first to discover the suicide.[1]

---

[1] Anonymous, *Die Zeit*, #1420, Abendblatt, September 7, 1906, p. 1, col. 3.

#129

Planck to Frau Boltzmann Gmund a/Tegersee, September 8, 1906

I was deeply shaken by the newspaper report of the sudden passing away of your husband, and I beg to be allowed to give expression to my inner feelings on the tragic end of this rich life, about which I pledge to you and your honored family my heartiest, warmest sympathy. What must the pitiable sick man have suffered, before he decided to take this final liberating step!

And what a blessed life so rich with successes has departed through him. I have always looked up to him as a model that one should strive after, even if one cannot reach his standard; for he was ahead of us all mainly in his love of truth with a genuinely inquiring mind in which he showed his conscien-tiousness in small as well as great tasks. Never will what he has done for our science be replaced or even approached.

To you most highly honored lady remains the consolation that you were able to give the best to this man that he ever received from his fellow human beings; for without you he would have gone through life in full loneliness. May this dark time leave no lasting hurt to your health behind.

With the very greatest and highest respect and honor, your true servant greets you...

#130

Seeliger[1] to Mach Munich, September 30, 1906

On returning to Munich I discover your friendly lines from September 16 and the 2nd edition of your *Erkenntnis und Irrtum* [2] which is naturally well-known to me. I thank you deeply for these rich gifts. It is of great value to possess your works including your essay on boundary shadows which I will read with great interest and profit.[3]

Boltzmann's tragic death has deeply affected us all. I was on very friendly terms with Boltzmann and both of our families were quite close to each other. For a long time we have thought that such an end was not improbable. But now that he has done it, we stand before what still seems inconceivable, in any case before an indescribably sad fact. With the expression of greatest respect...

[1] Hugo Ritter von Seeliger (1849-1924) was an astronomer at the University of Munich and Boltzmann's colleague there from 1890 until 1894.
[2] Ernst Mach, *Erkenntnis und Irrtum - Skizzen zur psychologie der forschung*, 2. Auflage, Leipzig, 1906. This is Mach's chief book on methodology of science and was largely derived from his philosophy lectures in Vienna circa 1895-1898.
[3] Presumably Ernst Mach's article: "Uber den Einfluss raumlich und zeitlich variierender Lichtreize auf die Gesichtswahrnehmung", *Sitzungsberichte der Kaiserl. Akad. Wien*, math.-naturw. Klasse, **115** (1906), Abt., 2A, pp. 633-648. The paper refers to what are now called "Mach Bands" which are inhibitory phenomena connected with the senses. Mach had discovered them in the 1860's but his papers were not noticed with the result that the phenomena had to be rediscovered around the turn of the century when Mach's priority was also discovered. This belated interest encouraged Mach to publish another article.

## #131

Jaumann to Mach [Brünn?], October 27, 1906

I take the liberty of openly disclosing to you that I would like to [occupy Boltzmann's chair and] become Professor of Theoretical Physics at the University of Vienna . . . .[1]

[1] Gustav Jaumann (1863-1924) had been Ernst Mach's longtime Assistant at the German University of Prague during the late 1880's and early 1890's finally accepting a post at the Technical University of Brünn which he retained until his death. Both Jaumann and Einstein published works in 1905 which caught the attention of other physicists and while neither man was able to replace Boltzmann in Vienna they both competed for a position in the German University of Prague which opened up in 1909-1910. With the help of Mach's recommendation and the fact that Jaumann unlike Einstein was Austrian, the position was officially offered to Jaumann but for unclear reasons, possibly financial, he rejected it. [See *The Collected Papers of Albert Einstein*, Vol. V, p. 256.] But while Einstein accepted it, he only stayed in Prague for less than two years and then went back to Switzerland.

## #131A

Meitner about Boltzmann's Institute [Concerning 1906-1907]

After Boltzmann's death, Planck was invited to take the post in Vienna - although he did not accept it - and I had occasion to see him when he came to Vienna to have a look around the Institute of Theoretical Physics. At that time, the Institute ... was located in a very primitive, converted apartment house in Türkenstrasse 3. The entrance looked like the entrance to a hen house, so that I often thought, 'If a fire breaks out here, very few of us will get out alive.'[1]

[1] Lise Meitner, "Looking Back", *Bulletin of the Atomic Scientists*, **20** (Novemeber 1964), p. 3. (Boltzmann's chair in physics went to his former student Friedrich Hasenöhrl who was born in 1874 and died fighting for Austria against Italy on the Tyrolean front in October 1915.)

For an interesting description of all three Türkenstrasse physical institutes, including those of Victor von Lang and Franz Exner as well as Boltzmann along with mention of many students who worked in them, see Karl Przibram, "Erinnerungen an ein altes physikalisches Institut", *Beiträge zur Physik und Chemie des 20 Jahrhunderts*, Verlag Vieweg: Braunschweig, circa 1960, pp. 1-6. (The current physical institutes on Boltzmanngasse were opened for use in 1913, though the Radium Institute next door had been opened as early as 1910 with Stephen Meyer as Director.)

#132

Frau Boltzmann to Brentano Vienna, November 28, 1906

Many, many thanks for your dear letter, and I will preserve in perpetual memory the wonderful profound poem which you have added alongside and will read it again and again. I have also repeatedly read with greatest interest your essay and hope that I may be allowed to keep it. I thank you very much for it. You have stimulated my belief in the immortality of the soul anew. What you say about the grounds for the existence of the soul is clearly represented. Although I do not have a background in philosophy, I think I have rather understood the treatise. But I am not clear why concept and judgment are heterogenous, since judgment is just a combination of concepts. Your beautiful idea that dimensional relations in God's world could be different than what is observed by the senses had previously been explained to me by my husband.

What you say in conclusion, which reduces the well-known arguments which you give at the start in favor of materialistic views to nothing, is a necessary consequence of the assumption of having one's own soul. . . . . How else could one otherwise hope to see one's loved ones again?

Since I do not believe that you possess a photograph of my husband, I am enclosing one. Please give my greetings to your respected wife and your son. I thank you again for sharing your sublime ideas with me which have stimulated me so much and given me so much to think about as well as the translation of the poem. With the expression of utmost respect...

#133

Frau Boltzmann to Lorentz Vienna, January 19, 1907

. . . . His discussions with you by mouth and pen Most Honored Professor belong to the most beautiful and happiest memories of his life . . . .[1]

[1] See Anne J. Kox, "The Correspondence between Boltzmann and H.A. Lorentz" in *Ludwig Boltzmann Gesamtausgabe*, Vol. VIII, edited by R. Sexl & J. Blackmore, Vieweg: Braunschweig, 1982, p. 82. The extant correspondence between the two scientists as listed in this article includes seventeen items and was from 1886 to 1897.

#134

Arthur Boltzmann to Hugo Buchholz Vienna, March 9, 1907

Regarding the lecture manuscripts of my dead father, I have taken a look at them but have only found the first beginning of the theory of elasticity worked out. But this prepared manuscript [*Druckmanuskript*] only includes the Introduction. I am astonshed myself how few manuscripts are at hand, but can remember that my father once told me, that he had preserved his first manuscripts carefully, but later had thrown away all typed and privately printed manuscripts [*gedruckte Manuskripte*] as well as proofs [*Korrekturen*]. Outside of what has already been mentioned no *Druckmanuskript* is at hand.[1]

[1] This letter can be found in Hugo Buchholz (ed), "Vorwort", *Ludwig Boltzmanns Vorlesungen über die Prinzipe der Mechanik, 3. Teil: Elastizitätstheorie und Hydrodynamik*, Barth: Leipzig, 1920, p. vii.

#135

Meyer to Sommerfeld Vienna, May 13, 1907

I have received your friendly letter of May 11, and since I have been entrusted with the Institute while Boltzmann's professorship remains unfilled I am the competent authority. I have known for some time that Boltzmann's model-bicycle belongs to you people in Munich [where Boltzmann taught from 1890 to 1894]. But I had hoped that you had already forgotten about it or that after more than ten years in Vienna that it would be regarded as our possession.

But now that you request to have it back, I will have it sent to you in the next few days and request on its arrival that you be friendly enough to send me Boltzmann's promissory note [*Schuldschein*] in return.[1]

I must remark, however, that two trips to America have not improved the operation of the instrument. Indeed, I have myself in its time straightened out the main axle, but completely straight it will never become again, so that the model does not always function free from problems.

With best regards...

[1] According to Boltzmann's letter to Professor Groth in Munich dated October 19, 1895, Boltzmann had received money not from normal laboratory funds or usual university sources but directly from the *Verwaltungsaussschusse* of Munich University to buy or make the machine in 1890 or 1891. After he left for Vienna, he asked Groth to send it to him, which Groth did. Boltzmann mentions in a later undated letter to Groth that he planned to pay for it when he could afford to do so. But apparently that never occurred or he forgot about it. The authorities in Munich, however, did not forget, and like Jean Valjean's nemesis, Inspecter Javert, were still on Boltzmann's trail eleven years later. The purpose of the instrument is not self-evident, especially since Stefan Meyer uses the odd expression *Bicykel-Modell* and not *Fahrrad*, which has become the normal German word for bicycle. On the other hand, there are extant humorous drawings by Carl Przibram of Boltzmann riding a bicycle. But if we look at article #99 "Beschreibung einiger Demonstrationsapparate" (1892) in Volume **III** of Boltzmann's *Wissenschaftliche Abhandlingen*, as prepared by Fritz Hasenöhrl and originally published in 1909, two possibilities stand out, namely as described in sections A & D. The first was a completed machine and the second was a proposed electrical model which used the expression (Bicycle). But regardless of which one it was, we have been informed by a knowledgable physicist that the instrument was intended as a mechanical analogy to help understand electrical induction, a purpose compatible with what Boltzmann taught in California in 1905 and which may help explain Meyer's contention that it had been taken to America twice.

#136

Brentano to Mach Florence, January 31, 1908

Also concerning Boltzmann I had a similar feeling, when he suddenly presented himself as my guest here in Belloguardo for a stay of several weeks. He was always eager, receptive, while partly developing his idiosyncratic arithmosophy (which is how I christened his child). This scientifically highly gifted man lacked neither interest in philosophy nor undiluted love of truth. And yet, what wonderful speculations did he not come to? You yourself certainly know much about him, although not so much as I, if what he said is correct which he made known to me, that I was the first human being who had the patience to hear him out.

That was as he confessed no small thing. Yet at other times our conversation was truly interesting and fruitful, namely when we turned to the principles of probability theory or to the problem whether or how one could with relative certainty determine the fact of true continuity (even if only on a phenomenal basis).

If someone had overheard us at that time, he might have awakened a very different impression of my philosophy than *Die Zeit* [1] which makes me a Scholastic or the good [Oskar] Kraus[2] who, without saying anything basically false, has allowed me to appear as a mystic.

But he himself is nothing like that, but rather a truly able person having as brave a character as ... von Krasnopolsky,[3] for example, while giving me evidence of having been persecuted, like others who elsewhere have been disconcerted by anti-Semitism. I know how very much your noble sense of justice regrets those conditions . . . . [4]

---

[1] This is the title of a Viennese newspaper.

[2] Oskar Kraus became a professor of Law at the German University of Prague in 1916. He was a former student of Brentano and is known to many historians of science for his strong opposition to Einstein's theory of relativity during the 1920's. See Klaus Hentschel, *Interpretationen und Fehlinterpretationen der speziellen und der allgemeinen Relativitätstheorie durch Zeitgenossen Albert Einsteins*, Birkhäuser: Basel, 1990, pp. 541-549.

[3] It is likely that Brentano is referring to Mach's Jewish colleague at the University of Prague, Horaz Krasnopolski, who was a professor of law.

[4] Franz Brentano's elliptical way of writing many of his letters is the despair of translators. (See also letter #93 to Boltzmann.) Nevertheless, we have tried to do the best we can. For the extant Mach-Brentano correspondence, see Franz Brentano, *Über Ernst Machs 'Erkenntnis und Irrtum'*, edited by Roderick M. Chisholm and Johann C. Marek, Rodopi, Amsterdam, 1988, pp. 203-227.

# Appendix A

## BOLTZMANN'S BATTALION

### A. Company: Wolfgang Stiller's Noteworthies[1]

| Teachers | Colleagues | Opponents |
|---|---|---|
| Josef Stefan<br>(1835-1893) | Willard Gibbs<br>(1839-1903) | Rudolf Clausius<br>(1822-1888) |
| Josef Loschmidt<br>(1821-1895) | Heinrich Streintz<br>(1848-1892) | P.G. Tait<br>(1831-1907) |
| James Clerk Maxwell<br>(1831-1879) | Franz Streintz<br>(1855-1922) | Ernst Mach<br>(1838-1916) |
| August Toepler<br>(1836-1912) | Franz Serafin Exner<br>(1849-1926) | Wilhelm Ostwald<br>(1853-1932) |
| Hermann von Helmholtz<br>(1821-1894) | Felix Klein<br>(1849-1925) | Georg Helm<br>(1851-1923) |
| Gustav Robert Kirchhoff<br>(1824-1887) | Max Planck<br>(1858-1947) | Henri Poincaré<br>(1854-1912) |
| Leo Königsberger<br>(1837-1921) | Woldemar Voigt<br>(1850-1919) | Ernst Zermelo<br>(1871-1953) |

| More Colleagues | Students | |
|---|---|---|
| Hendrik Anton Lorentz<br>(1853-1928) | Svante Arrhenius<br>(1859-1927) | Walther Nernst<br>(1864-1921) |
| Otto Heinrich Wiener<br>(1862-1927) | Gustav Jäger<br>(1865-1938) | Hantaro Nagaoka<br>(1865-1950) |
| Philipp Lenard<br>(1862-1947) | C. E. Curry<br>(1868-1935) | M. v. Smoluchowski<br>(1872-1917) |
| George Hartley Bryan<br>(1864-1928) | Lise Meitner<br>(1878-1968) | George Cecil Jaffé<br>(1880-1965) |
| Clemens Schaefer<br>(1878-1968) | Paul Ehrenfest<br>(1880-1933) | Ludwig Flamm<br>(1885-1964) |

Boltzmann's Successors

| in Munich: | in Leipzig: | in Vienna: |
|---|---|---|
| Arnold Sommerfeld<br>(1868-1951) | Theodor des Coudres<br>(1862-1926) | Friedrich Hasenöhrl<br>(1874-1915) |

---

[1] Wolfgang Stiller, *Ludwig Boltzmann - Altmeister der klassischen Physik - Wegbereiter der Quantentheorie und Evolutionstheorie*, Verlag Harri Deutsch: Frankfurt am Main, 1989. Stiller gives biographical information about the people on his list plus remarks or quotations from many of them about Boltzmann

## B. Company

Berta Karlik's & Erich Schmid's List of Some Austrian Physicists Who Were Students or Young Colleagues in Vienna in Boltzmann's Time.[2]

| | |
|---|---|
| Franz Aigner (1882-1945)<br>PHD 1906, Habil. World War I | Hans Benndorf (1870-1953)<br>PHD 1895, Habilitation 1899 |
| Viktor Conrad (1876-1962)<br>PHD 1900, Habil. 1905 | Felix Ehrenhaft (1879-1952)<br>PHD 1903, Habilitation 1905 |
| Felix Exner (1876-1930)<br>PHD 1900, Habilitation 1904 | Ludwig Flamm (1885-1964)<br>(Married Boltzmann's daughter) |
| Eduard Haschek (1875-1947)<br>PHD 1897, Habilitation 1903 | Friedrich Hasenöhrl (1874-1915)<br>PHD 1897, Habilitation 1899 |
| Fritz Kohlrausch (1884-1953)<br>PHD 1907, Habilitation 1911 | Anton Lampa (1868-1938)<br>PHD 1892, Habilitation 1897 |
| Friedrich von Lerch (1878-1946)<br>PHD 1901, Habilitation 1905 | Erwin Lohr (1880-1951)<br>PHD 1904, Habilitation 1908 |
| Heinrich Mache (1876-1954)<br>PHD 1898, Habilitation 1901 | Stefan Meyer (1872-1949)<br>PHD 1896, Habilitation 1900 |
| Adalbert Prey (1873-1949)<br>PHD 1896, Habilitation 1900? | Karl Przibram (1878-1973)<br>PHD (in Graz) Habilitation 1905 |
| Wilhelm Schmidt (1883-1936)<br>PHD 1905, Habilitation 1912 | Egon von Schweidler (1873-1948)<br>PHD 1895, Habilitation 1899 |
| M. v. Smoluchowski (1872-1917)<br>PHD 1895, Habilitation 1898 | Arthur Szarvassy (1873-1919)<br>PHD 1898, Habilitation 1905 |

Josef August Tuma (1866-1938) PHD 1892
Ausserordentlich Prof.1903 TH Prague

---

[2] Berta Karlik and Erich Schmid, *Franz S. Exner und sein Kreis*, Verlag der Österreichischen Akademie der Wissenschaften, Vienna, 1982. The authors give details about the life and careers of each of the people on their list, but since the primary focus is on Franz Exner and his circle, virtually all of those on the list were primarily experimental physicists.

C. Company

A. Dick's & G. Kerber's List of Doctoral Dissertations Completed with Boltzmann as either first or second advisor [*Gutachter*].[3]

Boltzmann as First Advisor
(1876-1906)

Vienna 1876 - Franz Hocevar (1853-1919): "On Some Determinate Integrals"

Graz[4] 1879 - Ignaz Klemencic (1853-1901): "Observations on the Elastic Aftereffect on Glass" [Dissertation # 228]

Graz 1881 - Friedrich Wrzal (1853- ): "Heat Capacity of Water Steam at Constant Saturation" [Dissertation #256]

Graz 1882- Thomas Romich [Romik?] (1854?- ): "Experimental Examination of Dielectrical Bodies with Reference to Their Dielectrical Aftereffect" [Dissertation #264]

Graz 1884 - Josef Hoffmann (1854- ): "Geometry in its Dependence on the Mass Relations of Space" [Dissertation #277]

Graz 1885 - Victor Hausmaninger (1885-1907): "On the Theory of the Longitudinal Thrust of Cylindrical Bodies" [Diss. #297]

Graz 1885 - Paul Czermak (1857-1912): "On the Value of Integrals A1 and A2 in Maxwell's Gas Theory when Based on a Power Law" [Dissertation #298]

Graz 1886 - Eduard Aulinger (1854- ): "On the Relation of Weber's Theory of Electrodynamics to the Principle of the Unity of Electrical Forces as Set up by Hertz" [Dissertation #306]

Graz 1888 - Anton Lampel (1854-1890): "On the Vibration of a Rotating Ball in the Face of Wind Resistance" [Dissertation #315]

Graz 1888- Richard Hiecke (1864-1948): On the Deformation of Electric Oscillations through Proximity of Unbroken Conductors" [Dissertation #324]

Graz 1890? Franz Streintz (1855-1922) [Because Boltzmann was called to Munich, Streintz finished his dissertation under Toepler.]

Graz 1891- Michael Radakovits (1866-1934) studied under Boltzmann in Graz 1884-1889. but received his PHD from other hands in 1891. He taught at Innsbruck, Czernowitz, & Graz. Much of his work was on determining the physical characteristics of molecules.

Munich 1893- Julius Werther (1870- ): "Contribution to the Theory of Apparatuses for the Manufacture of Micrometer Screws"

Munich 1893/94 Charles Emerson Curry (1868-1935): "Transmission of Compression Thrusts in a Gas"

Munich 1894- Samuel Preston Tolver (1844-1917): "On the Reciprocal Relation of some Hypotheses Which have been Proposed for the Explanation of Gravitation"

Munich 1894- Rudolf Camerer ( - ): "On the Total Reflection of Light on Thick (Course) Crystaline Substances"
Munich 1894- Curt Heinke (1864-1942): "Contributions to the Measurement of Inductive Coefficients and Capacities"

Vienna 1894-1900 Nothing listed, but See Boltzmann as second advisor below.

Leipzig 1900-1902 Nothing Listed

Vienna 1904- Karl Stanzel (1879- ) "On Diffusion in Itself [in sich selbst]"
Vienna 1904- Paul Ehrenfest (1880-1933): "The Movement of Fixed Bodies in Liquids and the Mechanics of Hertz"
Vienna 1906- Otto Frank (1883- ): "On the Potential Difference in the Place of Contact of 2 Different Electrolytes for the Case of the Cathode-ions Together [*untereinander*] and the anode-ions which possess the same Valence"
Vienna 1906- Johann Radakovits (1877- ): "Ionization of Gases by Galvanic Incandescent Wires"

---

[3] A. Dick and G. Kerber, *Ludwig Boltzmann - Katalog zur Ausstellung an der Zentralbibliothek für Physik in Wien,* Vienna, 1982, pp. 42-46. Where Boltzmann was the first advisor, the journals or monographs in which the dictoral dissertations were subsequently published are also cited.

[4] The list of Graz dissertations has been augmented by some mentioned by Paul Urban in his article "Ludwig Boltzmann in Graz" in *Ludwig Boltzmann - Ausgewählte Abhandlungen,* edited by Roman Sexl & John Blackmore, Friedrich Vieweg und Sohn: Braunschweig/-Wiesbaden, 1982, p. 68 and by Christa Jungknickel and Russell McCormmach in their book *Intellectual Mastery of Nature,* Volume II, University of Chicago Press: Chicago, 1986, pp. 67-68. Presumably, Boltzmann was also first advisor in the added cases, but that is not certain. But none of the lists appears to be complete, even the University of Graz "Rigorosenbuch 1866-1898". Boltzmann may have been the second or third advisor for other dissertations, to whit: Johann Svetina Jan. 17, 1881 and Franz Lampe Nov. 16, 1884. Many of Boltzmann's comments on the dissertations are mentioned in the Jungnickel and McCormmach account. There are some differences in dates and spelling of student names among the various incomplete and partly overlapping accounts of Graz dissertations.

Concerning the Munich dissertations, different sources give the order of middle and last names differently. In one it is Charles Emerson Curry and in another Charles Curry Emerson. Similarly one source lists Samuel Preston Tolver and the other Samuel Tolver Preston.

Boltzmann as Second Advisor[5]

(1896-1906)

| | | |
|---|---|---|
| Friedrich Erben<br>1896 | Stefan Meyer<br>1896 | Alfred Hauke<br>1897 |
| Fritz Hasenöhrl<br>1897 | Eduard Haschek<br>1897 | Rudolf Ludwig<br>1897 |
| Oskar Singer<br>1898 | Arthur Szarvassy<br>1898 | Leopold Kann<br>1898 |
| Heinrich Mache<br>1898 | Julius Hock<br>1899 | August Lehofer<br>1902 |
| Moriz Topolansky<br>1903 | Olga Steindler<br>1903 | Felix Ehrenhaft<br>1903 |
| Johann Knific<br>1904 | Vinzenz F. Blaha<br>1904 | Josef Zölss<br>1904 |
| Erwin Lohr<br>1904 | Josef Rozman<br>1905 | Wilhelm Schmidt<br>1905 |
| Friedrich Fasching<br>1905 | Elise Meitner<br>1906 | Selma Freud<br>1906 |
| Richard Dobner<br>1906[6] | | |

[5] The large number of doctoral dissertaions in Graz with Boltzmann as first advisor probably has to do with Boltzmann's long stay there, attractive personality, and especially with the fact that he held the chair in *experimental* physics. The seemingly odd fact that Boltzmann was very often second advisor in Vienna but rarely first advisor may be related to the fact that most of the students did their doctoral dissertations as in Graz in experimental physics. But there is also a possibility that as in Graz none of the lists of doctoral dissertations are complete. Nor may all extant lists have been consulted. As for the list of students who passed their *Habilitation* in Vienna, that is definitely not complete, since it only includes Franz Exner's students (but for whom there is evidence that they heard at least some lectures from Boltzmann. Other students of Exner for whom there is no evidence to this effect have not been included). One must also allow for some students who received a *Habilitation* under Victor von Lang as first advisor and Exner as second advisor but who also took courses under Boltzmann. A complete list of students who attended Boltzmann's lectures but who received a PHD or *Habilitation* under other professors would be long indeed.

[6] Exner and Boltzmann were the advisors for the dissertation, but Victor von Lang replaced Boltzmann for the actual examination on April 23, 1906. Dick and Kerber have also added a note on page 46 that Boltzmann was the third advisor for some dissertations in mathematics, meteorology, and chemistry.

## D. Company

### Other Notables

| Science Students | Science Students Interested in Philosophy | Philosophy Colleagues |
|---|---|---|
| Hugo Buchholz (1866-1921) | Hans Hahn (1879-1934) | Ernst Mach (1838-1916) |
| Josef Nabl (1876-1953) | Arthur Haas (1884-1941) | Franz Brentano (1838-1917) |
| Gerhard Kowalewsky[1] (1876-1950) | Philipp Frank (1884-1966) | Friedrich Jodl (1849-1914) |
| Arthur Boltzmann (1881-1952) | Hermann Broch (1886-1951) | Alois Höfler (1853-1922) |
| Lily Hahn (neé Minor) (1885 - ) | Gabriele Rabel (1886-1961) | Heinrich Gomperz (1873-1942) |

Vienna Youth Who Seem to Have Barely Missed Hearing Boltzmann:

| | | |
|---|---|---|
| Hans Thirring (1888-1976) | Erwin Schrödinger (1887-1961) | Ludwig Wittgenstein[2] (1889-1951) |

Three Little-Known Sources of Boltzmann's Philosophical Ideas:

| | | |
|---|---|---|
| Robert Zimmermann (1824-1898) | Oskar Schmitz-Dumont ( ? - ? ) | Branislav Petronijevic (1875-1924) |

[His books appeared from 1875 to 1896.]

---

[1] Technically, he was not Boltzmann's student but a *Privatddozent* at Leipzig during the 1900-1901 schoolyear while Boltzmann was there, but given their reported conversations it seems that Boltzmann acted as an advisor, friend, and teacher.

[2] Given Wittgenstein's strong identification with the linguistic phase in Boltzmann's thought circa 1904-1905 which is only present in published form in three short pieces in Boltzmann's *Populäre Schriften*, this translator has the suspicion that Wittgenstein must have had some other contact with Boltzmann's ideas, either through newspaper accounts, a friend who attended Boltzmann's lectures, or his own presence at some of them. It is also known that a fellow professor, Alois Höfler, gave lectures in Vienna on the similarities and differences between the ideas of Mach and Boltzmann. He had difficulty obtaining official permission for the lectures, but by 1909-1910 he did give them and possibly unofficial ones earlier, hence, there were other ways in which Wittgenstein could have become attracted to a linguistic philosophy which had so many resemblances to what Boltzmann had been thinking.

# Appendix B

## BEETHOVEN IN HEAVEN
## A Choral Jest

by Ludwig Boltzmann[1]

I

Done with dying and deeds unsung
My soul found solace on a higher rung.
Soaring, swooping, serenely swift
I caught a draft, an upward drift,
And there beyond Earth's spinning madness
I found God's world of winning gladness.

Glued to enter what most inspires
I heard instead his glorious choirs,
And angels fair sang songs so sweet
In words of joy none can repeat.
But still a flaw my ears did find:
Monotonous tone to bore the mind.

A gentle angel laughed to allay:
'Thou must be German of musical clay.
We envy thy skill in song and sound,
But God is eternal, forever bound.
The muse of time doth never end
And the tone of goodness can never bend.'

But voices vaunt a grand chorale
Of vintage Beethoven theatrical.
But clear and true, its completely new,
A work unnamed and below unfamed.
On God's command he wrote and planned;
The notes took shape appropriate, but bland.

---

[1] This mimeographed copy was found in the Institute of Physics in New York. According to handwritten remarks at the top of the copy, it was first given by one of Boltzmann's daughters to Paul Urban and then to F. Rohrlich, and then to the Institute. It has been freely translated by the editor in order to concentrate more strongly on what appears to be the author's main philosophical point.

II

The piece was played at every Fest
And always sung as if the best,
but woe is me and woe twice again;
I came to Heaven to hear this man,
Whose music on Earth still lives and lends
What Beethoven's Spirit no longer sends.

An angel firm then cast me far
To sup with beauty in Heaven's car.
I saw God's home, his sparkling realm,
His happy helpers, and He the helm:
'But grant me leave Beethoven to address,
The master muse and nothing less.'

He saw me and said: 'Give me your hand.
Welcome stranger from Mortal Land.
You hear the tones of my angel chorus.
We love them here, God and all of us.
Teutonic born your taste is true;
Tell me your thought, your honest view.'

Still as stone I dared not whisper.
'Flatter me not! The truth is crisper!'
'Oh my hero, my unmatchable master,
I do fear lo, there are signs of disaster.
Thy chorus is serene but the music mean,
Less than great. Oh fate! A dull sheen.'

III

'I must admit I do agree
In Heaven here I am not free.
To compose my work as best I can
I need a richer, better plan.
My angel friends love trumpets loud,
But the muse withers from music proud.

'If asked what robs creative fire
It is the tone *beyond* God's choir.
The deepest, strongest tone of all
Is pain and anguish when angels fall
And God's own Son on a Cross well-nailed
Who suffered and died for the sinful and failed.

What moves a mother's heart to her child
That nameless bond of feeling wild
Is love's own fear her child is lost
As she prays to heaven to Christ on the Cross:
'Oh please, oh God, help him revive!
Have mercy and love, keep him alive!'

'If thou hath ever wept with thy wife
Or heard lutéd tones cut like a knife,
Then the heart in pain is not strange
And its main range is musical as well.
Heaven is True Heaven, and far from Hell,
But the deepest tone, intones, a tragic knell.

IV

'Yes! My music was greater down below.
I felt the feeling of the suffering so.
I never mastered my ways or weeping.
I never won self-possession or keeping
A balanced view of Heaven and Earth,
But now I love angelic mirth. Ha! Ha!

'Yes! God can feel as king of poor and rich.
Human pain is still His pain, Christ's pitch.
But if suffering makes music, relief can leaven,
And higher than hearing is a happy Heaven.
To lust for depth of a tragic kind
Can sicken the soul, make angels blind.

'Yes! Grim Leonore was great, but greater still
Was Florestan's joy and generous will.[2]
Angels are kind, simple, happy, and brave;
Lesser music can their virtues save.
In this most happy, beautiful land
I let the lute *leave* my hand.'

'No! No! Must my words shock thee hard?
I love Earth and Beethoven *The Bard.*
I honor suffering and Leonore too!'
'Keep thy pain! Return home! Do!
To suffer in Heaven *can rebuff God!*
And prove thy ways, these days, *are still odd.*'

---

[2] In Beethoven's only opera, *Fidelio*, Lady Leonore found a skillful way to free Florestan her husband from a deep dungen ruled by the cruel Governor Pizarro.

# BIBLIOGRAPHY

## I

## Collections

1890 *The Scientific Papers of James Clerk Maxwell*, Cambridge University Press. Cambridge. [reprinted by Hermann: Paris, 1927 & Dover: New York, 1952]

1890-1900 P.G. Tait, *Scientific Papers*, two volumes, Cambridge University Press: Cambridge.

1891 *Gustav Kirchhoff: Gesammelte Abhandlungen*, edited and with a supplement added by Ludwig Boltzmann, J.A. Barth: Leipzig.

1894-95 [letters sent to the editor of *Nature* about Boltzmann's gas theory] from: L Boltzmann (3), G.H. Bryan (6), S.H. Burbury (5), E.P. Culverwell (6), G.F. Fitzgerald (1), J. Larmor (1), A. Schuster (1), & H.W. Watson (2).

1899 *Clark University 1889-1899, Decennial Celebration*, Worcester, Mass. [with Lectures by Picard, Boltzmann, Ramón y Cajal, Mosso, & Forel]

1904 *Boltzmann Festschrift*, edited by Stefan Meyer, J.A. Barth: Leipzig.

1904 *Congress of Arts and Science - Universal Exposition, St. Louis 1904*. Volume I - Philosophy and Mathematics. edited by H.J. Rogers. Houghton, Mifflin, and Company, Boston and New York, 1905. with lectures from Newcomb, Ostwald, Erdmann, Woodward, Nichols, Barus, Kimball, Nipher, Brace, Langevin, Rutherford, Boltzmann, & Poincaré. [reprinted by Tomash: Los Angeles: 1986]

[*POPULÄRE 1905*] Ludwig Boltzmann, *Populäre Schriften* , J.A. Barth: Leipzig. Contains 9 lectures on popular science, 6 on methodology, 4 on philosophy, 3 biography, one book review, and one long article on his lectures in Berkeley and trip to America. [reprinted in German in 1919 &1925 with the same pagination. The 1979 version is not complete. The English 1974 also not complete is below. The Japanese work, *Gendai no Kagaku* , (Science of Our Era), Chuokoron: Tokyo, 1979 includes three of the methodological articles.]

[*KLEIN 1907*] Arnold Sommerfeld (ed.) *Physik*, (Volume **5** of Part **I** of Felix Klein's *Encyklopädie der mathematischen Wissenschaften* as published from 1901 to 1908), B.G. Teubner: Leipzig, 1907.

[*WISS-AB 1909*] Ludwig Boltzmann: *Wissenschaftliche Abhandlungen*, edited by Fritz Hasenöhrl, three volumes, J.A. Barth: Leipzig. [reprinted by Chelsea Pub. Co., 1968] The outside cover says *Abhandlungen*, the inside *Wissenschaftliche Abhandlungen*, and the bottom of many pages *Gesammelte Wissenschaftliche Abhandlungen*. These volumes include Boltzmann's scientific articles as published in standard scientific journals and transactions. [for English translations see *KINETIC 1966*. Six articles are translated into Japanese in volumes **I**, **V**, and **VI** of *Butsurigaku Koten Ronbun Soosho* (Series of Classical Articles in Physics), Tokai Daigaku: 1971.]

1914 *Vorträge über die kinetische Theorie der Materie und Elektrizität*, edited by Peter Debye, Teubner: Leipzig. [includes articles by Planck, Debye, Smoluchowski, Nernst, Sommerfeld, and Lorentz]

1934-39 H.A. Lorentz: *Collected Papers*, 9 volumes, edited by P. Zeeman & A.D. Fokker, Martinus Nijhoff, The Hague.

1951-1954 Henri Poincaré, *Oevres*, Gauthiers-Villars: Paris.

1959 Paul Ehrenfest: *Collected Scientific Papers*, edited by Martin J. Klein, North-Holland: Amsterdam.

1961 Max von Laue: *Gesammelte Schriften*, 8 volumes, Vieweg: Braunschweig.

[*KINETIC 1966*] Stephen Brush, *Kinetic Theory*, Volume **II**, Pergamon Press: New York. [Volume **II** includes several articles by Boltzmann translated into English.]

[*EQUATION 1972*] *The Boltzmann Equation - Theory and Application*, edited by E.G.D. Cohen & W. Thirring, (Proceedings of the International Symposium: "100 Years of the Boltzmann Equation" in Vienna, 4th-8th September 1972), Acta Physica Austriaca Supplementum X, Springer Verlag: Vienna & New York, 1973. [This book includes articles by D. Flamm, E. Broda, M. Klein, G. Uhlenbeck, C. Cercignani, J. Foch, G. Ford, F.G.R. Cohen, J, Sengers, J. Dorfman, L. Waldmann, S. Hess, J. Befnackker, R. Kubo, K. Seeger & H. Pötzl, M. Kac, I. Prigogine, W. Wood, I. Kuscer, S. DeGroot, I. Khalatnikov, Ya. Sinai, D. Ruelle, O. Lanford, K. Przybram.]

1974 *Philosophical Foundations of Science*, edited by R.J. Seeger & R.S. Cohen (Vol. **XI** of Boston Studies in the Philosophy of Science), Reidel: Dordrecht. [with articles by Wigner, Bronowski, Klein, Bergmann, Agassi, Jarvie, Toulmin, McMullin, Sklar, Schaffner, and others]

[*THEO & PHIL 1974*] Ludwig Boltzmann: *Theoretical Physics & Philosophical Problems*, edited by Brian McGuinness, & translated by Paul Foulkes, (Vol. **V** of the Vienna Circle Collection), Reidel: Dordrecht, 1974. [This book includes translations of 14 scientific, methodological, and philosophical articles from *Populäre Schriften* plus selections from the prefaces to his two volume work *Vorlesungen über die Principe der Mechanik* plus his article "Model" from the *Encyclopaedia Britannica*.]

[*PISA PROB 1978*] *Probabilistic Thinking, Thermodynamics and the Interaction of the History and Philosophy of Science*, (Volume **2** of the Pisa Conference on the History and Philosophy of Science Proceedings), Reidel: Dordrecht, 1981, pp. 147-238. [with articles on Boltzmann by L. Krüger, E. Hiebert, V. Kartsev, & O. Lezhneva]

[*TAGUNG 1981*] *Ludwig Boltzmann* Internationale Tagung anlässlich des 75. Jahrestages seines Todes 5.-8. September 1981 - *Ausgewählte Abhandlungen*, edited by R. Sexl & J. Blackmore, (Ludwig Boltzmann Gesamtausgabe, Band **8**), Akademische Druck u. Verlagsanstalt, Graz / Friedrich Vieweg & Sohn: Braunschweig-Wiesbaden, 1982. [This book includes articles by D. Flamm (2), P. Urban, A. Kox, G. Fasol, K.v. Meyenn, E. Broda, F. Wallner, J. Blackmore, H. Hörz, W. Kaiser, A.I. Miller, M. Curd, S.G. Brush, A. Wehrl, S. Wagner, and H. Motz.]

[*ATOMS 1984*] Mary Jo Nye (ed.), *The Question of the Atom from the Karlsruhe Congress to the first Solvay Conference, 1860-1911.* (Vol. 4 of: A Series in the History of Modern Physics, 1800-1950), Tomash: Los Angeles & San Francisco, 1984. [with articles by Cannizzaro, Kelvin, Williamson, Mach, Perrin, Rutherford, Lockyer, Maxwell, Marignac, Berthelot, Helmholtz, Arrhenius, Mendeleev, Ostwald, Boltzmann, Thomson, Poincaré, Rutherford & Soddy, Einstein, Duhem]

[*WVF 1985*] Engelbert Broda: *Wissenschaft Verantwortung Frieden - Ausgewählte Schriften* edited by Paul Broda, Gitta Deutsch, Peter Markl, Thomas Schönfeld, and Helmuth Springer-Lederer, Franz Deuticke: Vienna, 1985. [contains five articles by Engelbert Broda on Boltzmann]

[*DEUTSCHES MU 1985*] *Ludwig Boltzmann 1844-1906. Vollender der klassischen Thermodynamik - Eine Dokumentation,* edited by W. Höflechner & A. Hohenester, (Ausstellung im Deutsches Museum 15. 11 - 29. 12. 1985), Deutsches Museum, 1985. [with articles by H. Mitter & P. Urban, W. Höflechner (107 pages), A. Dick & G. Kerber, A. Hohenester (2), and A. Hohenester & O. Lichtenauser & L. Mathelitsch]

[*AUSTRIA 1986*] J.C. Nyiri (ed.), *From Bolzano to Wittgenstein - The Tradition of Austrian Philosophy,* Verlag Hölder-Pichler-Tempsky, 1986. [Ten articles are in English and eight in German. The contributors are: W. Grassi & B. Smith, G. Weiler, R. Haller, W.G. Stock, R.M. Chisholm, E. Morscher, K. Mulligan, P.M. Simon, R. Lütthe, L. Aron, D. Münch, L. Pasquerella, W. Röd, J.C. Nyiri, F. Stadler, H. Rutte, E. Leinfellner-Rupertsberger, A.G. Gargani, and E. Weibl.]

[*HALLE 1987*] *Ludwig Boltzmann's Werk in naturwissenschaftlicher und philosophischer Sicht,* edited by S. Kirschke & W. Eisenberg, (Martin-Luther-Universität Halle-Wittenberg - Wissenschaftliche Beiträge), Halle (Saale), 1987. [contains articles by W. Eisenberg, E. Christke, U. Spretke, U. Niedersen, M. Koch, P. Wermes, H. Herwig, W. Kummer, W. Gründler, and K. Stöber]

[*STILLER 1989*] Wolfgang Stiller, *Ludwig Boltzmann - Altmeister der klassischen Physik, Wegbereiter der Quantenphysik und Evolutionstheorie,* Verlag Harri Deutsch: Thun & Frankfurt am Main, 1989. [This book includes a large nmuber of short selections and long quotations from Boltzmann's colleagues, students, friends, and opponents.]

[*EINSTEIN 1989*] *The Swiss Years: Writings, 1900-1909,* edited by J. Stachel, with D. Cassidy, J. Renn, and R. Schulmann as associate editors, D. Howard as assistant editor, A.J. Kox as contributing editor, and Ann Lehar as editorial assistant, (*The Collected Papers of Albert Einstein*, Volume 2), Princeton University Press: Princeton. [This book contains three articles by Einstein which were presumably sent to Boltzmann, two being apparently mentioned in the 1905 work by Boltzmann and Nabl which was published in Klein's encyclopedia in 1907.]

[*LB PRINCIPIEN 1990*] *Ludwig Boltzmann Principien der Naturfilosofi - Lectures on Natural Philosophy 1903-1906 with Two Essays by S.G. Brush and G. Fasol,* edited, translated from shorthand, and with an introduction by Ilse M. Fasol-Boltzmann, Springer Verlag: Berlin et al. [This book is primarily a collection of what survives from Boltzmann's lectures and lecture notes in philosophy from 1903 to 1906, but in addition it also includes some undated short philosophical manuscripts plus articles by Stephen Brush and Gerhard Fasol plus Ilse Fasol-Boltzmann's comments.

## II

## Obituaries of Boltzmann 1906-1908

### A. Vienna Newspapers: The First Week:

Sept. 7. (1) E. Ludwig, *Neue Freie Presse*, (Abendblatt), p. 3.
" " (2) Anon. " " " " " "
" " (3) Anon. " " " " " "
" " (4) Ernst Mach, *Die Zeit*, (Abendblatt), p. 1.
" " (5) Anon. " " " p. 1
" " (6) (W. Ostwald) " " " p. 2.
" " (7) Anon. " " " p. 2.
" 8 (8) Anon. *Neue Freie Presse*, (Morgenblatt), p. 1;
" " (9) Ernst Mach, " " "
" " (10) Franz Exner. " " "
" " (11) Anon. *Die Zeit*, (Morgenblatt), p. 5.
" 9 (12) *Neue Freie Presse*, p. 8.
" 11(13) *Neue Freie Presse*, (Morgenblatt), p. 8.
" 12(14) *Kleine Zeitung*, p. 2.
" 13(15) *Neue Freie Presse*, (Abendblatt), p. 19.

### B. Later Obituaries by Author.

Blaserna, P., *R. Accademia dei Lincei. Rendiconti*, **15** (1906) 570.
Bryan, G.H., *Nature*, **74** (Oct. 4, 1906) 569-570.
Bryan, G. H., *Proceedings of the Royal Society*, **A 80** (1908) xi-xiii.
Carus, P.,*The Open Court*, **20** (1906) 759.
Des Coudres, T., *Berichte über die Verhandlungen der Königliche Sachsischen Gesellschaft der Wissenschaften* (Leipzig), **58** (1906) 617-627.
Ehrenfest, P., *Mathematisch. - naturwissenschaftlich. Blätter*, **3** (1906) 205-209. [This obituary is also included in his 1959 *Collected Scientific Papers* .]
Höfler, A., *Süddeutsche Monatshefte*, **3** (October 1906) 418-422.
Höfler, A., *Zeitschrift für den phys. & chem. Unterricht*, 12. Jahrgang, Band **19**, Heft VII.
Jäger, G., *Akademie der Wissenschaften. Wien, Sitzungsberichte*, **109** (1906) 74.
Jäger, G., *Monatsheft für Mathematik und Physik*, **18** (1907) 3-8.
Lampa, A., *Naturwissenschaftliche Rundschau*, **21** (1906) 532-533.
Lampa, A., *Biographisches Jahrbuch und Deutscher Nekrolog*, Berlin, **11** (1908) 96-104.
Lang, Viktor von, *Almanach der Akademie der Wissenschaften*, Wien, **57** (1907) 307-09.
Lorentz, H.A., *Mitteilungen aus der mathematischen Gesellschaft in Hamburg*, **18** (1907) 3-7.
Lorentz, H.A.,*Verhandlungen der deutschen physikalischen Gesellschaft* , **9** (1907) 206-238. [also in *Collected Papers*, Vol. **9**, 359-91.]
Meyer, S., *Österreichische Rundschau*, **8** (1906) 479-483.
Smoluchowski, M. v., *Bull. Acad. Sci.*, Cracovie, 1906, p. 202.
Voigt, W., *Physikalische Zeitschrift*, **7** (1906) 649-650.
Voigt, W., *Göttingen Nachrichten*, **8** (1907) 69-82.
Voit, C., *Sitzungsberichte der bayrischen Akademie*, **37** (1907) 262-267.

## Section III

## Articles and Books

Adams, Henry, *The Education of Henry Adams*, privately published, 1907. [It was copyrighted by the Massachusetts Historical Society in 1918 and was republished by Modern Library in 1931. See especially Chapter XXXI, "The Grammar of Science".]

Adolph, K.-H., M. Lienert, & P. Ziesche, "Briefe von Ludwig Boltzmann an A. Toepler aus der Zeit 1871-1888", *Wissenschaftliche Zeitschrift der Technischen Universität Dresden*, **36** (1987), Heft I, pp. 1-12.

Bach, Alexander, "Boltzmann's Probability Distribution of 1877", *Archive for History of Exact Sciences*, **41** (1990) 1-40.

Bach, D.J., "Ernst Mach", *Frankfurter Zeitung*, Feb, 18, 1908, #49, p. 2, col. 1.

Bellone, Enrico, *A World on Paper: Studies on the Second Scientific Revolution*, MIT Press: Cambridge, 1980.

Bellone, Enrico, "Boltzmann e Lagrange: Quanti 'classici' e credenze sull'irreversibilita", in *Scienza e Filosofia: Saggi in onore di Ludovico Geymonat*, edited by Corrado Margiore, Garzanti: Milan, 1985.

Bernhardt, Hannelore, "Der Umkehreinwand gegen das H-Theorem und Boltzmanns statistische Deutung der Entropie", *NTM - Zeitschrift für Geschichte der Naturwissenschaft, Technik, und Medizin*, (Leipzig), **5** (1968) 35.

Bernhardt, Hannelore, "Der Wiederkehreinwand gegen Boltzmanns H-Theorem und den Begriff der Irreversibilität, *NTM*, **6** (1969), 27.

Bernhardt, Hannelore, "Über die Entwicklung und Bedeutung der Ergodenhypothese in den Anfängen der statistischen Mechanik", *NTM*, **8** (1971) 13-25.

Bierhalter, Günter, "Boltzmanns mechanische Grundlegung des zweiten Hauptsatzes der Wärmelehre aus dem Jahre 1866", *Archive for History of Exact Sciences*, **24** (1981) pp. 195-205.

Bierhalter, Günter, "Ziklische Zeitvorstellung, Zeitrichtung, und die frühen Versuche einer Deduktion des Zweiten Hauptsatzes der Thermodynamik", *Centaurus*, **33** (1990) pp. 345-367.

Bierhalter, Günter, "Von L. Boltzmann bis J.J. Thomson: Die Versuche einer mechanischen Grundlegung der Thermodynamik (1866-1890)", *Archive for History of Exact Sciences*, **44** (1992), 25-75.

Blackmore, John T., *Ernst Mach - His Life, Work, and Influence*, 2 volumes, doctoral dissertation, UCLA, 1970. University Microfilms, Ann Arbor Michigan, 1971. [See particularly Chapter 15, pp. 751-843.]

Blackmore, John T., *Ernst Mach - His Life, Work, and Influence*, University of California Press: Berkeley & Los Angeles, 1972. [See particularly Chapter 14, pp. 204-231.]

Blackmore, John, "Boltzmann's Concessions to Mach's Philosophy of Science", *Tagung 1981*, pp. 55-90.

Blackmore, John. (Co-editor of *Tagung 1981*. See Bibliography I, Collections for more information.)

Blackmore, John & Klaus Hentschel (eds.), *Ernst Mach als Aussenseiter - Machs Briefwechsel über Philosophie und Relativitätstheorie mit Persönlichkeiten seiner Zeit. Auszug aus dem letzten Notizbuch (Faksimile) von Ernst Mach*, Wilhelm Braumüller: Vienna, 1985. [includes 4 letters from Boltzmann to Mach.]

Blackmore, John, "Ludwig Boltzmann as a Major Source of 20th Century Philosophy", in *Logik, Wissenschaftstheorie, und Erkenntnistheorie*. (Akten des **11**. Internationalen Wittgenstein Symposiums), edited by P. Weingartner & G. Schurz, Hölder-Pichler-Tempsky: Vienna, 1987, pp. 105-112.

Blackmore, John (ed.), *Ernst Mach A Deeper Look - Documents and New Perspectives*. (Vol. **143** of Boston Studies in the Philosophy of Science), Kluwer: Dordrecht, 1992.

Bogolyubov, N.W. & Yu. v. Sannochkin, "Ludwig Boltzmann", *Uspekh Fiz. Nauk.*, **61** (1957) 7 [Translation AEC - Tr-3971, p. 7, U.S. National Science Foundation]

Bois-Reymond, Emil du, "Die sieben Welträtsel [1880]", *Vorträge über Philosophie und Gesellschaft*, Felix Meiner: Hamburg, 1974, pp. 159-187.

A. Selection of Boltzmann's Publications in German

Boltzmann, Ludwig, "Weitere Studien über das Wärmegleichgewicht unter Gasmolekülen", *Sitzungsberichte Akademie der Wissenschaften*, Vienna, **66** (1872) 275-370. [*WISS-AB 1909*, Vol. **I**, pp. 316-402]

Boltzmann, Ludwig, "Über Maxwells Elektrizitätstheorie", *Mitteilungen des naturwissenschaftlichen Vereins in Graz*, August 1873. [included in *POPULÄRE 1905*, pp. 11-24.]

Boltzmann, Ludwig, "Über die Beziehung eines allgemeine mechanischen Satzes zum zweiten Hauptsatze der Wärmetheorie", *Sitzungsberichte der Akademie der Wissenschaften*, **75** (1877) 67-73. [*WISS-AB 1909*, Vol. **II**, pp. 116-122]

Boltsmann, Ludwig, "Der zweite Hauptsatz der mechanischen Wärmelehre", A lecture given before the Imperial Academy of Sciences in Vienna on May 29 1886. [included in *POPULÄRE 1905*, pp. 25-50.

Boltzmann, Ludwig, *Festrede zur Feier des 301. Gründungstages der Karl-Franzens- -Universität zu Graz*, J.A. Barth: Leipzig, 1888.

Boltzmann, Ludwig, "Über die Bedeutung von Theorien", [Reply to the departing words of A. Tewes and H. Streintz on his leaving Graz for Munich on July 16, 1890. Included in *POPULÄRE 1905*, pp. 76-91.]

Boltzmann, Ludwig, *Vorlesungen über Maxwells Theorie der Elektricität und des Lichtes*, 2 volumes, J.A. Barth: Leipzig, 1891 & 1893. [2nd edition 1908. 3rd edition 1982. Spanish tranlation 1895 & 1899]

Boltzmann, Ludwig, "Über die Methoden der theoretischen Physik", *Katalog mathematischer und mathem.-physiikalischer Modelle, Apparate, und Instrumente*, herausg. im Auftrage des Vorstandes der Deutschen Mathematiker-Vereinigung von W. Dyck, Munich, 1892. [included in *POPULÄRE 1905*, pp. 1-10.]

Boltzmann, Ludwig, "Notes on Faraday and Maxwell" [Appendix to Boltzmann's translation of James Clerk Maxwell, *Über Faradays Kraftlinien*, Leipzig, 1895. The translation appeared in Ostwald's *Klassiker Nr.102*, Engelmann:Leipzig, 1898.]

Boltzmann, Ludwig, "Über Herrn Ostwalds Vortrag über den wissenschaftlichen Materialismus", Part III of "Ein Wort der Mathematik an die Energetik", *Annalen der Physik und Chemie*, Volume **57**, 1896. Part I starts on page 39. [included in *POPULÄRE 1905*, Part I p. 104-, Part III pp. 128-136.]

Boltzmann, Ludwig, "Entgegnung auf die wärmetheoretischen Betrachtungen des Hrn. E. Zermelo", *Annalen der Physik*, **57** (1896) 773-784. [*WISS-AB 1909*, Vol. **III**, pp. 567-578]

Boltzmann, Ludwig, *Vorlesungen über Gastheorie*, 2 volumes, J.A. Barth: Leipzig, 1896 and 1898. [2nd edition 1910-12, 3rd. 1923, republished 1981. French trans. 1902 & 1905. Russian trans. 1957. English translation by S. Brush 1966.]

Boltzmann, Ludwig, "Über die Frage nach der objektiven Existenz der Vorgänge in der unbelebten Natur", *Sitzungsberichte der Akademie der Wissenschaften* in Wien, Volume **106**, Abt. II, January 1897, p. 83. [*POPULÄRE 1905*, pp. 162-187]

Boltzmann, Ludwig, *Vorlesungen über die Principe der Mechanik*, 2 volumes, J.A. Barth: Leipzig, 1897 & 1904. [For Volume **III**, 1920, see Hugo Buchholz. In 1922 Barth published the 3rd edition of the 1897 book and the 2nd edition of the 1904 book as one volume. A reprint was published in 1974. For a long book review in English on the methodological parts of the book see *The Monist*, **9** (April 1899) 453-458.]

Boltzmann, Ludwig, "Über die Unentbehrlichkeit der Atomistik in der Naturwissenschaft", *Annalen der Physik* **60** (1897) 231. [*POPULÄRE 1925*, pp. 141-157]

Boltzmann, Ludwig, "Zu Hrn. Zermelos Abhandlung über die mechanische Erklärung irreversiler Vorgänge", *Annalen der Physik*, **60** (1897) 392-398.[*WISS-AB 1909*, Vol. **III**, pp. 579-586]

Boltzmann, Ludwig, "Nochmals über die Atomistik", *Annalen der Physik*, **61** (1897) 790 [*POPULÄRE 1905*, pp. 158-161].

Boltzmann, Ludwig, "Über die Entwicklung der Methoden der theoretischen Physik in neurer Zeit", given as a lecture at the Munich Natural Science Convention on Friday, September 22, 1899. [*POPULÄRE 1905*, pp. 198-227]

Boltzmann, Ludwig, "Über die Grundprinzipien und Grundgleichungen der Mechanik". [This was originally a series of four lectures given at the tenth anniversary of Clark University during the Summer of 1899. See *Clark University 1889-1899 Dicennial Celebration*, Worcester, pp. 261-309. *POPULÄRE 1905*, pp. 253-307]

Boltzmann, Ludwig, "Eugen von Lommel", *Jahresbericht der Deutschen Mathematiker-Vereinigung*, **8** (1900) 47-53. [*WISS-AB 1909*, Vol. **III**, pp. 673-681.

Boltzmann, Ludwig, "Über die Prinzipien der Mechanik". [This consists of two lectures. one given as an innaugural event in mechanics at Leipzig in November 1900 and the second as an innaugural speech in mechanics on his return to Vienna in October 1902. *Über die Prinzipien der mechanik, Zwei akademische Antrittsreden*, Leipzig, 1903. *POPULÄRE 1905*, pp. 308-337]

Boltzmann, Ludwig, "Ein Antrittsvortrag zur Naturphilosophie", Beilage zu Nr. 432 der Wiener Tageszeitung *Die Zeit*, December 11, 1903, pp. 1-2. [*POPULÄRE 1905*, pp. 338-344 and *LB PRINCIPIEN 1990*, pp. 150-156]. The published lecture was partly an attempt to correct versions of the actual lectures as reported earlier in *Neue Freie Presse*, October 27, 1903 and *Arbeiter Zeitung*, October 29, 1903. [Both articles are reproduced in *LB PRINCIPIEN 1990*, pp. 147-149 as well as Boltzmann's notes for what were actually two introductory lectures, October 26 & 27, 1903, pp. 77-82.]

Boltzmann, Ludwig, *Principien der Naturfilosofi - Lectures on Natural Philosophy 1903-1906*, Berlin, 1990. [For more information see Bibliography I, Collections, *LB PRINCIPIEN 1990*]

Boltzmann, Ludwig, "Über statistische Mechanik". This lecture was given under a different title in the Applied Mathematics Section of the Congress of Arts and Sciences at the Universal Exposition, St. Louis, Hall 7, September 24, 1904, 3 p.m. It was first published in English (See below) and then included at the end of 1905 in *Populäre Schriften*. [*POPULÄRE 1905*, pp. 344-363. Part of this lecture is also in *LB PRINCIPIEN 1990*, pp. 286-293]

Boltzmann, Ludwig, "Entgegnung auf einen von Prof. Ostwald über das Glück gehaltenen Vortrag". [This article was a reply to a lecture in Vienna by Wilhelm Ostwald given in November 1904. It was originally published in the journal *Umschau* but was extended when published again in *Populäre Schriften* at the end of 1905. *POPULÄRE 1905*, pp. 364-378

Boltzmann, Ludwig,"Über eine These Schopenhauers". [A lecture given before the Philosophical Society in Vienna on January 21, 1905. *POPULÄRE 1905*, pp. 385-402]

Bolztmann, Ludwig, "Reise eines deutschen Professors ins Eldorado". [This account of Boltzmann's trip to California in Summer 1905 seems to have been written especially for *Populäre Schriften. POPULÄRE 1905*, pp. 403-435.]

Boltzmann, Ludwig, *Populäre Schriften*, J.A. Barth: Leipzig, 1905. [See Bibliography I, Collections, *POPULÄRE 1905* for more information.]

Boltzmann, Ludwig & Josef Nabl, "Kinetische Theorie der Materie", *Enzyklopädie math. Wissenschaften*, Vol. **V**, Part 1, Teubner: Leipzig, 1907, pp. 493-557.

Boltzmann, Ludwig, *Wissenschaftliche Abhandlungen*. Vienna, 1909. [See Bibliography I, Collections, *WISS-AB 1909* for more information.]

B. Boltzmann's Publications in English[1]

Boltzmann, Ludwig, "Boiling-Points of Organic Bodies", *Philosophical Magazine and and Journal*, **42** (1871) 393. [*WISS-AB* 1909, Vol. **I**, p. 199]

Boltzmann, Ludwig, "Further Studies on the Thermal Equilibrium of Gas Molecules [1872]", in Stephen Brush, *KINETIC 1966*, Vol. **2**, pp. 88-175. [*WISS-AB 190*,. Vol. **I**, pp. 237-258].

Boltzmann, Ludwig, "On the Relation of a General Mechanical Theorem to the Second Law of Thermodynamics [1877]", in Stephen Brush, *KINETIC 1966*, Vol. **2**, pp. 188-. 193. [*WISS-AB 1909*, Vol. **II**, pp. 164-223]

Boltzmann, Ludwig, "The Second Law of Thermodynamics [1886]", *THEO & PHIL* 1974, pp. 13-32. [*POPULÄRE 1905*, pp. 25-50]

Boltzmann, Ludwig, "On the Significance of Theories [1890]", *THEO & PHIL 1974*, pp. 33-36. [*POPULÄRE 1905*, pp. 76-80]

Boltzmann, Ludwig, "On the Methods of Theoretical Physics [1892]", *THEO & PHIL 1974*, pp. 5-12. [*POPULÄRE 1905*, pp. 1-10]

Boltzmann, Ludwig, "On the Application of the Determinantal Relation to the Kinetic Theory of Polyatomic Gases", Appendix C to an article on thermodynamics Nr. 2, by G.H. Bryan in the *Report of the British Association for the Advancement of Science* at Oxford, 1894, pp. 102-106. [*WISS-AB 1909*, Vol. **III**, pp. 520-525]

Boltzmann, Ludwig, "On Maxwell's Method of Deriving the Equations of Hydrodynamics from the Kinetic Theory of Gases", *Report of the British Association*, Oxford, 1894, p. 579. [WISS-AB 1909, Vol. **III**, pp. 526-527}

Boltzmann, Ludwig, "On Certain Questions of the Theory of Gases", *Nature*, **51** (1895) pp. 413-415. Also in *THEO & PHIL 1974*, pp. 201-209. [*WISS-AB 1909*, Vol. **III**, PP. 535-544]

Boltzmann, Ludwig, "Reply to Culverwell", *Nature*, **51** (1895) 581. [*WISS-AB 1909*, Vol. **III**, p. 545]

Boltzmann, Ludwig, "On Energetics [1896]", *THEO & PHIL*, pp. 37-40. [*POPULÄRE* 1905, pp. 137-140]

Boltzmann, Ludwig, "On the Minimum Theorem in the Theory of Gases", *Nature*, **52** (1895) 221. [*WISS-AB 1909*, Vol. **III**, p. 546.

Boltzmann, Ludwig, "Reply to Zermelo's Remarks on the Theory of Heat [1896]", *KINETIC 1966*, Vol. **2**, pp. 218-228. [*WISS-AB1909*, Vol. **III**, pp. 567-578]

[1] Boltzmann also published a number of articles in French. See *WISS-AB 1909*.. There are also many articles translated into Russian in the anthology edited by U.I. Frankfurt, Verlag Nauka: Moscow, 1970.

Boltzmann, Ludwig, *Lectures on Gas Theory* [1896 & 1898], translated by Stephen Brush, University of California Press: Berkeley & Los Angeles, 1964.

Boltzmann, Ludwig, "Preface" to *Lectures on the Principles of Mechanics* [1897], Part One, *THEO & PHIL 1974*, pp. 223-254. [This is from *Vorlesungen über die Principe der Mechanik, Teil I*, a book which has apparently not yet been translated into English though there is a long book review in *The Monist*, **9** (1899) 453-456.]

Boltzmann, Ludwig, "Some Errata in Maxwell's Paper 'On Faraday's Lines of Force'", *Nature*, **57** (1897) 77-79.

Boltzmann, Ludwig, "On Zermelo's Paper 'On the Mechanical Explanation of Irreversible Processes' [1897]", *KINETIC 1966*, Vol. **2**, pp. 238-245. [*WISS-AB 1909*, Vol. **III**, pp. 579-586]

Boltzmann, Ludwig, "On the Necessity of Atomic Theories in Physics [1897]", *The Monist*, translated by Thomas J, McCormack, **12** (1901) 65-79. [Reprinted in *ATOMS 1984*, pp. 357-371. Translated by Paul Foulkes for *THEO & PHIL 1974*, pp. 41-53. [*POPULÄRE 1905*, pp. 141-157]

Boltzmann, Ludwig, "More on Atomism [1897]", *THEO & PHIL 1974*, pp. 54-56. [*POPULÄRE 1905*, pp. 158-161]

Boltzmann, Ludwig, "On the Question of the Objective Existence of Processes in Inanimate Nature [1897]", *THEO & PHIL 1974*, pp. 57-76. [*POPULÄRE 1905*, pp. 162-187]

Boltzmann, Ludwig, "Preface" in Charles Emerson Curry, *Theory of Electricity and Magnetism*, Macmillan: London, 1897.

Boltzmann, Ludwig, "The Recent Development of Method in Theoretical Physics [1899]", *The Monist*, **11** (1901) 226-257, also *THEO & PHIL 1974*, pp. 77-100 with a different title and translation. [*POPULÄRE 1905*, pp. 198-227]

Boltzmann, Ludwig, "Acceptance Speech on Becoming Doctor of Laws, honoris causá", *Clark University 1889-1899, Decennial Celebration*, Worcester, Mass., 1899, pp. 20 & 22. [Boltzmann gave four lectures at Clark University which were published in German in this book, pp. 261-309. For P. Foulkes' translations of the first two lectures see the following entry. For R. Weingartner's translation of part of the first lecture under a different title see the entry after that.]

Boltzmann, Ludwig, "On the Fundamental Principles and Equations of Mechanics [1899]", *THEO & PHIL 1974*, Lectures I & II, pp. 101-128. [*POPULÄRE 1905*, pp. 253-307]

Boltzmann, Ludwig, "Theories as Representations [1899]", in *Philosophy of Science*, edited by Arthur Danto & Sidney Morgenbesser, Meridian Books: Cleveland and New York, 1964, pp. 245-252. [*POPULÄRE 1905*, pp. 257-265]

Boltzmann, Ludwig, "On the Principles of Mechanics [1900 & 1902]", *THEO & PHIL* 1974, pp. 129-152. [*POPULÄRE 1905*, pp. 308-337]

Boltzmann, Ludwig, "Models", *Encyclopaedia Britannica*, 10th edition, 1902. Reprinted in the 11th edition, Vol. **10**, Cambridge, England, 1911, pp. 638-640. Reprinted again in *THEO & PHIL 1974*, pp. 213-220.

Boltzmann, Ludwig, "An Inaugural Lecture on Natural Philosophy [1903]", THEO & PHIL, pp. 153-158. [*POPULÄRE 1905*, pp. 338-344. This was the first of two inaugural lectures. The actual lecture given on October 26 as apparently remembered by listeners and commented on in Vienna newspapers seems to have been somewhat different. It began with the line "How have I come to teach philosophy?" which let loose a chain of laughter from his huge audience.]

Boltzmann, Ludwig, "Preface" to *Lectures on the Principles of Mechanics* [1904]. Part II, *THEO & PHIL 1974*, pp. 255-265. [This is from *Vorlesungen über die Principe der Mechanik*, Vol. **II**, a book which does not yet seem to have been translated into English.]

Boltzmann, Ludwig, "The Relations of Applied Mathematics", in *Congress of Arts and Science - Universal Exposition*, St. Louis, 1904, edited by Howard J. Rogers and translated by S. Epsteen, Houghton, Mifflin and Company: Cambridge, Mass., 1905, pp. 591-622. Reprinted in *Physics for a New Century*, edited by K.R. Sopka, Tomash: San Francisco & Los Angeles, 1986, pp. 267-279. [The same article with the title "On Statistical Mechanics", the order of paragraphs somewhat rearranged, and with a different translation can be found in *THEO & PHIL 1974*, pp. 159-172.]

Boltzmann, Ludwig, "Reply to a Lecture on Happiness given by Prof. Ostwald [1904]", *THEO & PHIL 1974*, pp. 173-184. [*POPULÄRE 1905*, pp. 364-378.]

Boltzmann, Ludwig, "On a Thesis of Schopenhauer's [1905]", *THEO & PHIL 1974*, pp. 185-198. [*POPULÄRE 1905*, pp. 385-402]

Boltzmann, Ludwig, "A German Professor's Journey into El Dorado [1905]", translated by Margaret Malt, *Annals of Nuclear Energy*, **4** (1977) 147-159. [Republished in *Theory and Statistical Physics* with a new translation by Walter Kutschera, **20** (1991) issues 5 & 6, pp. 499-523, and again in *Physics Today* (abbreviated) with a translation by Bertram Schwarzschild, **45** (Jan. 1992) 44-51. There is also a French translation as a separate work, Paris, 1987. *POPULÄRE 1905*, pp. 403-435]

Boltzmann, Ludwig, *Theoretical Physics and Philosophical Problems*, Boston, 1974. [For more information see Bibliography I, Collections, *THEO & PHIL 1974*]

See also: Fritz Rohrlich, "A Poem by Ludwig Boltzmann", *American Journal of Physics*, **60** (1992) 972-973.

Boring, Edwin A., "Franz Brentano", *A History of Experimental Psychology*, 2nd edition, Appleton-Century-Crofts: New York, 1957, pp. 356-361 & 379-380.

Bouveresse, Jacques, "Hertz, Boltzmann, et le probleme de la vérité des théories", in *La Vérité est-elle scientifique?*, edited by André Lichnérowicz & Gilbert Godofere, Editions Universitaires: Paris, 1991, pp. 107-141.

Brentano, Franz, "Nativistische, empiristische und anoetistische Theorie unserer Raumvorstellung [1906]", in *Philosophische Untersuchungen zu Raum, Zeit und Kontinuum*, edited by A. Kastil, S. Körder, & R. Chisholm, Felix Meiner Verlag: Hamburg, 1976, pp. 164-177.

Brentano, Franz, *Über Ernst Machs "Erkenntnis und Irrtum"*, edited by R. Chisholm & J. Marek, Rodopi: Amsterdam, 1988. [This book includes the Mach-Brentano letters.]

Broch, Hermann, "The So-Called Philosophical Rudiments of Empiric Knowledge", in *The Annotated Works of Hermann Broch*, Vol. **10/1**, edited by P.M. Lötzeller, Suhrkamp: Frankfurt am Main, 1974-1981, pp. 131-145. [Brock made many criticisms of the empirical philosophies of Mach and Boltzmann.]

Broda, Engelbert, *Ludwig Boltzmann - Mensch - Physiker - Philosoph*, mit einem Geleitwort von Hans Thirring, Franz Deuticke: Wien, 1955. reprinted in Berlin 1957, 2nd Vienna edition 1986, Japanese translation 1957, American translation 1983. There is also a Romanian translation.

Broda, Engelbert, "Die philosophische Auffassungen Ludwig Boltzmanns", *Weg und Ziel*, **14** (1956) 525-534.

Broda, Engelbert, "Philosophical Biography of Ludwig Boltzmann", *EQUATION 1972*, pp. 17-52.

Broda, Engelbert, "Erklärung des Entropiesatzes und der Liebe aus den Prinzipien der Wahrscheinlichkeitsrechnung", *Physikalische Blätter*, **32** (1976) 337-341.

Broda, Engelbert, "Der Einfluss von Ernst Mach und Ludwig Boltzmann auf Albert Einstein" in *Einstein Centenarium*, edited by H.J. Treder, Akademie Verlag: Berlin, 1979, pp. 227-237.

Broda, Engelbert, "Ludwig Boltzmann, der Mann und das Werk (zum 75. Todestag), *Mitteilungen der Österreicischen Gesellschaft für Geschichte der Naturwissenschaften*, **1** (1981) 21-25. Also in *WVF 1985*, pp. 79-87.

Broda, Engelbert, "The Intellectual Quadrangle: Mach-Boltzmann-Planck-Einstein", *Cern Report 81-10*, Geneva, January, 1981, v + 20 pp. republished in 1983 with added material. Translated into German in *WVF 1985*, pp. 129-150.

Broda, Engelbert, "Boltzmann and Darwin", *TAGUNG 1981*, pp. 129-142.

Broda, Engelbert, "The Interaction of Boltzmann with Mach, Ostwald, and Planck and His Influence on Nernst and Einstein", *Science and Technology, Humanism and Progress*, Academy of the Socialist Republic of Romania: Bucharest, 1982, pp. 35-57.

Broda, Engelbert, "Boltzmann als evolutionistischer Philosoph", *Berichte zur Wissenschaftsgeschichte*, **6** (1983) 103-114. Also in *WVF 1985*, pp. 88-100.

Broda, Engelbert, "Ludwig Boltzmann, Albert Einstein, und Franz Josef", *Wiener Geschischtsblätter*, **38** (1983) 100-119. Also in *WVF 1985*, pp. 151-161.

Broda, Engelbert, "Ludwig Boltzmann - Die Mechanik und der Vitalismus", *WVF 1985*, pp. 101-128.

Broda, Engelbert, *Wissenschaft, Verantwortung, Frieden*, Vienna, 1985 [See Bibliography I, Collections, *WVF 1985* for more information.]

Brush, Stephen, "Translator's Introduction", Ludwig Boltzmann, *Lectures on Gas Theory*, University of California Press: Berkeley & Los Angeles, 1964, pp. 1-17. [translated by Roman Sexl and updated for Ludwig Boltzmann's *Vorlesungen über Gastheorie*, (Ludwig Boltzmann Gesamtausgabe, Band I), Vieweg: Braunschweig, 1981, pp. 7-24.]

Brush, Stephen, *Kinetic Theory*, 3 volumes, Pergamon Press: New York, 1965-1972. [Volume **II** includes four articles by Boltzmann and is designated as *KINETIC 1966* in Bibliography I, Collections.] Volume **II** was translated into German in 1970.

Brush, Stephen, "Thermodynamics and History", *The Graduate Journal*, **7** (1966) 477-565.

Brush, Stephen, "Mach and Atomism", *Synthese*, **18** (1968) 192-215. [Included in *The Kind of Motion We Call Heat*, Vol. **I**, North Holland: Amsterdam, 1976, pp. 299.]

Brush, Stephen, "Ludwig Bo;tzmann", *Dictionary of Scientific Biography*, Vol. **II**, edited by Charles Coulston Gillispie, Charles Scribner's Sons: New York, 1970, pp. 260-268. [Reprinted in *The Kind of Motion We Call Heat*, Vol. **I**, North Holland: Amsterdam, 1976, pp. 231-248.]

Brush, Stephen, "The Development of the Kinetic Theory of Gases: Randomness and Irreversibility", *Archive for History of Exact Sciences*, **12** (1974) 1- 88. [In somewhat revised form this article appeared in *The Kind of Motion We Call Heat*, Vol. **II**, North Holland: Amsterdam, 1976, pp. 543-654.]

Brush, Stephen, *The kind of Motion We Call Heat*, 2 volumes, North Holland: Amsterdam, 1976. Republished in 1986. [We have listed a few of the included articles separately, but there are others.]

Brush, Stephen, "Statistical Mechanics and the Philosophy of Science" in *Proceedings of the Philosophy of Science Association*, edited by F. Suppe & P.D. Asquith, 1977, pp. 551-584. [Included in *Statistical Physics and the Atomic Theory of Matter, from Boyle and Newton to Landau and Onsager*, Princeton University Press: Princeton, 1983, pp. 259-275.]

Brush, Stephen, "Nietzsche's Recurrence Revisited" The French Connection", Journal of the History of Philosophy, **19** (1981) 235-238. [Included in *Temperature of History*, New York, 1978/79.]

Brush, Stephen, *Statistical Physics and the Atomic Theory of Matter, from Boyle and Newton to Landau and Onsager*, Princeton University Press: Princeton, 1983.

Brush, Stephen, *The Temperature of History: Phases of Science and Culture in the Nineteenth Century*, Burt Franklin: New York, 1978/79. Translated into German, 1987.

Brush, Stephen, "Ludwig Boltzmann and the Foundation of Natural Science", in *LB Principien 1990*, pp. 43-61.

Buchheim, G., "Zur Geschichte der Elektrodynamic: Briefe Ludwig Boltzmanns an Hermann von Helmholtz", *NTM - Schriftenreihe Geschichte d. Naturwiss., Technik, Medizin*, (Leipzig), **5** (1968) 125-131.

Buchholz, Hugo, *Das mechanische Potential, nach Vorlesungen von L. Boltzmann bearbeitet, und die Theorie der Figur der Erde*. J.A. Barth: Leipzig, 1908. [This book includes a copy of the last known letter from Boltzmann, March 1906.]

Buchholz, Hugo, *Angewandte Mathematik ... mit einem Anhang über das elastische und das hydronamische Potential auf grund von Vorlesungen L. Boltzmanns*, Leipzig,

1916. [According to A. Dick & G. Kerber this book contained 820 pages including 454 pages reworked from Boltzmann's Lectures, which were placed in an appendix. The publishers then brought out the appendix as a separate work in the book below.]

Buchholz, Hugo, *Ludwig Boltzmanns Vorlesungen über die Prinzipe der Mechanik, 3. Teil, Elastizitätstheorie und Hydromechanik*, J.A. Barth: Leipzig, 1920.

Bunge, Mario, "Ludwig Boltzmann (1844-1906), defensor del materialismo", *Minerva*, (Buenos Aires), **2** (1944) 70-72.

Burtt, Edwin Arthur, *The metaphysical Foundations of Modern Science*, 1924. [This classic work has subsequently been published in 1932, 1949, 1952, 1954, and 1980.]

Burtt, Edwin Arthur, "Introduction", *The English Philosophers from Bacon to Mill*, edited by E.A. Burtt, Random House: New York, 1939, pp. ix-xxii.

Campogalliani, Paolo, "Irreversibiliita e principio probabalistico nella costruzione teorica di Ludwig Boltzmann", *Scienza e storica: Bolletino del centro internazionale di storia dello spazio e del tempo*, **8** (1991) 25-30.

Campogalliani, Paolo, "Frecca del tempo e principio probabilistico nella construzione teorica di Ludwig Boltzmann", *Physis*, **29** (1992) 447-463.

Carus, Paul, "Professor Ostwald's Philosophy", *The Monist*, **17** (1907) 516-540.

Cekic, Miodrag, "The Influence of Mach's Philosophy in Yugoslavia", in *Ernst Mach A Deeper Look - Documents and New Persprectives*, edited by J. Blackmore, Kluwer: Dordrecht, 1992, pp. 379-401. [Cekic describes the philosophy of Branislav Petronijevic who seems to have influenced Boltzmann's late system of philosophical classification. His theory of multi-dimensions and discontinuity also seems to have interested Boltzmann.]

Chisholm, Roderick, "Franz Brentano", *The Encyclopedia of Philosophy*, Volume **I**, New York, 1974, pp. 365-368.

Chisholm, Roderick, (Co-editor of a book, 1976. See Franz Brentano for more data.)

Chisholm, Roderick & Johann C. Marek, (Co-editors of a book, 1988. See Franz Brentano above for more information.)

Christke, Ernst, "Ludwig Boltzmann als Vollender der 'klassischer' und Wegbereiter der modernen Physik", *HALLE 1987*, pp. 16-30.

Clark, Peter, "Atomism Versus Thermodynamics" in *Method and Appraisal in the Physical Sciences*, edited by C. Howson, Cambridge University Press: Cambridge, 1976, pp. 41-105.

Clark, Peter, "Review of the 'Kind of Motion We Call Heat'", by Stephen Brush, *British Journal for the Philosophy of Science*, **33** (1982) 165-186.

Classen, Johannes W., "Die Prinzipien der Mechanik bei Boltzmann und Hertz", *Jahresbericht der Hamburg Wiss. Ansichten*, Sonderdruck, Hamburg, 1898, 13 pages.

Classen, Johannes, W., *Vorlesungen über moderne Naturphilosophen*, C. Boysen: Hamburg, 1908. [It includes lectures on Hertz, Poncaré, Mach, Boltzmann, and many others.]

Cohen, E.G.D. & Walter Thirring (eds.), *The Boltzmann Equation -Theory and Applications*, Springer Verlag: Vienna, 1973. [See Bibliography I, Collections, *EQUATION 1972* for more details.]

Cohen, R.S., (Co-editor of a book, 1974. See R.J. Seeger and Bibliography I, Collections.)

Comte, Auguste, *Cours de philosophique positiv*, Paris, 1830-1842.

Cook, John W., *Wittgenstein's Metaphysics*, Cambridge University: Cambridge, 1994.

Cornforth, Maurice, *Science Versus Idealism - An Examination of 'Pure' Empiricism and Modern Logic*, Lawrence & Wishart: London, 1946.

Crombie, Alistair, *Styles of Scientific Thinking in the European Tradition*, London, 1994.

Curd, Martin, *Ludwig Boltzmann's Philosophy of Science: Theories, Pictures, and Analogies*, University of Pittsburgh PHD, 1978. [University Microfilms, 78-16829]

Curd, Martin, Review of Ludwig Boltzmann, Theoretical Physics and Philosophical Problems, Selected Writings, edited by B. McGuinness in *Philosophy of Science*, **45** (1978) 148.

Curd, Martin, "Popper on Boltzmann's Theory of the Direction of Time", *TAGUNG 1981*, pp. 263-304.

D'Agostino, Salvo, "Boltzmann and Hertz on the Bild Conception of Physical Theory", *History of Science*, **28** (1990) 381-398.

Daub, Edward, E., "Probability and Thermodynamics - The Reduction of the Second Law", *Isis*, **60** (1969) 318-330. [Discusses the controversy between Boltzmann and Loschmidt over the issue of reversibility during the 1870's while pointing out that Loschmidt was actually much more concerned with something else.]

De Groot, S.R., "Relativistic Boltzmann Theory", *EQUATION 1972*, pp. 529-562.

De Groot, S.R., "Foreword", *THEO & PHIL 1974*, pp. ix-xiii,

Deltete, Robert J., *The Energetics Controversy in Late Noneteenth-Century Germany; Helm, Ostwald, and Their Critics*. 2 vols., 1983, Yale Dissertation, UMI, Dissertation Information Service, A Bell Howell Company, Ann Arbor, 1984. [This excellent work also contains a first-rate bibliography.]

Dick, Auguste & Gabriele.Kerber, *Ludwig Boltzmann - Katalog zur Ausstellung an der Zentralbibliothek für Physik in Wien*, Vienna, 1982. [This catalog includes copies of many letters and documents by and about Boltzmann. The physics library also contains a large amount of unsorted Nachlass and correspondence from deceased physics professors going back to early in the 20th century. A well-intended effort is being made to organize and publish the material using computers, but even at the current advanced rate of progress, it will almost certainly take many years.]

Dick, Auguste & Gabriele Kerber, "Ludwig Boltzmann's Nachleben", *DEUTSCHES MU 1985,* pp. 138-141.

Dilwirth, Craig, "Empiricism vs. Realism - The Perennial Debate", *The Fabric of Science,* Kluwer: Dordrecht, 1994, pp. 3-27.

Dugas René, *La Théorie physique au sens de Boltzmann et ses prologements modernes,* Edition du Griffon: Neuchatel, 1959.

Duhem, Pierre, *Thermodynamique et chemie. Lecons élémentaires a l'usage des chemistes,* Hermann: Paris, 1902. 2nd edition 1910.

Duhem, Pierre, *Traité d'energétique ou de thermodynamique générale,* 2 volumes, Gautier-Villars: Paris, 1911.

Eckert, Michael & Willibald Pricha, "Boltzmann, Sommerfeld, und die Berufungen auf die Lehrstelle für theoretische Physik in Wien und München 1890-1917", *Österreichische Gesellschaft für Geschichte der Naturwissenschaften,* **4** (1984) 101-119.

Eggarter, T.P., "A Comment on Boltzmann's H-Theorem and Time Reversal", *American Journal of Physics,* **41** (1973) 874-877.

Ehrenfest, P. & T., "Über zwei bekannte Einwände gegen das Boltzmannsche H-Theorem", *Physikalische Zeitschrift,* **8** (1907) 311-314. [Also in *Collected Scientific Papers,* pp. 146-149.]

Ehrenfest, P & T, "Begriffliche Grundlagen der statistischen Auffassung in der mechanik", *Encyklopädie der mathematischen Wissenschaften,* **IV**, pt. 32, pp. 1-90, Leipzig, 1911. [The English translation by M.J. Moravcsik was published in book form as *The Conceptual Foundations of the Statistical Approach in Mechanics,* Ithaca, 1959]

Ehrenfest, Paul, *Collected Scientific Papers,* edited by Martin J. Klein, Amsterdam, 1959.

Ehrenhaft, F., H. Molisch, H. Thirring, and J. Tandler, "Ludwig Boltzmann zur Enthüllung des Denkmales auf dem Ehrengrabe am 4. Juli 1933", *Elektrotechnik und Maschinenbau,* (Vienna), **51** (1933) 690-696.

Einstein, Albert, "Folgerungen aus den Capillaritätserscheinungen", *Annalen der Physik,* **4** (1901) 513-523.

Einstein, Albert, "Kinetische Theorie des Wärmegleichgewichtes und des zweiten Hauptsatzes der Thermodynamik", *Annalen der Physik,* **9** (1902) 417-433.

Einstein, Albert, "Eine Theorie der Grundlagen der Thermodynamik", *Annalen der Physik,* **11** (1903) 170-187.

Eisenberg, Wolfgang, (co-editor and contributor to *HALLE 1987.* See Bibliography I, Collections.)

Elkana, Yehuda, "Boltzmann's Scientific Program and its Alternatives", in Elkana (ed.), *The Interaction Between Science and Philosophy*, Humanities press: Atlantic City, 1974, pp. 243-274.

Exner, Franz, *Naturwissenschaften*, **9** (1921) 177-180. [Exner discusses Loschmidt and Boltzmann.]

Fasol, Gerhard, "Comments on Some Manuscripts by Ludwig Boltzmann", *TAGUNG 1981*, pp. 87-96.

Fasol, Gerhard, "Summary - Lectures on Natural Philosophy", in *LB PRINCIPIEN*, 1990, pp. 65-74.

Fasol-Boltzmann, Ilse M., *Ludwig Boltzmann Principien der Naturfilosofi - Lectures on Natural Philosophy 1903-1906*, Springer: Berlin, 1990. [For more information see Bibliography I, Collections *LB PRINCIPIEN 1990*.]

Fasol-Boltzmann, Ilse M., "Introduction", *LB PRICIPIEN 1990*, pp. 1-39. [in English and German]

Ferrier, James Frederick, *Institutes of Metaphsic*, 1854. 2nd edition 1856. [Ferrier is widely credited with introducing the term "epistemology" into English.]

Feyerabend, Paul, "Ludwig Boltzmann", *Encyclopedia of Philosophy*, Vol. I, New York, 1969, pp. 334-337.

Flamm, Dieter, "Life and Personality of Ludwig Boltzmann", *EQUATION 1972*, pp. 3-16. [Professor Flamm is Ludwig Boltzmann's grandson.]

Flamm, Dieter, "Der Entropiesatz und das Leben: 100 Jahren Boltzmannsches Prinzip", *Naturwissenschaftliche Rundschau*, **32** (1979) 225-239.

Flamm, Dieter, "Die Persönlichkeit Ludwig Boltzmanns" & "Aus dem Leben Ludwig Boltzmanns", *TAGUNG 1981*, pp. 11-20 & 21-56. [The second article includes many selections from the large correspondence between Boltzmann and his future wife from 1872 until their marriage in 1876.]

Flamm, Dieter, "Ludwig Boltzmann and His Influence on Science, *Studies in the History and Philosophy of Science*, **14** (1983) 255-278.

Flamm, Ludwig, "Die Persönlichkeit Boltzmanns", *Wiener Chemiker Zeitung*, **47** (1944) 27-30. Also in *Österreichische Naturforscher und Techniker* (Vienna Gesellschaft für Kultur und Technik), 1951. [Professor Ludwig Flamm was the son-in-law of Ludwig Boltzmann and the father of Dieter Flamm mentioned above. This article commemorated the 100th anniversary of Boltzmann's birth.]

Flamm, Ludwig, "Zum 50. Todestag von Ludwig Boltzmann", *Physikalische Blätter*, **12** (1956) 408 to 412. [The following articles also largely commemorated the 50th anniversary of Boltzmann's death: *Anzeiger der Österreichischen Akademie der Wissenschaften*, 1956, pp. 141-145; *Universum*, **11** (1956) 565-567; *Chemiker Zeitung*, **58** (1957) 61-65; and *Uspekh Fiz. Nauk*, **61** (1957) 3.]

Frank, Philipp, *Einstein - His Life and Tmes,* edited by Shuichi Kusaka & translated by George Rosen, Alfred A. Knopf: New York, 1947.

Gardner, M.R., "Realism and Instrumentalism in 19th Century Atomism", *Philosophy of Science,* **46** (1979) 1-34.

Gearhart, Clayton A., "Einstein Before 1905, The Early Papers on Statistical Mechanics", *American Journal of Physics,* **58** (1990) 68-80.

Gerlach, G. Walther, "Ludwig Boltzmann und die Atomistik", *Geist und Gestalt,* Band **2**, (Bayrische Akademie), C.H. Beck: Munich, 1959, pp. 88-91.

Gibbs, J. Willard, *Elementary Principles in Statistical Mechanics,* E. Scribner's Sons: New York, 1902.

Gibbs, J. Willard, *Scientific Papers,* Longmans: London & New York, 1906. [Republished in 1928 with a slightly different title. Reprinted in 1948 and 1960.]

Grünbaum, Adolf, "Popper's Views on the Arrow of Time", in *The Philosophy of Karl Popper,* edited by Paul Arthur Schilpp, Volume II, Open Court: La Salle, 1974, pp. 775-797.

Grunwald, M., "Die philosophischen Ansichten Ludwig Boltzmanns", *Wissenschaftliche Zeitschrift der Friedrich Schiller Universität,* Jena/Thüringen, Math. - Natur. Classe, **11** (1962) 279-300.

Grunwald, M., "Boltzmanns Verteidigung des Materialismus in der anbrechenden 'Krise der Physik'" in *Naturwissenschaft - Tradition - Fortschritt: Beiheft zur Zeitschrift für Geschichte der Naturwissenschaften, Technik, und Medizin,* Berlin, 1963.

Haas, Arthur Erich, *Die Entwicklungsgeschichte des Satzes von der Erhaltung der Kraft,* Vienna, 1909.

Haller, Rudolf, *Fragen zu Wittgenstein und Aufsätze zur Österreichischen Philosophie,* Rodopi: Amsterdam, 1986.

Haller, Rudolf & Friedrich Stadler, (eds.) *Ernst Mach - Werk und Wirkung,* Hölder-Pichler-Tempsky: Vienna, 1988,.

Haller, Rudolf & Friedrich Stadler (eds.), *Wien-Berlin-Prag - Der Aufstieg der wissenschaftlichen Philosophie,* Hölder-Pichler-Tempsky: Vienna, 1993. [Even though the title is deceptive since the book only covers the rise of the Vienna Circle and parallel movements in Berlin and Prague and ignores non-positivist philosophy of science, still, Boltzmann's influence, especially his linguistic ideas, is detectible.]

Hasenöhrl, Friedrich (ed.), Ludwig Boltzmann, *Wissenschaftliche Abhandlungen,* 3 Volumes, J.A. Barth: Leipzig, 1909. [See Bibliography I, Collections, *WISS AB 1909* for more information.]

Hasenöhrl, Friedrich, *Bericht über die Errichtung eines Denkmals für Boltzmann,* Selbstverlag: Vienna, 1913.

Heilbron, J.L., "Fin-de-Siecle Physics", in C.G. Bernhard et al. (eds.), *Science, Technology, and Society in the Time of Alfred Nobel*, Pergamon Press: Oxford, 1982, pp. 51-73.

Heilbron, J.L., *The Dilemmas of an Upright Man - Max Planck as Spokesman for German Science*, University of California Press: Berkeley, etc., 1986.

Helmholtz, Anna von, *Ein Leben in Briefen*, edited by Ellen von Siemens-Helmholtz, 2 volumes, Berlin, 1929.

Hentschel, Klaus (Co-editor of *Ernst Mach als Aussenseiter*, 1985. See J. Blackmore.)

Hentschel, Klaus, *Interpretationen und Fehlinterpretationen der speziellen und der allgemeinen Relativitätstheorie durch Zeitgenossen Albert Einsteins*, Birkhäuser: Basel, 1990.

Herneck, Friedrich, "Wiener Physik vor 100 Jahren", *Physikalische Blätter*, **17** (1961) 455-461.

Herneck, Friedrich, *Bahnbrecher des Atomzeitalters*, Buchverlag der Morgen: Berlin, 1966.

Herrick, Clarence L. "The Passing of Scientific Materialism", *The Monist*, **15** (1905) 46-86. [This article is about how energeticism, phenomenalism, and "the electromadnetic world picture" were widely believed to be replacing the atomic theory in physics..]

Hertz, Heinrich, *Die Prinzipien der Mechanik*, J.A. Barth: Leipzig, 1894. English translation, 1899.

Herwig, Helge, "Ludwig Boltzmanns progressive philosophische Ansichten zum Character physikalischer Theorien", *HALLE 1987*, pp. 93-99.

Hiebert, E.N., *The Conception of Thermodynamics in the Scientific Thought of Mach and Planck*, Ernst-Mach-Institut: Freiburg im Breisgau, 1968.

Hiebert, Erwin, "The Energetics Controversy and the New Thermodynamics", D.H.D. Roller (ed.), *Perspectives in the History of science and Technology*, University of Oklahoma Press: Norman, 1971, pp. 67-86.

Hiebert, Erwin, "The State of Physics at the Turn of the Century" in *Rutherford and Physics at the Turn of the Century*, edited by Mario Bunge & William Shea, Watson: New York, 1979, pp. 1-28.

Hiebert, Erwin, "Boltzmann's Conception of Theory Construction: The Promotion of Pluralism, Provisionism, and Pragmatic Realism", *PISA PROB 1978*, Vol. **II**, 1980, pp. 175-198.

Hilbert, David, "Foundations of the Kinetic Theory of Gases [1912]", *KINETIC 1972*, pp. 89-101.

Hintikka, J. et al. (eds.), *PISA PROB 1978*, Vol. **II**, Reidel: Dordrecht, 1980.

Hirst, R.J. (ed.), *Perception and the External World*, Macmillan: New York, 1965.

Hoffmann, Dieter, "Ludwig Boltzmann. Pionier der Atomphysik", *Urania*, **57** (1981) Heft 10, p. 58-.

Hoffmann, Dieter & Hubert Laitko (eds.), *Ernst Mach - Studien und Dokumente zu Leben und Werk*, Deutscher Verlag der Wissenschaften: Berlin, 1991.

Höflechner, Walter, "Ludwig Boltzmann. Sein akademischer Werdegang in Österrreich - dargestellt nach archivalischen Materiallen", *Mitteilungen der Österreichischen Gesellschaft für Geschichte der Naturwissenschaften*, **2** (1982) 43-62.

Höflechner, Walter, "Zum Leben Ludwig Boltzmanns", *DEUTSCHES MU 1985*, PP. 31-137. [This biography of Boltzmann is the most detailed and useful one for scholars interested in his academic career and transfers.]

Höflechner, Walter & Adolf Hohenester eds.), *Boltzmann - Vollender der klassischen Thermodynamik, 1844-1906*, Deutsches Museum: Munich, 1985. [For more information see Bibliography I, Collections, *DEUTSCHES MU 1985*. ]

Höfler, Alois, "Zur Geschichte und Wurzel der Machschen Philosophie", *Zeitschrift für den physikalischen und chemischen Unterricht*, **23** (1910) 1-16.

Hohenester, Adolf, "Dreidimensionale Ausstellungsgegenstände", *DEUTSCHES MU 1985*, pp. 138-142. [This article describes four machines designed, built, or used by Boltzmann, three in Graz and one in Munich. The most famous was his "bicycle" made in Munich and transferred to Vienna, which is now in the Deutsches Museum in Munich.]

Hohenester, Adolf, (co-editor of book with Walter Höflechner above. For more data see Bibliography I, Collections, *DEUTSCHES MU 1985*.]

Holton, Gerald, *Science and Anti-Science*, Harvard University Press: Cambridge, 1993.

Hörz, Herbert, "Helmholtz and Boltzmann", *TAGUNG 1981*, pp. 191-206.

Hörz, Herbert & Andreas Laass, *Ludwig Boltzmanns Wege nach Berlin*, Akademie Verlag: Berlin, 1989. [This book includes Boltzmann's letters to Helmholtz and much official correspondence between Boltzmann and Berlin officials during the late 1880's when Boltzmann *almost* transferred from Graz to Berlin.]

Hoyer, Ulrich, "Von Boltzmann zu Planck", *Archive for History of Exact Sciences*, **23** (1980) 47-86.

Hoyer, Ulrich, "Ludwig Boltzmann und das Grundlagenproblem der Quantentheorie", *Zeitschrift für allgemeine Wissenschaftstheorie*, **15** (1984) 201-210.

Hoyer, Ulrich, "Thermodynamics and Philosophy - Ludwig Boltzmann", *Journal of Non-Equilibrium Thermodynamics*, **12** (1987) 11-26.

Hull, David, *Science as Process*, University of Chicago Press: Chicago, 1988.

Jaffé, George Cecil, "Recollections of Three Great Laboratories", *Journal of Chemical Education*, **29** (1952) 230-238.

Jäger, Gustav, "Ludwig Boltzmann", *Neue Österreichische Biographie*, Vol. **II**, Amalthea Verlag: Vienna, 1925, pp. 117-137.

Jodl, Margaret, *Friedrich Jodl, Sein Leben und Wirken, dargestellt nach Tagebücher und Briefen*, Stuttgart/Berlin, 1920.

Johnston, William M., *The Austrian Mind*, University of California Press: Berkeley and Los Angeles, 1972.

Jones, W.T., *Kant to Wittgenstein & Sartre*, 2nd edition, Harcourt Brace: New York, 1969.

Jost, R., "Boltzmann and Planck", *Einstein Symposium Berlin*, ed. by H. Nelkowski et al., Berlin, 1979.

Jungnickel, Christa & Russell McCormmach, *Intellectual Mastery of Nature - Theoretical Physics from Ohm to Einstein*, 2 volumes, University of Chicago Press: Chicago and London, 1986.

Kaiser, Walter, *Atomistik und Positivismus am Ende des 19. Jahrhunderts*, Magister Arbeit, Stuttgart, 1971.

Kaiser, Walter, "Boltzmanns mechanische Darstellung von Thermodynamik und Elektrodynamik",. *TAGUNG 1981*, pp. 207-230.

Kaiser, Walter, "Einleitung", Ludwig Boltzmann, *Vorlesungen über Mazwells Theorie der Elektricität und des Lichtes*, (Ludwig Boltzmann Gesamtausgabe, Band II), Vieweg: Braunschweig, 1982 [1891-93], pp. 5-32..

Kaizik, Jürgen, *Versuch einer Begegnung - Mach gegen Boltzmann*, Franz Deuticke: Vienna, 1986. [This book is a reflective discussion which becomes an imaginary dialogue between Mach and Boltzmann with Mach generally holding the upper hand.]

Kant, Immanuel, *Prolegomena zu einer künftigen Metaphysik* . . . Felix Meiner Hamburg, Hamburg, 1953. [1783]

Kant, Immanuel, *Kritik der Urteilskraft*, Felix Meiner: Hamburg, 1959. [1790]

Kant, Immanuel, *Kritik der reinen Vernunft*, Felix Meiner: Hamburg, 1960. [1781 & 1787]

Karlik, Berta & Erich Schmid, *Franz Exner und sein Kreis*, Verlag der Österreichischen Akademie der Wissenschaften: Vienna, 1982.

Kerber, Gabriele, (Co-author of book., 1982. See Auguste Dick.)

Kerber, Gabriele, (Co-author of article, 1985. See Auguste Dick.)

Kerber, Gabriele, (Co-author of a book, 1992. See Brigitte Zimmel.)

Keyserling, Arnold, *Der Wiener Denkstil. Mach, Carnap, Wittgenstein*, Graz/Vienna, 1965.

Kienzl, Wilhelm, *Meine Lebenswanderung - Erlebtes und Erschautes*, J. Engelhorn: Stuttgart, 1926. 2nd edition 1953. [He was an opera composer and friend of both Mach and Boltzmann.]

Kirchhoff, Gustav, *Gesammelte Abhandlungen*, J.A. Barth: Leizig, 1882. [Kirchhoff died in 1887 and in 1891 Boltzmann published a supplement.]

Kirschke, Siegrfried & Wolfgang Eisenberg eds.), *Ludwig Boltzmanns Werk in naturwissenschaftlicher und philosophischer Sicht*, Halle, 1987 [For more information see Bibliography I, Collections, *HALLE 1987*.]

Kirsten, Christa & Hans-Gunter Körber, *Physiker über Physiker, Wahlvorschläge zur Aufnahme von Physikern in die Berliner Akademie 1870-1929*, Berlin, 1975.

Kiss, Endre, *Der Tod der K.U.K. Weltordnung in Wien*, Hermann Böhlaus Nachfolger: Vienna-Cologne-Graz, 1986.

Klein, Martin J. (ed.), *Paul Ehrenfest, Collected Scientific Papers*, Amsterdam, 1959.

Klein, Martin J., *Paul Ehrenfest*, Vol. I, , American Elsevier Pub. Co.: New York, 1970.

Klein, Martin, "Mechanical Explanation at the End of the Nineteenth Century", *Centaurus*, **17** (1972) 58-82.

Klein, Martin J, "The Development of Boltzmann's Statistical Ideas", *EQUATION 1972*, pp. 53-107.

Klein, Martin J, "The Maxwell-Boltzmann Relationship", in J. Kestin (ed.), *Transport Phenomena 1973*, American Institute of Physics: New York, 1973, pp. 297-308.

Klein, Martin J., "Boltzmann, Monocycles, and Mechanical Explanation", in *Philosophical Foundations of Science*, edited by R.J. Seeger & R.S. Cohen, (Boston Studies in the Philosophy of Science, Volume **11**), Reidel: Dordrecht, 1974, pp. 155-175.

Klein, Martin J., et al. (eds.), *The Collected Papers of Albert Einstein*, Vols. **III** & **V**, Princeton University Press: Princeton, both volumes 1993. [For more information see Bibliography I, Collections, *EINSTEIN 1989*. The full list of editors, associate editors, assistant editors, contributing editors, and editorial assistants seems to shift from volume to volume in this very important series. Many will be referred to in the index.]

Kleinpeter, Hans, *Der Phänomenalismus - Eine naturwissenschaftliche Weltanschauung*, J.A. Barth: Leipzig, 1913. [This book by a follower of Ernst Mach seems to have helped end the use of the term *Phänomenologie* as a general expression for presentist philosophies, replacing it with *Phänomenalismus* (or phenomenalism in English) while reserving the term *Phänomenologie* for Husserl's philosophy and similar approaches Strangely, only physicists and a few psychologists seem to have kept the general use of the word "Phänomenologie". Mach and Boltzmann writing *before* Höfler (1910) and Kleinpeter (1913) also still used "Phänomenologie" as the general philosophical label for presentist or empirical world views such as those of Berkeley and Hume and Mach himself, though Boltzmann sometimes called them forms of *Idealismus*.]

Koch, Martin, "Ludwig Boltzmann. Ein Vorläufer der Quantentheorie? Bemerkungen zum Verhältnis von klassischer und nichtklassischer Physik", *Wissenschaftlicher Zeitschrift der Humboldt Universität zu Berlin*. Math. Naturw. Reihe, **32** (1983) 317-320.

Koch, Martin, "Naturdialektik und Atomistik, Ludwig Boltzmann und die Philosophie", *HALLE 1987*, pp. 56-74.

Koch, Martin, "From Boltzmann to Planck: On Continuity in Scientific Revolutions", in *World Views and Scientific Discipline Formation*, edited by W.R. Woodward & R.S. Cohen, Kluwer: Dordrecht, 1991, pp. 141-150.

Koenigsberger, Leo, *Hermann von Helmholtz*, Vol. **I**, Braunschweig 1902. [An abridged English translation by F. S. Welby with preface by Lord Kelvin appeared in 1906. The book includes parts of several letters from Boltzmann to Helmholtz.]

Koenigsberger, Leo, *Mein Leben*, Winter: Heidelberg, 1919.

Köhler, W., "Boltzmann und 2. Hauptsatz", *Erkenntnis*, 1932, pp. 336-353. Also said to be included in *Annalen der Philosophie* (Leipzig).

Körber, Hans-Günter (ed.), *Aus dem wissenschaftlichen Briefwechsel Wilhelm Ostwalds*, Vol. **I**, Akademie Verlag: Berlin, 1961.

Körber, Hans-Günter (Co-editor of book, 1975. See Christa Kirsten.)

Kowalewski, Gerhard, *Bestand und Wandel - Meine Lebens Erinnerungen*, Oldenbourg: Munich, 1950.

Kox, Anne, "The Correspondence Between Boltzmann and H.A. Lorentz", *TAGUNG 1981* pp. 73-86.

Kreuzer, Franz, "Im Gespräch mit Engelbert Broda und Rupert Riedl" in *Ich bin also denke ich. Die Evolutionäre Erkenntnistheorie aus Anlass des 75. Todestages von Ludwig Boltzmann*, Franz Deuticke: Vienna, 1981. [This book also includes two articles from Boltzmann, "Der zweite Hauptsatz der mechanischen Wärmetheorie" (1886) and "Über die Frage nach der objektiven Existenz der Vorgänge in der unbelebten Natur" (1897), pp. 80-97 & 89-120.]

Kubo, Ryogo, "Statistical mechanics: A Survey of the One Hundred Years", in *Scientific Culture in the Contemporary World*, edited by Vıtorio Mathieu & P. Rossi, Scientia: Milan, 1979, pp. 131-157.

Kuhn, Thomas S., *Black-Body Theory and the Quantum Discontinuity 1894-1912*, Clarendon Press: Oxford, 1978.

Laass, Andreas, (Co-author of book, 1989. See Herbert Hörz.)

Laitko, Hubert, (Co-editor of book, 1991, See Dieter Hoffmann.)

Lange, F.A., *Geschichte des Materialismus und Kritik seiner Bedeutung in der Gegenwart*, 2 Bände, Reclam: Iserlohn, 1866. [2nd edition 1873-1875]

Larmor, Sir Joseph, *Aether and Matter,* Cambridge, 1900.

Laue, Max von, "Zu Ludwig Boltzmanns 100. Geburtstag", *Forschungen und Fortshritte,* **20** (1944) Nr. 4, 5, & 6, pp. 46-47. [Also included in *Gesammelte Schriften,* 1961.]

Lenard, Philipp, *Grosse Naturforscher,* J.F. Lehmanns Verlag: Munich, 1929.

Lenin, Vladimir, *Materialism and Empirio-Criticism - Critical Comments on a Reactionary Philosophy,* International Publishers: New York, 1927. [1909] Vladimir Lenin praises Boltzmann and criticizes Mach, an attitude repeated many times in leftwing literature since 1909.]

Leplin, Jarret (ed.), *Scientific Realism,* University of California Press: Berkeley & Los Angeles, 1984.

Lestienne, Rémy, "A la mémoire de Ludwig Boltzmann: l'entropie est-elle objective?", *Fundamenta Scientiae,* **8** (1987) 173-184.

Lewes, G.H., *A Biographical History of Philosophy,* George Routledge & Sons: Broadway, Ludgate Hill, London, 1902. [1845-46]. Reprinted as well in 1859 and 1970. While some Positivists think their philosophy is the only truly scientific one, Lewes' book takes the opposite approach as if Positivism were not a philosophy at all but merely a methodology opposed to all philosophy. This approach appealed to Boltzmann in 1903.

Lienert, M. (Co-author-editor of an article, 1987. See K. -H. Adolph.)

Linke, Paul F., "Franz Brentano", *Neue Deutsche Biographie,* Vol. **II**, Duncker & Humboldt: Berlin, 1971, pp. 593-595.

Loeb L., *Kinetic Theory of Gases,* McGraw-Hill: New York, 1922. [2nd edition 1934]

Lorentz, H.A., "Ludwig Boltzmann (Gedächtnisrede)", *Collected Papers,* Vol. **IX**, The Hague, 1939, pp. 359-390.

Loschmidt, Josef, *Über die Natur des Aethers,* Gerold & Sohn: Vienna, 1862.

Loschmidt, Josef, "Über den Zustand des Wärmegleich gewichtes eines System von Körpern", *Die Sizungsberichte der Akademie der Wissenschaften, Wien,* Parts 1 & 2, **73** (1876) 128-142 & **73** (1876) 366-372.

Lovejoy, Arthur, *The Revolt Against Dualism,* Open Court: La Salle, 1960 [1930]

Mach, Ernst, *Die Geschichte und die Wurzel des Satzes von der Erhaltung der Arbeit,* Calve: Prague, 1872. 2nd edition 1909. English 1911.

Mach, Ernst, *Beiträge zur Analyse der Empfindungen,* G. Fischer: Jena, 1886. five editions by 1906, and English, Russian, and Italian editions by 1904. [The title is slightly changed in later editions by dropping the words *Beiträge zur.*]

Mach, Ernst, *Die Principien der Wärmelehre.* Historisch-kritisch entwickelt. J.A. Barth: 1896. 2nd edition 1900. [English translation: *Principles of the Theory of Heat,* edited by Brian McGuinness with an introduction by Matin Klein, Reidel: Dordrecht, 1986.]

Mach, Ernst, *Erkenntnis und Irrtum. Skizzen zur Psychologie der Forschung*. J.A. Barth: Leipzig, 1905. 2nd edition 1906. [English translation: *Knowledge and Error*, edited by Brian McGuinness, Reidel: Dordrecht, 1976.]

Mache, Heinrich, "Österreichische grossen Physiker und ihre Spitzenleistungen", in *Schriften des Pedagog. Inst. der Stadt Wien*, Heft 13, 1936.

Marek, Johann C., (Co-editor of a book, 1988. See Franz Brentano.)

Martin, H. de, *Josef Loschmidt*, Doctoral Dissertation for the University of Vienna, Vienna, circa 1950. [This dissertation includes four letters from Boltzmann to Loschmidt, all from the 1890's.]

Mason, Stephen F., "German Nature-Philosophy", *History of the Sciences*, Collier: New York, 9th printing 1975 [1956], pp. 349-362.

Maxwell, James Clerk, *The Scientific Papers of James Clerk Maxwell*, Cambridge University Press: Cambridge, 1890. [Reprinted 1952 and in French 1927.]

Maxwell, James Clerk, *Über Faradays Kraftlinien*, translated by Ludwig Boltzmann with 31 pages of additional notes, Leipzig, 1895.

Mayerhöfer, Josef, "Der Kampf um die Atomistik und die Relativitätstheorie in Wien", *Österreichische Gesellschaft für Geschichte der Naturwissenschaften - Mitteilungen*, **12** (1992) 23-50.

McCormack, Thomas J., (book review), Ludwig Boltzmann, *Vorlesungen über die Principe der Mechanik*, Vol. **I**, J.A. Barth: Leipzig, 1897 in *The Monist*, **9** (April 1899) 453-458.

McCormmach, Russell, (Co-author of a book, 1986. See Christa Jungnickel.)

McGuinness, Brian, (Editor of two books. See Ernst Mach.)

Meitner, Lise, "Looking Back", *Bulletin of the Atomic Scientists*, **20** (Nov. 1964) 2-7.

Meyenn, Karl von, "Boltzmann als Kritiker und Rezensent", *TAGUNG 1981*, pp. 97-128. [This article includes 11 letters from Boltzmann to Felix Klein, 1896-1899.]

Meyer, Stefan, & Egon von Schweidler, "Über das Verhalten von Radium und Polonium im magnet. Felde", *Physikalische Zeitschrift*, Jahrgang 2, Volume **I**, 1900.

Meyer, Stefan (ed.), *Boltzmann Festschrift*, J.A. Barth: Leipzig, 1904.

Meyer, Stefan, "Die Vorgeschichte der Gründung und das erste Jahrzehnt des Institutes für Radiumforschung", *Sitzungsberichte der Österreichische Akademie der Wissenschaften*, **159** (1950) 1-26.

Meyer, Stefan, *Festschrift zum 40 Jährigen Bestand des Wiener Instituts für Radiumforschung*, Springer: Vienna, 1950.

Miller, Arthur I., "On the Origins, Methods, and Legacy of Ludwig Boltzmann's Mechanics", *TAGUNG 1981*, pp. 231-262.

Millikan, Robert, *The Autobiography of Robert A. Millikan*, New York, 1950.

Mitter, H. & Paul Urban, "Ludwig Boltzmann und die Entwicklung der statistischen Physik", *DEUTSCHES MU 1985*, pp. 11-27.

Molisch, H., (Co-author of an article, 1933. See Felix Ehrenhaft.)

Moore, Walter, *Schrödinger - Life and Thought*, Cambridge University Press: Cambridge, 1992. [This biography originally published in 1989 shows significant research and investigates Schrödinger's attraction toward the ideas of Boltzmann and Mach.]

Motz, Hans, "Did the Germ of General relativity Come from Boltzmann?", *TAGUNG 1981*, pp. 355-362.

Moyer, Albert E., *American Physics in Transition - A History of Conceptual Change in the Late Nineteenth Century*, (A Series in the History of Modern Physics Volume **III**), Tomash: Los Angeles & San Francisco, 1983. [This book gives an interpretation of the atmosphere in America and especially St. Louis when Boltzmann lectured there in 1904.]

Moyer, Albert, "Foreword", Physics for a New Century - Papers Presented at the 1904 St. Louis Congress, edited by Katherine R. Sopka, Tomash: Los Angeles and San Francisco, 1986, pp. xiii-xx.

Müller, A.I., "Ludwig Boltzmann als Philosoph", *Philosophisches Jahrbuch der Görresgesellschaft*, **20** (1907) 85-90.

Müller, R., "Ludwig Boltzmann", *Familie und Volk - Zeitschrift für Genealogie und Bevölkerungskunde*, **5** (1956) 213-216.

Nabl, Josef, (Co-author of article, 1907. See Ludwig Boltzmann.)

Niedersen, Uve, "Ludwig Boltzmann in seinen Beziehungen zu Wilhelm Ostwald, *HALLE 1987*, pp. 49-55.

Nye, Mary Jo, *Molecular Reality: A Perspective on the Scientific Work of Jean Perrin*, American Elsevier: New York, 1972.

Nye, Mary Jo (ed.), *The Question of the Atom - From the Karlsruhe Congress to the First Sovay Conference, 1860-1911*, (A Series in the History of Modern Physics 1800-1950, Volume **IV**), Tomash: Los Angeles & San Francisco, 1984. [This book includes selections from various supporters of atomism such as Boltzmann as well as several opponents including Mach, Ostwald, and Duhem.]

Nyhof, John, *Instrumentalism and Beyond*, PHD Dissertation, University of Otago, Dunedin, New Zealand, 1981. [Boltzmann is much discussed.]

Nyhof, John, "Philosophical Objections to the Kinetic Theory", *British Journal for the Philosophyof Science*, **39** (1988) 81-109.

Nyiri, J.C. (ed.), *Von Bolzano zu Wittgenstein - Zur Tradition der Österreichischen Philosophie*, Verlag Hölder-Pichler-Tempsky: Vienna, 1986. [For more information see Bibliography I, Collections, *AUSTRIA 1986*.]

Ostwald, Grete, *Mein Vater*, Berliner Union: Stuttgart, 1953.

Ostwald, Wilhelm, "Die Überwindung des wissenschaftlichen Materialismus", *Verhandlungen Ges. Deutscher Naturforscher und Aertzte*, (Leipzig), Part I, 1895, p. 155-168.

Oswald, Wilhelm, "Zur Energetik", *Annalen der Physik*, **58** (1896) 154-167.

Ostwald, Wilhelm, *Abhandlungen und Vorträge allgemeines Inhaltes (1887-1903)*, Veit: Leipzig, 1904.

Ostwald, Wilhelm, "A Theory of Happiness", *The International Quarterly*, Volume XI, July 1905, p. 316-.

Ostwald, Wilhelm, "Energische Theorie des Glücks", *Annalen der Naturphilosophie*, **4** (1905) 457-74.

Ostwald, Wilhelm, "The Modern Theory of Energetics", *The Monist*, **17** (1907) 481-515.

Ostwald, Wilhelm, *Grosse Männer - Studien zur Biologie des Genies*. Akademische Verlagsgesellschaft: Leipzig, 1909. [5th edition 1919]

Ostwald, Wilhelm, *Lebenslinien. Eine Selbstbiographie*, 3 vols., Klasing: Berlin, 1926-27.

Petronievics, Branislav, *Principien der Metaphysik*, Volume I, Heidelberg, 1904.[It seems that in Serbo-Croatian the name is spelled "Petronijevic", which is how we have normally spelled it. But his German and English articles use the other spelling.]

Petronievics, Branislav, "A Multi-Dimensional Space Conception", *The Monist*, **17** (1907) 630-631. [This is a friendly reply to a review pp. 316-317 of his book on metaphysics mentioned above. Petronijevic is defending a theory of mathematical discontinuity as well as a theory of multi-dimensions, which may be partly why Boltzmann (and Mach) were interested in his ideas.]

Planck, Max, "Gegen die neuere Energetik", *Annalen der Physik*, **57** (1896) 72-78.

Planck, Max, "Die Einheit des physikalischen Weltbildes", *Physikalische Zeitschrift*, **10** (1909) 62-75

Planck, Max, "Persönliche Erinnerungen aus alten Zeiten", *Naturwissenschaften*, **33** (1946) 220-235.

Planck, Max, *Vorträge und Erinnerungen . Fünfte Auflage der Wege zur physikalischen Erkenntnis*, Stuttgart, 1949.

Plato, Jan von, "Boltzmann's Ergodic Hypothesis", *Archive of the History of Exact Sciences*, **42** (1991) 71-89.

Poincaré, Henri, "Sur le probleme des trois corps et les equations des dynamique", *Acta Mathematica*, **13** (1889) 1-270.

Poincaré, Henri, "Mechanism and Experience [1893]", in *KINETIC 1966*, pp. 203-207.

Poincaré, Henri, *Science and Hypothesis*, Dover [1903]: New York, 1952.

Poincaré, Henri, *Ouvres*, Gauthier-Villars: Paris, 1951-1954.

Popper, Sir Karl, "The Arrow of Time", *Nature*, **177** (1956) 538. See also: *Nature*, **178** (1956) 383, **179** (1957) 1297, and **181**(1958) 402-403.

Popper, Sir Karl, "Irreversibility or Entropy since 1905", *The British Journal for the Philosophy of Science*, **8** (1957) 151-155.

Popper, Sir Karl, "Probability Magic or Knowledge out of Ignorance", *Dialectica*, **11** (1957) 354-372.

Popper, Sir Karl, "Time's Arrow and Entropy", *Nature*, **207** (1965) 233-234. See also **213** (1967) 320, **214** (1967) 322.

Popper, Sir Karl, "Autobiography of Karl Popper" in *The Philosophy of Karl Popper*, edited by P.A. Schilpp, Volume I, Open Court: La Salle, 1974, p. 3-181.

Popper, Sir Karl, "Boltzmann and the Arrow of Time" in *Unended Quest - An Intellectual Biography*, Fontan/Collins, Glasgow, 5th impression, 1980, pp. 156-162. [This book is basically the same as "Autobiography of Karl Popper" above.]

Pozdena, R., *Österreichische Jahrbuch*, **34** (1916) 1-27.

Pricha, Willibald, "Briefe von Ludwig Boltzmann an Arnold Sommerfeld, Carl Runge, und Leo Graetz", *Mitteilungen der Österreichischen Gesellschaft für Geschichte der Naturwissenschaften*, **3** (1983) 14-23.

Pricha, Willibald, (Co-author of an article, 1984. See Michael Eckert.)

Prigogine, Ilya, *From Being to Becoming - Time and Complexity in the Physical Sciences*, W.H. Freeman: San Francisco, 1980.

Prigogine, Ilya & Isabelle Stengers, *Dialog mit der Natur - Neue Wege naturwissenschaftliche Denkens*, Piper: Munich, 1981.

Prigogine, Ilya & Isabelle Stengers, *Order Out of Chaos*, Banatam Books: Toronto, 1984.

Przibram, Karl, "Erinnerungen an ein altes physikalisches Institut" in *Beiträge zur Physik und Chemie des 20 Jahrhunderts*, Verlag Vieweg: Braunschweig, Year?, pp. 1-6. [This article describes the old Türkenstrasse laboratories of Lang, Exner, and Boltzmann, that is, before the large physical institute was completed in 1913.]

Przibram, Karl, "Erinnerungen an Boltzmanns Vorlesungen", *EQUATION 1972*, pp. 641-642.

Rabel, Gabriele, "Mach und die 'Realität der Aussenwelt'", *Physikalische Zeitschrift*, **21** (1920) 433-437.

Ramunni, Girolamo, "Peut-on analyser le langage scientifique sans se soucier de l'outil mathematique? Le cas de Boltzmann confronté `a celui de Ostwald", *Documents pour l-histoire du voocabulaire scientifique*, (Institute national de la langue Francaise, CNAS), **8** (1986) 121-132.

Reichenbach, Hans, *The Direction of Time*, University of California Press: Berkeley and Los Angeles, 1956. 2nd edition 1971.

Renn, Jürgen & Robert Schulmann (eds.), *Albert Einstein & Mileva Maric - The Love Letters*, translated by Shawn Smith, Princeton University Press: Princeton, 1992.

Röd, Wolfgang, "Alois Riehl und der Herbartianismus in Österreich", *AUSTRIA 1986*, pp. 132-140. [This article gives some information on Boltzmann's philosophy teacher, Robert Zimmermann, at the University in Vienna during the 1860's.]

Rohrlich, Fritz, "A Poem by Ludwig Boltzmann", *American Journal of Physics*, **60** (1992) 972-973.

Sannochkin, Yu. v., (co-author of article, 1957. See N.N. Bogolyubov.)

Schaefer, Clemens, "Ludwig Boltzmann", *Naturwissenschaften*, **33** (1946) 33.

Scheibe, E., "The Physicists' Conception of Progress", *Studies in History and Philosophy of Science*, **19** (1989) 141-159.

Schmid, Erich, (Co-author of book, 1982. See Berta Karlik.)

Schmitz-Dumont, Oskar, *Zeit und Raum in ihren Denknotwendigen Bestimmungen abgeleitet aus dem satze d. Widerspruche*, (Volume **I** of *Zur Filosofia der Wissenschaften*), Koschny: Leipzig, 1875.

Schmitz-Dumont, Oskar, *Bedeutung der Pangeometrie*, Koschny: Leipzig, 1877.

Schmitz-Dumont, Oskar, *Die mathematische Elemente der Erkenntnistheorie*, C. Duncker: Berlin, 1878.

Schmitz-Dumont, Oskar, *Lichtäther und elektrische Welle. Eine Weiterführung der Maxwellschen Wellentheorie*, Hückner: Dresden, 1890.

Schmitz-Dumont, Oskar, *Naturphilosophie als exakte Wissenschaft mit bes. Berücksicht. der matemat. Physik*, Duncker und Humblot: Leipzig, 1896.

Schreter, Jorg, *Zur Kritik der philosophischen Grundposition des Wienerkreises*, Berlin, 1977.

Schrödinger, Erwin, "Irreversibility", *Proceedings of the Royal Irish Academy*, **53A** (1950) 189-195.

Schulmann, Robert, (Co-author of book, 1992. See Jürgen Renn.)

Scott, W.T., *Erwin Schrödinger*, University of Massachusetts Press: Amherst, 1967. [Schrödinger alleges that Boltzmann's teachings were "his first love in science".]

Seeger, R.J. & R.S. Cohen (eds.), *Philosophical Foundations of Science*, Dordrecht, 1974. [For more information see Bibliography I, Collections.]

Sexl, Roman & John Blackmore (eds.), *TAGUNG 1981*, Vienna, 1982. [See Bibliography I, Collections for more information.]

Smekal, Adolf. "Adiabaten Hypothese v. Boltzmannsche Prinzip", *Physikalische Zeitschrift*, **19** (1918) 137-142.

Smekal, Adolf. G., *Mitteilungen des naturwissenschaftlichen Vereins für Steiermark*, (Graz), **87** (1957) 12-19.

Smoluchowski, Marian von, "Gültigkeitsgrenzen des zweiten Hauptsatzes der Wärmelehre" in *Vorträge über die kinetische Theorie der Materie und Elektrizität*, edited by Peter Debye, Teubner: Leipzig, 1914.

Sommerfeld, Arnold, "Dem Andenken Ludwig Boltzmanns", *Enzyklopädie Math. Wissenschaften*, Vol. V., Part 1, Teubner: Leipzig, 1907, between pp. 492 & 493.

Sommerfeld, Arnold, "Das Werk Boltzmanns", *Wiener Chemiker Zeitung*, **47** (1944) 25-28.

Sommerfeld, Arnold, "Ludwig Eduard Boltzmann", *Neue Deutsche Biographie*, zweiter Band, Duncker & Humblot: Berlin, 1971, pp. 436-437.

Sopka, Katherine R. (ed.), *Physics for a new Century - Papers Presented at the 1904 St. Louis Congress*, Tomash: Los Angeles & San Francisco, 1986. [Reprinted from *Congress of Arts and Science*, edited by H.J. Rogers, Houghton, Mifflin: Boston and New York, 1905. This book contains lectures by Boltzmann, Ostwald, and many others. See Bibliography I, Collections, for more information.]

Spretke, Ulf, "Die Kontroverse zwischen Planck und Ludwig Boltzmann - ein fruchtbringender wissenschaftlicher Meinungsstreit", *HALLE 1987*, pp. 31-44.

Stachel, John (ed.), *The Collected Papers of Albert Einstein*, Volumes **I** & **II**, Princeton University Press: Princeton, 1987 & 1989. [For more informations see Bibliography I, Collections, and *EINSTEIN 1989*.]

Stadler, Friedrich, *Vom Positivismus zur wissenschaftliche Weltauffassung - Am Beispiel der Wirkungsgeschichte von Ernst Mach in Österreich von 1895 bis 1934*, Löcker: Vienna & Munich, 1982.

Stadler, Friedrich (ed.), *Scientific Philosophy: Origins and Developments*, (Vienna Circle Institute Yearbook [1993], Volume **I**), Kluwer: Dordrecht, 1993. [Even though the title is misleading since it only includes articles influenced by the Positivist-Analytic tradition, nevertheless, many of them are useful and some reflect traces of Boltzmann's ideas.]

Stadler, Friedrich, (Co-editor of two books, 1988 & 1993. See Rudolf Haller.)

Steckline, Vincent S., "Zermelo, Boltzmann, and the Recurrence Paradox", *American Journal of Physics*, **51** (1983) 894-897.

Stengers, Isabelle, (Co-author of two books, 1981 & 1984. See Ilya Prigogine.)

Stiller, Wolfgang, "Ludwig Boltzmann - Pioneer of Atomistics and Evolution", *Isotopenpraxis*, **22** (1986) 257.

Stiller, Wolfgang, *Ludwig Boltzmann - Altmeister der klassischen Physik - Wegbereiter der Quantenphysik und Evolutionstheorie*, Verlag Harri Deutsch: Thun & Frankfurt, am Main, 1989. [This book includes much information on Boltzmann's colleagues, students, and opponents. It also tries to make his physics understandable.]

Stöber, Konrad, "Populationsdenken bei Boltzmann, Darwin, und Mendel", *HALLE 1987*, pp. 118-128.

Streintz, Franz, "Ludwig Boltzmanns 60. Geburtstag", *Grazer Tagblatt*, Nr. 51, February 20, 1904.

Tait, P.G., *Scientific Papers*, Cambridge University Press: Cambridge, 1898-1900.

Tandler, Julius, (Co-author of an article, 1933. See Felix Ehrenhaft.)

Teske, Armin., *Marian Smoluchowski*, Wroclaw, 1977. [This book includes information from Gustav Jäger about Boltzmann.]

Thiele, Joachim, "Briefe von Gustav Theodor Fechner und Ludwig Boltzmann an Ernst Mach", *Centaurus*, **11** (1966) 222-235.

Thiele, Joachim, *Wissenschaftliche Kommunikation - Die Korrespondenz Ernst Machs*, A. Henn Verlag: Kastellaun, 1978.

Thiele, Joachim, "Briefe Hugo von Seeliger an Ernst Mach" in *Philosophia Naturalis*, **17** (1979) 391-399.

Thirring, Hans, (Co-author of an article, 1933. See Felix Ehrenhaft.)

Thirring, Hans, "Ludwig Boltzmann", *Physik und Chemie*, **37** (1937) 1-6. See also: *Journal of Chemical Education*, **29** (1952) 298-300; *Acta Physica Austriaca*, **11** (1956) 1-10; *Arbeiter Zeitung*, Nr. 256, Sept. 5, 1956, p. 5; and *Naturwissenschaftliche Rundschau*, **11** (1957) 411-415.

Thirring, Hans, "Ludwig Flamm", *Almanach für das Jahr 1965 / Österreichische Akademie der Wissenschaften*, **115** (1966) 387-398.

Thirring, Walter, (Co-editor of a book, 1973. See E.G.D. Cohen and Bibliography I, Collections, *EQUATION 1972*.)

Tomás, Milan, "Philosophical Portrait of Ludwig Boltzmann with Stress on His Philosophical Thought in the Exact Sciences" [in Czech], *Rozpravy Ceskoslovenské Akademie ved rada spolecenskych ved, roc*, ses 4., Akademie: Prague, 1970.

Treder, Hans-Jürgen, "Boltzmanns Kosmologie und die räumliche und zeitliche Unendlichkeit der Welt", *Wiss. z. Humboldt Universität*, naturwiss. Reihe, **26** (1977) 13-15.

Treder, Hans-Jürgen, *Grosse Physiker und ihre Probleme*, Akademie Verlag: Berlin, 1983.

Uebel, Thomas (ed.), *Rediscovering the Vienna Circle - Austrian Studies on Otto Neurath and the Vienna Circle*, (Boston Studeies in the Philosophy of Science, Volume **133**.) Kluwer: Dordrecht, 1991.

Urban, Paul, "Ludwig Boltzmann in Graz", *TAGUNG 1981*, pp. 57-72.

Urban, Paul, (Co-author of an article, 1985. See H. Mitter and Bibliography I, Collections, *DEUTSCHES MU 1985*.)

Visser, Henk, "Wittgenstein's Debt to Mach's Popular Scientific Lectures", in *Mind*, **91** (1982) 102-105.

Visser, Henk, "Mach's Method in Wittgenstein's Later Philosophy", in *8. Internationalen Wittgenstein Symposiums* [1983], edited by R, Haller & W.L. Gombocz, Hölder-Pichler-Tempsky: Vienna, 1984.

Wagner, Siegfried, "Ludwig Boltzmann and the Special Theory of Relativity", *TAGUNG 1981*, pp. 341-354.

Wald, Frantisek, *Die Energie und ihre Entwertung, Studien über den zweiten Hauptsatz der mechanischen Wärmetheorie*, Engelmann: Leipzig, 1889.

Wallner, Friedrich, "Boltzmann, Hertz, and Wittgenstein", *TAGUNG 1981*, pp. 143-154.

Wedberg, Anders, *A History of Philosophy*, Volume **3**, Translated by Bergljot Wedberg, Clarendenon Press: Oxford, 1984.

Wehrl, Alfred, "Entropie seit Boltzmann", *TAGUNG 1981*, pp. 329-340.

Wermes, Peter, "Atomismus und naturwissenschaftlicher Materialismus bei Ludwig Boltzmann", *HALLE 1987*, pp. 75-92.

Wiesner, A.R., *Die Geschichte des Faches Philosophie an der Universität Wien 1948-1938*, Dissertation (unpublished), Vienna, 1950.

Wilson, Andrew D., "Hertz, Boltzmann, and Wittgenstein Reconsidered", *Studies in the History and Philosophy of Science*, **20** (1989) 245-263.

Wilson, Andrew D., *Representing Reality: Ludwig Boltzmann and the Nature and Purpose of Theoretical Physics*, Cornell U..Dissertation, 1990. University Microfilms Order No. 89-24534, 457 pages. [The reader should be warned, however, that the author has apparently forbidden University Microfilms to release copies, something which is sometimes done at the behest of a potential publisher.]

Wilson, Andrew D., "Mental Representation and Scientific Knowledge: Boltzmann's *Bild* Theory of Knowledge in Historical Context", *Physis*, **28** (1991) 769-795.

Winteler-Einstein, Maja, "Albert Einstein, Beitrag für sein Lebensbild", Typescript, February 15, 1924. [She mentiones that Einstein corresponded with Boltzmann.]

Wood, Ledger, "Recent Epistemological Schools", in *A History of Philosophical Systems*, edited by Vergilius Ferm, The Philosophical Library: New York, 1950, pp. 516-539.

Zermelo, Ernst, "Über einen Satz der Dynamik und die mechanische Wärmetheorie", *Annalen der Physik*, **57** (1896) 485-494. [Translated into English by Stephen Brush in *KINETIC 1966*, pp. 208-217.]

Zermelo, Ernst, "Über mechanische Erklärungen irreversibler Vorgänge - Eine Antwort auf Hrn. Boltzmanns Entgegnung", *Annalen der Physik*, **59** (1896) 793-801. [Translated into English by Stephen Brush in *KINETIC 1966*, pp. 229-237.]

Ziesche, P. (Co-editor of an article, 1987. See K. -H. Adolph.)

Zimmel, Brigitte & Gabriele Kerber, *Hans Thirring, ein Leben für Physik und Frieden*, Böhlau: Vienna, 1992.

Zimmermann, Robert, *Leibniz Monodologie* (1847). For other books see: *Leibniz und Herbart* (1849); *Philosophische Propadeutik* (1852 & 1867); *Leibniz und Lessing* (1855); *Über das Tragische und die Tragödie* (1856); *Schiller als Denker* (1859); *Philosophie und Erfahrung* (1861), *Allgemeine Ästhetik* (1865); and *Kant und die positive Philosophie* (1874). [Zimmermann was Boltzmann's philosophy professor in Vienna during the middle 1860's.]

Zuckerkandl-Szeps, Berta, *Österreich - Intim. Erinnerungen, 1892-1942*, Hrsg. Reinhard Federmann: Frankfurt am Main, 1970.

# NAME INDEX

(**Boldface** is for correspondence and *italics* for bibliography.)

68. W. Krajewski (ed.): *Polish Essays in the Philosophy of the Natural Sciences.* Translated from Polish and edited by R.S. Cohen and C.R. Fawcett. 1982 ISBN 90-277-1286-7; Pb 90-277-1287-5
69. J.H. Fetzer: *Scientific Knowledge.* Causation, Explanation and Corroboration. 1981 ISBN 90-277-1335-9; Pb 90-277-1336-7
70. S. Grossberg: *Studies of Mind and Brain.* Neural Principles of Learning, Perception, Development, Cognition, and Motor Control. 1982 ISBN 90-277-1359-6; Pb 90-277-1360-X
71. R.S. Cohen and M.W. Wartofsky (eds.): *Epistemology, Methodology, and the Social Sciences.* 1983. ISBN 90-277-1454-1
72. K. Berka: *Measurement.* Its Concepts, Theories and Problems. Translated from Czech. 1983 ISBN 90-277-1416-9
73. G.L. Pandit: *The Structure and Growth of Scientific Knowledge.* A Study in the Methodology of Epistemic Appraisal. 1983 ISBN 90-277-1434-7
74. A.A. Zinov'ev: *Logical Physics.* Translated from Russian. Edited by R.S. Cohen. 1983 ISBN 90-277-0734-0
 *See also* Volume 9.
75. G-G. Granger: *Formal Thought and the Sciences of Man.* Translated from French. With and Introduction by A. Rosenberg. 1983 ISBN 90-277-1524-6
76. R.S. Cohen and L. Laudan (eds.): *Physics, Philosophy and Psychoanalysis.* Essays in Honor of Adolf Grünbaum. 1983 ISBN 90-277-1533-5
77. G. Böhme, W. van den Daele, R. Hohlfeld, W. Krohn and W. Schäfer: *Finalization in Science.* The Social Orientation of Scientific Progress. Translated from German. Edited by W. Schäfer. 1983 ISBN 90-277-1549-1
78. D. Shapere: *Reason and the Search for Knowledge.* Investigations in the Philosophy of Science. 1984 ISBN 90-277-1551-3; Pb 90-277-1641-2
79. G. Andersson (ed.): *Rationality in Science and Politics.* Translated from German. 1984 ISBN 90-277-1575-0; Pb 90-277-1953-5
80. P.T. Durbin and F. Rapp (eds.): *Philosophy and Technology.* [*Also* Philosophy and Technology Series, Vol. 1] 1983 ISBN 90-277-1576-9
81. M. Marković: *Dialectical Theory of Meaning.* Translated from Serbo-Croat. 1984 ISBN 90-277-1596-3
82. R.S. Cohen and M.W. Wartofsky (eds.): *Physical Sciences and History of Physics.* 1984. ISBN 90-277-1615-3
83. É. Meyerson: *The Relativistic Deduction.* Epistemological Implications of the Theory of Relativity. Translated from French. With a Review by Albert Einstein and an Introduction by Milić Čapek. 1985 ISBN 90-277-1699-4
84. R.S. Cohen and M.W. Wartofsky (eds.): *Methodology, Metaphysics and the History of Science.* In Memory of Benjamin Nelson. 1984 ISBN 90-277-1711-7
85. G. Tamás: *The Logic of Categories.* Translated from Hungarian. Edited by R.S. Cohen. 1986 ISBN 90-277-1742-7
86. S.L. de C. Fernandes: *Foundations of Objective Knowledge.* The Relations of Popper's Theory of Knowledge to That of Kant. 1985 ISBN 90-277-1809-1

# Boston Studies in the Philosophy of Science

87. R.S. Cohen and T. Schnelle (eds.): *Cognition and Fact.* Materials on Ludwik Fleck. 1986 ISBN 90-277-1902-0
88. G. Freudenthal: *Atom and Individual in the Age of Newton.* On the Genesis of the Mechanistic World View. Translated from German. 1986 ISBN 90-277-1905-5
89. A. Donagan, A.N. Perovich Jr and M.V. Wedin (eds.): *Human Nature and Natural Knowledge.* Essays presented to Marjorie Grene on the Occasion of Her 75th Birthday. 1986 ISBN 90-277-1974-8
90. C. Mitcham and A. Hunning (eds.): *Philosophy and Technology II.* Information Technology and Computers in Theory and Practice. [*Also* Philosophy and Technology Series, Vol. 2] 1986 ISBN 90-277-1975-6
91. M. Grene and D. Nails (eds.): *Spinoza and the Sciences.* 1986 ISBN 90-277-1976-4
92. S.P. Turner: *The Search for a Methodology of Social Science.* Durkheim, Weber, and the 19th-Century Problem of Cause, Probability, and Action. 1986. ISBN 90-277-2067-3
93. I.C. Jarvie: *Thinking about Society.* Theory and Practice. 1986 ISBN 90-277-2068-1
94. E. Ullmann-Margalit (ed.): *The Kaleidoscope of Science.* The Israel Colloquium: Studies in History, Philosophy, and Sociology of Science, Vol. 1. 1986 ISBN 90-277-2158-0; Pb 90-277-2159-9
95. E. Ullmann-Margalit (ed.): *The Prism of Science.* The Israel Colloquium: Studies in History, Philosophy, and Sociology of Science, Vol. 2. 1986 ISBN 90-277-2160-2; Pb 90-277-2161-0
96. G. Márkus: *Language and Production.* A Critique of the Paradigms. Translated from French. 1986 ISBN 90-277-2169-6
97. F. Amrine, F.J. Zucker and H. Wheeler (eds.): *Goethe and the Sciences: A Reappraisal.* 1987 ISBN 90-277-2265-X; Pb 90-277-2400-8
98. J.C. Pitt and M. Pera (eds.): *Rational Changes in Science.* Essays on Scientific Reasoning. Translated from Italian. 1987 ISBN 90-277-2417-2
99. O. Costa de Beauregard: *Time, the Physical Magnitude.* 1987 ISBN 90-277-2444-X
100. A. Shimony and D. Nails (eds.): *Naturalistic Epistemology.* A Symposium of Two Decades. 1987 ISBN 90-277-2337-0
101. N. Rotenstreich: *Time and Meaning in History.* 1987 ISBN 90-277-2467-9
102. D.B. Zilberman: *The Birth of Meaning in Hindu Thought.* Edited by R.S. Cohen. 1988 ISBN 90-277-2497-0
103. T.F. Glick (ed.): *The Comparative Reception of Relativity.* 1987 ISBN 90-277-2498-9
104. Z. Harris, M. Gottfried, T. Ryckman, P. Mattick Jr, A. Daladier, T.N. Harris and S. Harris: *The Form of Information in Science.* Analysis of an Immunology Sublanguage. With a Preface by Hilary Putnam. 1989 ISBN 90-277-2516-0

## Boston Studies in the Philosophy of Science

105. F. Burwick (ed.): *Approaches to Organic Form*. Permutations in Science and Culture. 1987 ISBN 90-277-2541-1
106. M. Almási: *The Philosophy of Appearances*. Translated from Hungarian. 1989 ISBN 90-277-2150-5
107. S. Hook, W.L. O'Neill and R. O'Toole (eds.): *Philosophy, History and Social Action*. Essays in Honor of Lewis Feuer. With an Autobiographical Essay by L. Feuer. 1988 ISBN 90-277-2644-2
108. I. Hronszky, M. Fehér and B. Dajka: *Scientific Knowledge Socialized*. Selected Proceedings of the 5th Joint International Conference on the History and Philosophy of Science organized by the IUHPS (Veszprém, Hungary, 1984). 1988 ISBN 90-277-2284-6
109. P. Tillers and E.D. Green (eds.): *Probability and Inference in the Law of Evidence*. The Uses and Limits of Bayesianism. 1988 ISBN 90-277-2689-2
110. E. Ullmann-Margalit (ed.): *Science in Reflection*. The Israel Colloquium: Studies in History, Philosophy, and Sociology of Science, Vol. 3. 1988 ISBN 90-277-2712-0; Pb 90-277-2713-9
111. K. Gavroglu, Y. Goudaroulis and P. Nicolacopoulos (eds.): *Imre Lakatos and Theories of Scientific Change*. 1989 ISBN 90-277-2766-X
112. B. Glassner and J.D. Moreno (eds.): *The Qualitative-Quantitative Distinction in the Social Sciences*. 1989 ISBN 90-277-2829-1
113. K. Arens: *Structures of Knowing*. Psychologies of the 19th Century. 1989 ISBN 0-7923-0009-2
114. A. Janik: *Style, Politics and the Future of Philosophy*. 1989 ISBN 0-7923-0056-4
115. F. Amrine (ed.): *Literature and Science as Modes of Expression*. With an Introduction by S. Weininger. 1989 ISBN 0-7923-0133-1
116. J.R. Brown and J. Mittelstrass (eds.): *An Intimate Relation*. Studies in the History and Philosophy of Science. Presented to Robert E. Butts on His 60th Birthday. 1989 ISBN 0-7923-0169-2
117. F. D'Agostino and I.C. Jarvie (eds.): *Freedom and Rationality*. Essays in Honor of John Watkins. 1989 ISBN 0-7923-0264-8
118. D. Zolo: *Reflexive Epistemology*. The Philosophical Legacy of Otto Neurath. 1989 ISBN 0-7923-0320-2
119. M. Kearn, B.S. Philips and R.S. Cohen (eds.): *Georg Simmel and Contemporary Sociology*. 1989 ISBN 0-7923-0407-1
120. T.H. Levere and W.R. Shea (eds.): *Nature, Experiment and the Science*. Essays on Galileo and the Nature of Science. In Honour of Stillman Drake. 1989 ISBN 0-7923-0420-9
121. P. Nicolacopoulos (ed.): *Greek Studies in the Philosophy and History of Science*. 1990 ISBN 0-7923-0717-8
122. R. Cooke and D. Costantini (eds.): *Statistics in Science*. The Foundations of Statistical Methods in Biology, Physics and Economics. 1990 ISBN 0-7923-0797-6

123. P. Duhem: *The Origins of Statics.* Translated from French by G.F. Leneaux, V.N. Vagliente and G.H. Wagner. With an Introduction by S.L. Jaki. 1991 ISBN 0-7923-0898-0
124. H. Kamerlingh Onnes: *Through Measurement to Knowledge.* The Selected Papers, 1853-1926. Edited and with an Introduction by K. Gavroglu and Y. Goudaroulis. 1991 ISBN 0-7923-0825-5
125. M. Čapek: *The New Aspects of Time: Its Continuity and Novelties.* Selected Papers in the Philosophy of Science. 1991 ISBN 0-7923-0911-1
126. S. Unguru (ed.): *Physics, Cosmology and Astronomy, 1300-1700.* Tension and Accommodation. 1991 ISBN 0-7923-1022-5
127. Z. Bechler: *Newton's Physics on the Conceptual Structure of the Scientific Revolution.* 1991 ISBN 0-7923-1054-3
128. É. Meyerson: *Explanation in the Sciences.* Translated from French by M-A. Siple and D.A. Siple. 1991 ISBN 0-7923-1129-9
129. A.I. Tauber (ed.): *Organism and the Origins of Self.* 1991 ISBN 0-7923-1185-X
130. F.J. Varela and J-P. Dupuy (eds.): *Understanding Origins.* Contemporary Views on the Origin of Life, Mind and Society. 1992 ISBN 0-7923-1251-1
131. G.L. Pandit: *Methodological Variance.* Essays in Epistemological Ontology and the Methodology of Science. 1991 ISBN 0-7923-1263-5
132. G. Munévar (ed.): *Beyond Reason.* Essays on the Philosophy of Paul Feyerabend. 1991 ISBN 0-7923-1272-4
133. T.E. Uebel (ed.): *Rediscovering the Forgotten Vienna Circle.* Austrian Studies on Otto Neurath and the Vienna Circle. Partly translated from German. 1991 ISBN 0-7923-1276-7
134. W.R. Woodward and R.S. Cohen (eds.): *World Views and Scientific Discipline Formation.* Science Studies in the [former] German Democratic Republic. Partly translated from German by W.R. Woodward. 1991 ISBN 0-7923-1286-4
135. P. Zambelli: *The Speculum Astronomiae and Its Enigma.* Astrology, Theology and Science in Albertus Magnus and His Contemporaries. 1992 ISBN 0-7923-1380-1
136. P. Petitjean, C. Jami and A.M. Moulin (eds.): *Science and Empires.* Historical Studies about Scientific Development and European Expansion. ISBN 0-7923-1518-9
137. W.A. Wallace: *Galileo's Logic of Discovery and Proof.* The Background, Content, and Use of His Appropriated Treatises on Aristotle's *Posterior Analytics.* 1992 ISBN 0-7923-1577-4
138. W.A. Wallace: *Galileo's Logical Treatises.* A Translation, with Notes and Commentary, of His Appropriated Latin Questions on Aristotle's *Posterior Analytics.* 1992 ISBN 0-7923-1578-2
Set (137 + 138) ISBN 0-7923-1579-0

139. M.J. Nye, J.L. Richards and R.H. Stuewer (eds.): *The Invention of Physical Science.* Intersections of Mathematics, Theology and Natural Philosophy since the Seventeenth Century. Essays in Honor of Erwin N. Hiebert. 1992 ISBN 0-7923-1753-X
140. G. Corsi, M.L. dalla Chiara and G.C. Ghirardi (eds.): *Bridging the Gap: Philosophy, Mathematics and Physics.* Lectures on the Foundations of Science. 1992 ISBN 0-7923-1761-0
141. C.-H. Lin and D. Fu (eds.): *Philosophy and Conceptual History of Science in Taiwan.* 1992 ISBN 0-7923-1766-1
142. S. Sarkar (ed.): *The Founders of Evolutionary Genetics.* A Centenary Reappraisal. 1992 ISBN 0-7923-1777-7
143. J. Blackmore (ed.): *Ernst Mach – A Deeper Look.* Documents and New Perspectives. 1992 ISBN 0-7923-1853-6
144. P. Kroes and M. Bakker (eds.): *Technological Development and Science in the Industrial Age.* New Perspectives on the Science–Technology Relationship. 1992 ISBN 0-7923-1898-6
145. S. Amsterdamski: *Between History and Method.* Disputes about the Rationality of Science. 1992 ISBN 0-7923-1941-9
146. E. Ullmann-Margalit (ed.): *The Scientific Enterprise.* The Bar-Hillel Colloquium: Studies in History, Philosophy, and Sociology of Science, Volume 4. 1992 ISBN 0-7923-1992-3
147. L. Embree (ed.): *Metaarchaeology.* Reflections by Archaeologists and Philosophers. 1992 ISBN 0-7923-2023-9
148. S. French and H. Kamminga (eds.): *Correspondence, Invariance and Heuristics.* Essays in Honour of Heinz Post. 1993 ISBN 0-7923-2085-9
149. M. Bunzl: *The Context of Explanation.* 1993 ISBN 0-7923-2153-7
150. I.B. Cohen (ed.): *The Natural Sciences and the Social Sciences.* Some Critical and Historical Perspectives. 1994 ISBN 0-7923-2223-1
151. K. Gavroglu, Y. Christianidis and E. Nicolaidis (eds.): *Trends in the Historiography of Science.* 1994 ISBN 0-7923-2255-X
152. S. Poggi and M. Bossi (eds.): *Romanticism in Science.* Science in Europe, 1790–1840. 1994 ISBN 0-7923-2336-X
153. J. Faye and H.J. Folse (eds.): *Niels Bohr and Contemporary Philosophy.* 1994 ISBN 0-7923-2378-5
154. C.C. Gould and R.S. Cohen (eds.): *Artifacts, Representations, and Social Practice.* Essays for Marx W. Wartofsky. 1994 ISBN 0-7923-2481-1
155. R.E. Butts: *Historical Pragmatics.* Philosophical Essays. 1993 ISBN 0-7923-2498-6
156. R. Rashed: *The Development of Arabic Mathematics: Between Arithmetic and Algebra.* Translated from French by A.F.W. Armstrong. 1994 ISBN 0-7923-2565-6

# Boston Studies in the Philosophy of Science

157. I. Szumilewicz-Lachman (ed.): *Zygmunt Zawirski: His Life and Work.* With Selected Writings on Time, Logic and the Methodology of Science. Translations by Feliks Lachman. Ed. by R.S. Cohen, with the assistance of B. Bergo. 1994 ISBN 0-7923-2566-4
158. S.N. Haq: *Names, Natures and Things.* The Alchemist Jābir ibn Ḥayyān and His *Kitāb al-Aḥjār* (Book of Stones). 1994 ISBN 0-7923-2587-7
159. P. Plaass: *Kant's Theory of Natural Science.* Translation, Analytic Introduction and Commentary by Alfred E. and Maria G. Miller. 1994 ISBN 0-7923-2750-0
160. J. Misiek (ed.): *The Problem of Rationality in Science and its Philosophy.* On Popper vs. Polanyi. The Polish Conferences 1988–89. 1995 ISBN 0-7923-2925-2
161. I.C. Jarvie and N. Laor (eds.): *Critical Rationalism, Metaphysics and Science.* Essays for Joseph Agassi, Volume I. 1995 ISBN 0-7923-2960-0
162. I.C. Jarvie and N. Laor (eds.): *Critical Rationalism, the Social Sciences and the Humanities.* Essays for Joseph Agassi, Volume II. 1995 ISBN 0-7923-2961-9
    Set (161–162) ISBN 0-7923-2962-7
163. K. Gavroglu, J. Stachel and M.W. Wartofsky (eds.): *Physics, Philosophy, and the Scientific Community.* Essays in the Philosophy and History of the Natural Sciences and Mathematics. In Honor of Robert S. Cohen. 1995 ISBN 0-7923-2988-0
164. K. Gavroglu, J. Stachel and M.W. Wartofsky (eds.): *Science, Politics and Social Practice.* Essays on Marxism and Science, Philosophy of Culture and the Social Sciences. In Honor of Robert S. Cohen. 1995 ISBN 0-7923-2989-9
165. K. Gavroglu, J. Stachel and M.W. Wartofsky (eds.): *Science, Mind and Art.* Essays on Science and the Humanistic Understanding in Art, Epistemology, Religion and Ethics. Essays in Honor of Robert S. Cohen. 1995 ISBN 0-7923-2990-2
    Set (163–165) ISBN 0-7923-2991-0
166. K.H. Wolff: *Transformation in the Writing.* A Case of Surrender-and-Catch. 1995 ISBN 0-7923-3178-8
167. A.J. Kox and D.M. Siegel (eds.): *No Truth Except in the Details.* Essays in honor of Martin J. Klein. 1995 ISBN 0-7923-3195-8
168. J. Blackmore (ed.): *Ludwig Boltzmann* His Later Life and Philosophy, 1900–1906. 1995 ISBN 0-7923-3231-8

*Also of interest:*

R.S. Cohen and M.W. Wartofsky (eds.): *A Portrait of Twenty-Five Years Boston Colloquia for the Philosophy of Science, 1960-1985.* 1985 ISBN Pb 90-277-1971-3

*Previous volumes are still available.*

KLUWER ACADEMIC PUBLISHERS – DORDRECHT / BOSTON / LONDON

GPSR Compliance

*The European Union's (EU) General Product Safety Regulation (GPSR) is a set of rules that requires consumer products to be safe and our obligations to ensure this.*

*If you have any concerns about our products, you can contact us on ProductSafety@springernature.com*

In case Publisher is established outside the EU, the EU authorized representative is:

Springer Nature Customer Service Center GmbH
Europaplatz 3
69115 Heidelberg, Germany

**Batch number: 10762211**

Printed by Printforce, the Netherlands